PLANE AND GEODETIC SURVEYING
FOR ENGINEERS

PLANE AND GEODETIC SURVEYING

FOR ENGINEERS

By the Late DAVID CLARK, M.A., B.Sc.

M.Inst.C.E.   M.Inst.C.E.I.   M.Am.Soc.C.E.
Professor of Civil Engineering, University of Dublin

VOLUME ONE
PLANE SURVEYING

SIXTH EDITION REVISED BY
J. E. JACKSON,
M.A., F.R.A.S., F.R.I.C.S.

With a chapter on Field Work for Aerial Surveys by
A. G. DALGLEISH,
M.A., A.R.I.C.S.

CONSTABLE LONDON

Published by Constable & Co Ltd
10 Orange Street, London WC2

First published	1923
Second edition	1932
Reprinted	1937
Third edition	1940
(revised by James Clendinning)	
Reprinted 3 times	
Fourth edition	1946
Reprinted 10 times	
Fifth edition	1958
Reprinted 7 times	
Sixth edition	1969
(revised by J. E. Jackson)	

Filmset by Photoprint Plates Ltd, Wickford
Printed in Great Britain by
Butler & Tanner Ltd, London & Frome
Bound by Leighton-Straker Bookbinding Co Ltd,
London

CONTENTS

Introduction

Page No.

10   CHAPTER I
   INSTRUMENTS—CONSTRUCTION AND
   ADJUSTMENT

12   Parts Common to Several Instruments
13   Optical Components
        Mirrors
        Prisms
        Parallel Plate
        Lenses
23   Telescopes
33   Bubble Tubes
37   The Vernier
39   The Magnetic Needle
40   The Tripod Stand
41   Chain Surveying Instruments
52   The Theodolite
67   Temporary Adjustments of the Theodolite
68   Angle Measurement by Theodolite
71   Testing and Adjustment of the Theodolite
74   Adjustment of the Transit Theodolite
78   Effects of Instrumental Errors of a Theodolite
80   The Compass
82   The Level
87   Levelling Staffs
93   Temporary Adjustments of the Level
93   Testing and Adjustment of the Level
95   Hand Levels and Clinometers
97   Precautions in Using Instruments
99   References

101   CHAPTER II
    LINEAR MEASUREMENTS

102   Ranging or Setting-out Chainage Lines

Page No.

103   Linear Measurements with the Chain
108   Errors in Ordinary Chaining
110   Linear Measurements with the Long Steel Band or Tape
112   Using the Tape on the Flat
118   Using the Tape in Catenary
121   Corrections to Measured Lengths
124   Standardising Steel Tapes, Spring Balances and
         Thermometers
125   Errors in Measurement with Steel Tapes
127   Propagation of Errors in Tape Measurements
127   Examples
129   References

130   CHAPTER III
       CHAIN SURVEYING

130   Chain Surveying—Principles
132   Field Work
135   Chain Survey Problems
140   Plotting a Chain Survey

144   CHAPTER IV
       TRAVERSING

145   Bearings
150   Field Work of Theodolite Traversing
154   Angular Measurements
155   Theodolite Observation of Angles
161   Theodolite Observation of Bearings
168   Linear Measurements
169   Sources of Error in Theodolite Traversing
173   Field Checks in Traversing
178   Office Tests for Locating Gross Errors in Traversing
179   Propagation of Error in Traversing
181   Miscellaneous Problems in Theodolite Surveying
186   Compass Traversing
190   References

Page
No.
191 CHAPTER V
OFFICE COMPUTATIONS

198 Circular Measure
199 Logarithms of the Sines and Tangents of very Small Angles
201 Rectangular Coordinates
208 Theodolite Traverse Computations
218 Computation of Compass Traverses
221 Miscellaneous Problems in Rectangular Coordinates
227 Plotting Traverses
232 Examples
236 References

237 CHAPTER VI
ORDINARY LEVELLING

237 Principles
245 Field Work
246 Longitudinal Sections
251 Cross Sections
258 Sources of Error in ordinary Levelling
267 Examples
268 References

269 CHAPTER VII
PLANE TABLE SURVEYING

269 The Plane Table and Accessory Instruments
273 Attachment of Paper to the Board
273 Setting Up the Plane Table
274 Techniques in Plane-Tabling
281 Accuracy of Plane-Tabling

282 CHAPTER VIII
CONTOURS AND CONTOURING

286 Methods of Contouring
295 Uses of Contour Plans and Maps
299 References

*Page
No.*

300 CHAPTER IX
MEASUREMENT OF AREAS AND VOLUMES

301 Measurement of Area
314 Measurement of Volume
328 The Mass Diagram
334 Examples
337 References

338 CHAPTER X
SETTING OUT WORKS

340 Curves
345 Setting Out Curves
347 Setting Out Simple Curves by Chain and Tape Only
351 Problems in Ranging Simple Curves
359 Transition Curves
366 Vertical Curves
368 Miscellaneous Operations in Setting Out
375 Examples
378 References

379 CHAPTER XI
TACHYMETRIC SURVEYING

380 The Stadia Method
391 The Tangential System
392 Subtense Method
395 Uses of Tachymetric Surveying
396 Examples
398 References

400 CHAPTER XII
HYDROGRAPHICAL SURVEYING

400 Tides and Sea Level
402 Mean Sea Level
407 Mean Sea Level as a Datum Surface for Heights
408 Shore Line Surveys
408 Position Fixing
413 Sounding

Page
No.
415 Plotting Soundings
420 The Survey of Tidal Currents
422 Stream Measurement
424 Area-Velocity Method of Discharge Measurement
435 Weir Method of Discharge Measurement
440 Chemical Method of Discharge Measurement
441 Continuous Measurements of Stream Discharge
442 References

443 CHAPTER XIII
     FIELD WORK FOR AERIAL SURVEYS

444 Provision of Control where the Photographs are already
        taken
456 Provision of Control in advance of the Photography
459 Bibliography
461 Answers to Examples
464 Appendix
465 Index

PREFACE TO THE FIRST EDITION

This textbook is designed to form a complete treatise on plane surveying with such parts of geodetic work as are of interest to the civil engineer. The author would emphasise at the outset that he does not claim that a knowledge of geodesy is a very essential part of the equipment of the engineer. The execution of surveys of such extent and character as to necessitate the general methods of geodetic surveying and levelling does, however, occasionally fall within his province. For this reason, a knowledge of its principles is required in the examinations of the Institution of Civil Engineers and of universities and colleges.

In a general text on surveying there is little room for originality, except in treatment. Although this work is intended to serve as a reference book for practising engineers and surveyors, the chief aim has been to cover ground suitable for a degree course, and, while it is hoped that the book will prove of value to those pursuing a college course, the needs of the self-taught student have been specially kept in view. In consequence, many explanatory notes and practical hints have been inserted, particularly with reference to the more common surveying operations. The latter are not meant to take the place of practice in the field, which, needless to say, is an essential part of a training in surveying, but are intended as a guide to the reader with limited opportunities for field practice, and are mainly suggested by the author's experience of the initial mistakes of young engineers in practice and of students undergoing field training in camp.

The subject-matter is presented in two volumes. The first covers in ten chapters the more common surveying operations of the engineer, and the second deals with astronomical and geodetic work and the methods employed in large surveys generally.

In the arrangement of the present volume it has been thought desirable, for convenience of reference, to group descriptions of the more commonly used instruments and their adjustments to form Chapter I. It is hoped that the detailed method of treating the subject of adjustments will afford a sound understanding of the geometrical principles in each case. In Chapters II to VI the subjects of Chain Surveying, Theodolite and Compass Traversing,

Ordinary Levelling, Plane Table Surveying, and Contouring are described as applied to cadastral and engineering surveys. Chapter VII deals with the office work of computing areas and volumes, the latter with particular reference to the measurement of earthwork quantities. The practice of setting out works is treated in Chapter VIII. The setting out of railways is the only branch of this subject meriting detailed description, and problems in connection with the setting out of simple, compound, reverse, and transition curves are treated. The principles and practice of Tacheometry are given in Chapter IX, and Hydrographical Surveying, including Marine Surveying and Stream Measurement, is dealt with in Chapter X. Owing to the number of texts available on mine surveying, no special reference has been made to that subject.

Sets of illustrative numerical examples and answers are given for practice. For permission to reproduce questions set by the Institution of Civil Engineers, the University of London, and the Royal Technical College, Glasgow, the author desires to express his thanks to the authorities concerned.

Lists of references have been inserted, after the appropriate chapters, on such subjects as readers might wish to pursue further. The author would gratefully acknowledge the assistance he has derived from these and other books and papers on Surveying.

D.C.

PREFACE TO THE SECOND EDITION

In the preparation of this edition it has been found desirable to enlarge the text and to re-write a considerable part of the book. The number of illustrations has also been increased. The eight years which have elapsed since the issue of the First Edition have witnessed extensive developments in the design and manufacture of surveying instruments, and an endeavour has been made to bring the book up to date in respect of these. For permission to reproduce questions set by the Institution of Civil Engineers, the University of London, the Royal Technical College, Glasgow, and Trinity College, Dublin, the author desires again to express his indebtedness. He would also express his thanks to Messrs. C. F. Casella and Co., Ltd., Cooke, Troughton and Simms, Ltd., George Russell and Co., Ltd., E. R. Watts and Son, Ltd., Henry Wild Surveying Instruments Supply Co. Ltd., and Carl Zeiss for placing at his disposal information regarding their recent instruments. He gratefully acknowledges his further indebtedness to Messrs. Cooke, Trough-

ton and Simms for permission to reproduce Figs. 71, 72, and 301, to
Messrs. C. F. Casella and Co. and Henry Wild Co. for Fig. 69, and
to Messrs. Carl Zeiss for Fig. 102.

D.C.

TRINITY COLLEGE,
DUBLIN, 1931.

PREFACE TO THE THIRD EDITION

In undertaking the revision of this book for a third edition, I have
added certain matter, some of which, perhaps, is of greater interest
and importance to engineers and surveyors working abroad
or in the Colonies than it is to those whose work is confined to
home practice. In Great Britain, the existence of the Ordnance
Survey often simplifies the work of the private engineer or surveyor
very considerably and makes it unnecessary for him to aim at
the same degree of accuracy that is sometimes required abroad.
Consequently, I have added a new chapter on linear measurements,
in which work with the long steel band or tape, an article that is
much more extensively used abroad than it is at home, is dealt
with in considerable detail, and I have also made fairly considerable
additions to the matter dealing with the theodolite traverse, for
which purpose I have divided the original chapter on traversing
into two, one dealing with the field work and the other with the
office computations. Hitherto, most engineers and surveyors have
regarded the limit of the standard of accuracy attainable by simple
theodolite traversing as lying somewhere between 1/2000 and
1/5000, but, with the more extensive use of modern small theodolites with micrometer readings, such as the small Tavistock theodolite, and with reasonably careful taping with the long steel band, an
accuracy of anywhere between 1/10,000 and 1/30,000 is now easily
attainable in ordinary engineering and cadastral work. This
makes it possible to substitute traversing for triangulation in cases
where accuracy is necessary but where triangulation is difficult
or unduly expensive.

In the chapters dealing with linear measurements and theodolite
traversing I have used some of the results of the theory of errors,
although a formal treatment of the theoretical aspects of this
subject is reserved for Chapter IV of Vol. II. This is because it is
of the utmost importance that the surveyor should have some
idea of the different sources of error inherent in his work, the
probable magnitude of these errors, and the manner in which they
are propagated. Otherwise, it is not possible for him to do his

work in the most economical manner or to select the best methods. This applies particularly to traversing, and hence, and mainly for purposes of reference, I have thought it advisable to include with the description of certain operations some discussion of the resulting effect of those errors that are likely to affect them.

Among other additions are a short description of road transition curves, following the discussion of transition curves on railways already included in the second edition, and a brief description of echo sounding, with special reference to its advantages and possibilities as applied to engineering problems. Road transition curves are becoming of increasing importance in the lay-out of modern roads, and echo sounding is a fairly recent development which is likely to replace the older methods of sounding in modern hydrographical surveying.

In conclusion, I should like to express my thanks to the following:— The Astronomer Royal, for providing me with data regarding the present values of the magnetic elements; The Director General, Ordnance Survey, for permission to include a short summary of the Ordnance Survey methods of detail survey, and Colonel G. Cheetham, D.S.O., M.C., R.E., Ordnance Survey, for looking over my draft on this subject and providing me with other information; Messrs. Carl Zeiss (London), Ltd., for lending the block of Fig. 74; Mr. A. D. Simms, of Messrs. Cooke, Troughton & Simms, for providing me with special information regarding instruments made by his firm; Messrs. E. R. Watts & Sons for providing details regarding the Connolly Standard Compass manufactured by them; Messrs. Henry Hughes & Son, Ltd., for lending the blocks used in printing Figs. 369 and 370 and for giving me information concerning echo sounding apparatus of their manufacture, and Dr. E. B. Worthington, of the Freshwater Biological Association, for allowing the use of a photograph, reproduced as Fig. 371, of an echo sounding record obtained during the course of an echo sounding survey of the English Lakes.

J. CLENDINNING

ANGMERING-ON-SEA,
SUSSEX.
21st June 1939.

PREFACE TO THE SIXTH EDITION

This sixth edition is a complete revision. It has been entirely reset thus providing the opportunity to discard obsolete material, re-write where necessary, or incorporate portions of the previous edition that seemed suitable, having regard to present-day requirements of the land surveyor and the civil engineer. The appendices that had been added to the fifth edition have disappeared, and new material is in its proper places in the chapters. This procedure may have resulted in a patchwork, and it is hoped that this structure will not be too obvious to the reader.

Measurements of angles and distances are now, as they always have been, the basic material of land-surveying, but the recent development of photogrammetry as a mapping tool has obliged authors of textbooks on surveying to consider including something on this rather special and highly mechanised technique. In practice, the camera is flown by aeronauts and the map is constructed by photogrammetrists, and it is unlikely that any of these practitioners is a trained land-surveyor. Neither is it desirable that one trained in land-surveying should occupy himself in piloting aircraft or operating plotting machines, even if he were able to do either of these jobs. The land-surveyor, as such, comes into the process mainly in the provision of ground control, the special requirements of which have been developed from the experiences of several decades of air-photographic mapping. Hence the appearance here of the additional Chapter XIII, written by Mr. Dalgleish, and intended to describe ground-control survey methods, rather than photogrammetric theory and techniques, which are more or less fully dealt with in many modern texts.

Referring in more detail to changes made in preparing this edition, mention may be made of the increased attention given to elementary geometrical optics, in consideration of modern developments of optical reading systems which make extensive use of reflecting prisms and the parallel-sided plate device. Some obsolete, or nearly obsolete, instrumental features are no longer included; it can be assumed that any modern telescope has internal focusing and zero additive constant for tachymetry. The accounts of the nature of propagation of errors, especially in traversing, seemed to need some re-writing. Mention is made of the use of mechanical computation, referring only to the simplest types of calculating machines. Plane-tabling occupies fewer pages: this method still has its value, but it is no longer a primary method for continuous topographical mapping. Fewer pages are devoted to descriptions of tidal phenomena in Chapter XII.

Mr. J. Clendinning, who prepared the previous edition of this volume, might have undertaken the present revision, but was prevented from doing so by ill health. As many readers will know, he died early in 1966. I am indebted to him for some notes and suggestions he gave me when he learned that I had agreed to make the sixth edition. I also wish to thank many others who have discussed the book with me, suggested the inclusion, or exclusion, of certain items, or provided information included in the new text. Particularly, I am grateful to Mr. D. F. Munsey M.A., Lecturer at the School of Military Science, Shrivenham, to my brother, Mr. F. S. Jackson M.A., who is a civil engineer, and to Mr. A. G. Dalgleish M.A. (now Deputy Director (Mapping) at the Directorate of Overseas Surveys) who also read through the whole of the original drafts.

Most of the exercises given at the ends of some of the Chapters are new: in some of the exercises, quantities are expressed in the metric system of units. References to literature have been brought up to date, though no doubt they are incomplete, and many of the older references have been omitted.

J.E.J.

FITZWILLIAM COLLEGE,
CAMBRIDGE.
February 1969.

INTRODUCTION

Surveying is the art of making such measurements of the relative positions of points on the surface of the Earth that, on drawing them to scale, natural and artificial features may be exhibited in their correct horizontal or vertical relationships.

Less comprehensively, the term, 'surveying,' may be limited to operations directed to the representation of ground features in plan. Methods whereby relative altitudes are ascertained are distinguished as 'levelling,' the results being shown either as vertical sections or by conventional symbols on a plan.

Plane and Geodetic Surveying. A plan is a projection upon a horizontal surface, and in its construction all linear and angular quantities used must be horizontal dimensions. It is impossible to give a complete representation of distances following the undulations of the ground other than by a scale model. Now a horizontal surface is normal to the direction of gravity as indicated by a plumb line, but, on account of the form of the Earth, the directions of plumb lines suspended at different points in a survey are not strictly parallel, and the plane horizontal at one point does not precisely coincide with that through any other point. It is not the irregular shape of the Earth's physical surface that is referred to here, but the almost regular curvature of a level surface which is necessarily perpendicular to the vertical everywhere.

In surveys of small extent the effect of curvature is quite negligible, and it is justifiable to assume that a level surface of the Earth is a horizontal plane within the area covered. Surveying methods based on this supposition are comprised under the head of *Plane Surveying*. The assumption becomes invalid in the accurate survey of an area of such extent that it forms an appreciable part of the Earth's surface. Allowance must then be made for the effect of curvature, and the operations belong to *Geodetic Surveying*.

No definite limit can be assigned for the area up to which a survey may be treated as plane, since the degree of accuracy required forms the controlling factor. The sum of the interior angles of a geometrical figure laid out on the surface of the Earth differs from that of the corresponding plane figure only to the extent of one second for about every 76 square miles, or 200 sq. km., of area,

1

so that, unless extreme accuracy is required, plane surveying is applicable to areas of several hundreds of square miles.

Plane Surveying. Plane surveying is of wide scope and utility, and its methods are employed in the vast majority of surveys undertaken for various purposes, such as engineering, architectural, legal, commercial, scientific, geographical, exploratory, military, and navigational. As applied to civil engineering, all surveying methods are utilised in the various surveys required for the location and construction of the different classes of works within the province of the engineer. These surveys may be rapid reconnaissances of an exploratory character undertaken to facilitate the selection of an approximate site for the work. They are followed by more detailed surveys of the selected region, in which a much greater degree of accuracy is sought, and from which the best location is ascertained. The obtaining of various data required in the design of the proposed works forms part of the preliminary operations, and may involve surveying methods of a specialised character. Previous to and during construction, the surveyor's duties also include the routine of setting out the lines and levels of the works and the measurement of areas and volumes.

Aerial Surveying. An engineer may have air-photographs of a site on which he has to design or set out works. It is important that he should understand their uses and limitations.

Though an air-photograph will probably show all the details the engineer requires, and indeed probably very much more, it must not be treated as a map because it is not possible to ensure that the axis of an air-camera is exactly vertical. The tilt may be several degrees in a so-called vertical photograph, so there may be appreciable differences of scale between opposite sides of a photograph. In any case, even if the camera axis is vertical, the relative positions of points at different heights will not be correctly shewn.

The most valuable property of an air-photograph is that angles round a point at or close to its centre may be regarded as correct, and in making maps from air-photographs this property is exploited. In other words, a near-vertical air-photograph may be regarded as a record of angles taken at its centre point or at any point very close to the centre.

If air-photographs from two different positions cover an area of common ground, and they are viewed by means of a suitable stereoscope, a three-dimensional picture is seen and can be of very great value, especially in the preliminary stages of design of a project.

The production of accurate maps from air photographs involves the use of techniques and equipment which will not normally be available to an engineer-surveyor. These mapping processes, called photogrammetry, are outside the scope of this book. However, mapping from air-photographs requires some control by ground-survey methods, and this subject is discussed in Chapter XIII.

Geodetic Surveying. Geodetic surveys are usually of a national character, occasionally works of international co-operation, and they are undertaken as a basis for the production of accurate maps of wide areas, as well as for the furtherance of the science of *Geodesy*, which treats of the size and form of the Earth. The most refined instruments and methods of observation are employed, and the operations are directed to the determination of the positions on the Earth's surface of a system of points which serve as controls for all other surveys.

Use of Geodetic Data. As a general rule, in countries where a geodetic survey exists, much useful information concerning points established by it may be obtained, either free or on payment of a small fee, on application to the government survey authority. This information may include descriptions to enable the points to be found on the ground, coordinate values for each point, and, in many cases also, the height above sea level. Also, it often happens that the true bearings between fixed points, not too far apart, are available and can be used either to orientate the survey or to provide checks on the bearings. If extensive survey operations in any area are contemplated, it is always well, even when there is no legal obligation or reason to do so, to ascertain if information of this kind can be obtained, because, not only may it save the engineer or surveyor a great deal of work in providing his own control points, but it may also enable him to obtain most useful checks at various stages of his work.

Another advantage of using points established by the government survey is that it enables local surveys to be laid down and plotted, without any great difficulty, on the official printed maps and plans. Indeed, in many countries the law insists that certain classes of surveys must be tied in to the 'government datum points'. This is particularly the case where surveys of property, in which the resulting plans are to be used as part of legal documents conveying title to land, are involved, and the laws of some countries insist that plans of such things as the underground workings in mines should show the relation of the survey to some point or points established by the official survey department.

In Great Britain, there are certain cases where special maps or plans, based on the Ordnance Survey sheets, have to be prepared and submitted to the authorities concerned, while the larger scale plans indicate the positions of bench marks, this information being of great value to civil engineers and others concerned with levelling operations.

Field and Office Work. A record of measurements made on the ground is usually, in plane surveys at least, of little or no service until the dimensions are represented to scale on a drawing. The usual stages in the production of the finished drawing may be summarised as:

Field Work: (*a*) A preliminary examination, or reconnaissance, of the ground to discover how best to arrange the work:

(*b*) The making of the necessary measurements.

(*c*) The recording of the results in a systematic form.

Office Work: (*a*) The making of any calculations necessary to transform the field measurements into a form suitable for plotting.

(*b*) The plotting or drawing of those dimensions.

(*c*) The inking-in and finishing of the drawing.

(*d*) The calculation of quantities to be shown on the drawing for special purposes, such as areas of land, etc.

Principles of Surveying. The fundamental principles upon which various survey methods are based are themselves of a very simple nature, and may be stated here in a few lines.

It is always practicable to select two points in the field and to measure the distance between them. These can be represented on paper by two points placed in a convenient position on the sheet and at a distance apart depending upon the scale to which it is

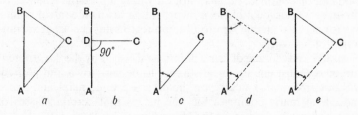

Fɪɢ. 1.

proposed to plot the survey. From these initial points others can be located by two suitable measurements in the field and plotted in their relative positions on the sheet by means of appropriate drawing instruments. Points so obtained serve in turn to fix the positions of others.

The more direct methods of locating a point C, with respect to two given points A and B, are illustrated in Fig. 1.

(*a*) Distances AC and BC are measured, and C is plotted as the intersection point of arcs with centres A and B and radii scaling the measured distances. This system is employed in linear or chain surveying (Chapter III); it may also be described as *trilateration*–a word recently brought into use through the introduction of the so-called 'electromagnetic' distance-measuring instruments.

(*b*) Perpendicular CD and the position on AB of its foot D are measured, and C is plotted by the use of a set-square. This method, termed offsetting, is used, in combination with other surveying systems, for locating subsidiary points not required for extending the survey but required for defining details.

(*c*) Distance AC and angle BAC are measured, and C is plotted either by means of a protractor or trigonometrically. The traverse method of surveying (Chapter IV) is founded on this principle.

(*d*) Angles BAC and ABC are measured, and C is plotted by solution of the triangle or by use of a protractor. Distance AB being known, C is located without further linear measurement, and, in consequence, the method, known as triangulation, is applicable to the most extended surveys, in which it is desirable to reduce linear measurement to a minimum.

(*e*) Angle BAC and distance BC are measured, and C is plotted trigonometrically or by protracting the angle and swinging an arc from B. This method is of minor utility, and is required only in exceptional cases. Angle C must not be near 90°.

The same methods could be employed in measuring relative altitudes as well as positions in plan, but for this purpose methods (*b*), (*c*), and (*d*) only are of practical utility. If the diagrams are regarded as elevations, with AB horizontal, ordinary spirit levelling (Chapter VI) is illustrated by (*b*). The elevation of C relatively to that of A is obtained by establishing instrumentally a horizontal line AB through A, such that AB and C are in the same vertical plane, and measuring the vertical distance CD. Trigonometrical levelling (Vol. II, Chapter V) is exemplified in (*c*) and (*d*).

Several of the above systems, both for horizontal and vertical location, may be employed in the same survey, and various types of instruments are available for the angular and linear measurements. A survey may therefore be executed in several ways by different combinations of instruments and methods, and some parts of the work may require different treatment from others. The principal factors to be considered are the purpose of the survey and the degree of accuracy required, the nature of the country, the extent of the survey, and the time available for both field and office work. To select the methods best suited to a particular case demands on the part of the surveyor a degree of skill which can be acquired only by experience.

Working from the Whole to the Part. In most types of survey the ruling principle is to work from the whole to the part. Thus, in fairly extensive surveys, such as those of a large estate or of a town, the first thing to be done is to establish a system of control points. The positions of these points are fixed with a fairly high standard of accuracy, but, between them, the work may be done by less accurate and, consequently, by less expensive methods. In a town survey, for instance, the 'primary horizontal control' will consist either of triangulation or of a traverse surrounding the whole town. If triangulation is adopted as the control, the larger 'main' or 'primary' triangles will be surveyed with the greatest care, but these will be 'broken down' by smaller 'minor' or 'secondary' triangles, which may be measured by less rigid methods and less elaborate instruments than those used in the survey of the larger triangles. Similarly, if the control consists solely of points established by traverses, other traverses, which will usually run along the main streets, will be used to connect points on the outer surround. Some of these radial traverses will probably be measured with the same degree of accuracy as the main outer surround traverse so as to stiffen it and provide a traverse 'network' rigidly held together. Between these radial traverses, however, other less important ones, surveyed by less precise methods, will be established, and, as a general rule, the survey of detail—that is of the positions and shapes of houses, streets, etc.—will be done by still less elaborate methods, using the minor traverses as a frame from which to work. The idea of working in this way is to prevent the accumulation of error, which, in some cases, tends to magnify itself very quickly. If the reverse process is adopted and the survey is made to expand outwards, it will generally be found that minor errors become so magnified in the process of expansion as to give rise to serious discrepancies at some stage in the work. On the other hand, if an accurate basic control is established in the first place, not only are large errors prevented and minor ones controlled and localised, but it will be found that the detail begins to fall almost automatically into its proper place, like fitting in the smaller pieces in a jig-saw puzzle.

These remarks regarding 'working from the whole to the part' apply also to such operations as levelling. Thus, in contouring on a fairly large scale, it will generally be found advisable to establish a system of bench marks, using an accurate high-class Level for the purpose. The actual survey of the contours or form lines can then be made by using an Abney Level or Indian Clinometer, or other instrument of minor accuracy, or even, when the scale of the plan or map is very small and the 'contour interval' large, by aneroid barometer.

Nature of the Subject. By virtue of the simplicity of the underlying principles of plane surveying, there is little of theory to be studied, and a training in the subject must be chiefly directed towards a thorough working knowledge of field methods and the associated instruments, as well as of office routine. Success in the field is the outcome not only of skill in solving the larger problems connected with the general organisation of surveys, but also of attention to the methodical performance of the numerous details of field work. Frequent practice in the field under expert guidance saves the beginner much memorising of these details, makes for skill and speed in manipulating instruments, and promotes systematic habits of work. Numerous minor problems requiring special treatment are likely to be encountered in field work, and to the beginner these are apt to assume a more difficult aspect on the ground than they do on paper. A little experience is necessary before one can entirely overcome the distractions of field conditions, especially when these are aggravated by physical fatigue and bad weather.

Even in favourable circumstances there are many ways in which errors may enter into the work, and it is important to realise that *absolute precision can never be attained.* Any desired degree of refinement of practical utility may, however, be secured by the use of suitable instruments and methods of observation. The surveyor must keep in view the uses to which his results will be put, and must select those methods which will yield sufficient accuracy for the purpose. Much time may be wasted in needless refinements, and the necessary judgment must be cultivated by a study of the nature and relative importance of the various sources of error affecting different surveying operations.

Errors. While a knowledge of the theory of errors as dealt with in Vol. II, Chapter IV is not required in connection with small surveys, we shall find it convenient to use one or two fundamental relations in the chapters dealing with linear measurements and traversing. This matter is of practical importance because traversing involves a combination of linear and angular measures; and, unless it is possible to form some sort of estimate of the errors likely to arise from each type of measurement, the manner in which they are propagated, and their probable effect on the final result, it is not possible to choose the best or most economical methods of working. Consequently, although reference must be made to Vol. II for an account of the principles and of the scientific basis of the theory of errors, those results that are of practical importance regarding the subject under discussion will be stated and used in Chapters II and IV of this volume, but, if necessary, the sections dealing with errors may be omitted on a first reading or left over until Chapter IV of the second volume has been studied in detail. Here, however, it may be well to describe the principal types of error and to give a very brief explanation of how they are propagated, and their total effect estimated.

The ordinary errors met with in all classes of survey work may be classified as mistakes, systematic or cumulative errors, and accidental or compensating errors.

Mistakes arise from inattention, inexperience, or carelessness. Since an undetected mistake may produce a very serious effect upon the final results, the surveyor must always arrange his work to be self-checking, or take such check measurements as will ensure the detection of mistakes. On discovering that a mistake has been made, it may be necessary to repeat the whole or part of the survey. It should always be possible for the surveyor to guarantee that his completed work is free from mistakes.

Systematic Errors are those which persist and have regular effects in the performance of a survey operation. Their character is understood, and their effects can be eliminated either by the exercise of suitable precautions or by the application of corrections to the results obtained. Such errors are of a constant character, and are regarded as positive or negative according as they make the result too great or too small. Their effect is therefore cumulative: thus, if, in making a measurement with a 50-ft. linen tape which is 1 in. too long, the tape is extended ten times, the error from this source will evidently be 10 in. The effects of constant errors may become very serious, and the precautions to be adopted against them in various field operations are treated in detail throughout the text.

Accidental Errors include all unavoidable errors which are present notwithstanding that every precaution may have been taken. They arise from various causes, such as want of perfection of human eyesight in observing and of touch in manipulating instruments, as well as from the lack of constancy in the conditions giving rise to systematic errors. Thus, in the example cited above the error of 1 in. in the tape may fluctuate a little on either side of that amount by reason of small variations in the pull to which it is subjected or even by changes in the humidity of the atmosphere. Such errors are usually of minor importance in surveying operations when compared with the systematic errors, because their chief characteristic is that they are variable in sign, plus errors tending to be as frequent as minus errors, so that they are of a compensating nature and thus tend to balance out in the final results. In consequence, it is needless to adopt elaborate precautions against the occurrence of errors of this type while possibly overlooking the propagation of serious cumulative errors.

As regards the method of propagation of the different kinds of error, the effect of the cumulative errors is additive since each one tends always to be of one particular sign. The effect of the accidental errors is also additive in the sense that, in any particular observation, the total error is the sum of all the errors made during that observation. However, we do not know the signs of the individual errors, but we do know that, while one error may be positive, another one is equally likely to be either positive or negative. Accordingly, the ordinary addition law cannot be applied, and all that is possible is to form a general estimate of the probable effect of combining all errors, when each one is as likely to have a positive as it is to have a negative sign. In this case, the mathematical law of probability shows that the best way of combining errors of this kind is to take the square root of the sum of the squares of the individual probable errors. Mathematically, this can be expressed as:

$$r = \pm(\varepsilon_1{}^2 + \varepsilon_2{}^2 + \varepsilon_3{}^2 + \ldots + \varepsilon_n{}^2)^{\frac{1}{2}}$$

where r is the probable error resulting from the combination of the probable errors $\pm\varepsilon_1$, $\pm\varepsilon_2$, $\pm\varepsilon_3 \ldots \pm\varepsilon_n$, the plus or minus sign in all cases indicating the equal probability of either sign. Quantities calculated in this way must be regarded as estimates of averages taken over a large number of cases: the actual error in any particular measurement is unknown.

CHAPTER I

INSTRUMENTS–CONSTRUCTION
AND ADJUSTMENT

1. Before proceeding to consider the actual work of surveying it is desirable that a knowledge of the instruments employed should be obtained, and in this chapter are described the instruments in most regular use by the engineer and surveyor in everyday work of linear and angular surveying, and levelling. Those of a more specialised character are treated later under the branches of surveying with which they are related.

While the necessity of a thorough understanding of instruments is self-evident, it is not suggested that the surveyor need at first make an exhaustive study of the construction of all the fittings which go to make up his instruments. The more detailed his knowledge, however, the better qualified will he be to effect temporary repairs–an important matter when working far from headquarters. In general, these individual fittings can be treated in groups forming definite and essential parts of the instrument, and the fundamental principles underlying the arrangement, use and adjustment of the instrument are to be studied by reference to the functions and relationships of those parts.

2. The Nature of Adjustments. Adjustments are of two kinds– *Temporary and Permanent.* The former are those which have to be made at each setting up of the instrument, and, as they are part of the regular routine in using the instrument, need no preliminary explanation. The so-called permanent adjustments, on the other hand, are directed to eliminating derangements or defects which may develop either through wear or accident, and in general consist in setting essential parts into their proper positions relatively to each other.

The method of making these adjustments in any instrument is entirely dependent upon the arrangement of its essential parts, and upon the geometrical relationships they are designed to bear to each other. The adjustment of an instrument is simply a practical problem in geometry. Unless so regarded, the operation can only be performed mechanically, and the methods will be much more readily forgotten.

The examination of an instrument from the geometrical standpoint should lead the student to an understanding of: (*i*) The nature of the errors of measurement that will be caused by erroneous relationships between adjustable parts or by defects in non-adjustable parts; (*ii*) the manner in which the instrument can be tested to discover whether it is in adjustment or not; (*iii*) the nature of the adjustment necessary to eliminate an error discovered; and (*iv*) the appropriate method of manipulating the instrument so that the effect of any remaining error may be minimised or eliminated.

3. The Principle of Reversal. In testing for instrumental errors, considerable use is made of the method of reversal. Such tests are usually directed to examining whether a certain part is truly parallel, or perpendicular, to another, and on reversing one part relatively to the other, any erroneous relationship between them is made evident.

To take the simple instance of examining the perpendicularity between two edges of a set square, let the line BC (Fig. 2) be drawn with the set square in the first position, and BC' after reversal. The reversal constitutes the test, for if C and C' coincide, the

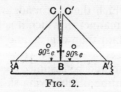

Fig. 2.

required condition, that ABC = 90°, obtains. If not, let *e* be the angular error. Then clearly CBC' = 2*e*, or the apparent error on reversal is *twice* the actual error, and therefore, if correction is possible, the error to be eliminated is half the observed discrepancy. This example also shows that good results are obtainable from a defective instrument, by reversing and taking the mean of the two erroneous results.

As a second example of the effect of reversal, the bubble tube (section 36) may be studied.

4. Notes on Adjustments. Most instruments have several relationships to be established by adjustment, and it is important that adjustments should be performed in such sequence and in such a manner that (*a*) those already executed will not be disturbed in making subsequent adjustments, (*b*) in performing any adjustment, the effects of possible errors awaiting adjustment will be balanced out if they are such as to influence the accuracy of that under examination.

As it is sometimes troublesome to ensure that a particular adjustment will be quite independent of the others, it is well to repeat the adjustments from first to last and so gradually perfect them all, especially if serious errors have been discovered.

In making an adjustment, it is difficult to eliminate an error completely at the first trial, so the test and correction may need to be repeated a few times, using greater refinement as the error decreases.

Instability of the instrument makes it almost impossible to adjust it satisfactorily, and the instrument should therefore be placed on firm ground. In adjustments involving the taking of test sights, parallax (section 28) must be carefully eliminated.

Adjustment screws must be left bearing firmly, so that they will not slacken on being accidentally touched, but they should never be forced. The screws are commonly provided with holes into which an adjusting pin, called a *tommy bar*, may be fitted, and, being rotated in the manner of a capstan, are usually referred to as *capstan screws*. Careful inspection may be required, to determine the correct way to turn a capstan screw so as to reduce an error discovered.

The length of time an instrument will maintain its adjustments to the accuracy with which they are made depends both upon the instrument and on the manner in which it is handled. Some adjustments are more important than others, and should be tested every day when precise work is being done. However, to obtain the best results in surveying, it should be assumed that no instrument is free from error, and the routine of observing should be arranged so as to eliminate the effects of any residual errors, or of any defects in non-adjustable parts.

PARTS COMMON TO SEVERAL INSTRUMENTS

5. Before the design and use of particular instruments is considered, some features common to various instruments will be treated separately. Parts thus dealt with will be optical components, telescope, bubble tube, vernier, micrometer systems, magnetic needles, and tripod stand.

In the majority of surveying instruments, the small fittings may be broadly classed according to their functions as optical parts or as measuring parts. In many instruments, such as the theodolite and the level, a telescope assists the eye in obtaining distinct vision of distant objects and furnishes a definite line of sight of which the direction and inclination are either known or to be measured. The provision of a telescope in such instrument is not absolutely essential, but the utility and accuracy of the instrument would be greatly impaired without it. Some instruments, however, such as the tachymeter, the photo-theodolite and the sextant, depend for their action upon certain laws of optics, and in these instruments the optical parts are essential and assume the character of measuring parts. Optical systems for reading divided scales are now widely used, not only in precise theodolites, but in some instruments of quite modest orders of accuracy, because these systems can facilitate, and greatly speed up, the observer's work.

OPTICAL COMPONENTS

6. Almost all surveying instruments contain components such as lenses and mirrors. Many modern instruments contain dozens of pieces of glass which perform various functions that can be understood in terms of optical principles.

7. Mirrors. The simplest optical components of instruments are plane mirrors, whose function is to change the direction of a beam of light. A polished metal surface will do this, as indicated in Fig. 3, (a) and (b). The direction of the reflected beam is determined by the well-known simple rule that the incident and reflected rays make equal angles with the surface. This also leads to the rule relating to the image point of a source of light as at A in Fig. 3 (b):

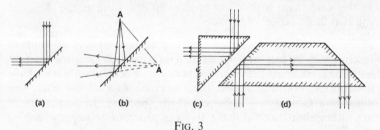

FIG. 3

after reflection, the rays of light are proceeding as if they had come from the image point A', which is situated so that AA' is perpendicular to the mirror surface and bisected by it. Polished mirrors are used in some simple stereoscopes, for instance. However, a metal surface exposed to the air may rapidly become tarnished and lose much of its reflecting power, so such mirrors are not likely to be found in instruments for use in the open air.

To overcome this disadvantage, the deflection of beams of light is usually done by internal reflection in a block of glass, as indicated, in a very simple way, in Fig. 3 (c). Blocks of glass with suitably positioned plane polished surfaces are very extensively used in modern theodolites; a typical component is illustrated in Fig. 3 (d).

8. Prisms. A prism is a block of glass with two plane faces that are not parallel: it deflects a ray of light as indicated in Fig. 4 (a).

Prisms suffer from the disadvantage arising from the well-known fact that light of different colours is differently deflected. For this reason, prisms of large deflection are not normally found in surveying instruments, though prisms of very small deflection have some uses.

9. Parallel Plate. An optical component that is widely used nowadays is the parallel plate, which can be employed to displace a beam

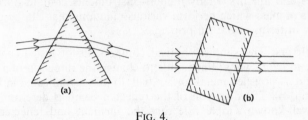

FIG. 4.

of light, without changing its direction, as indicated in Fig. 4 (b). In this case there is no effect of colour dispersion, as the effects at the two faces cancel out.

10. Lenses. These are of course very extensively used in all kinds of instruments. A *lens* may be defined as a portion of a transparent substance enclosed between two surfaces of revolution which have a common normal, termed the *Principal Axis* of the lens. The curved surfaces are of spherical shape, because the formation of any other shape would make the lens much more expensive: one face of a lens may actually be plane.

Lenses are classified as convex or concave. The various forms are shown in Fig. 5, these being distinguished as: (*a*) double convex or

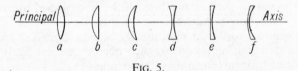

FIG. 5.

biconvex; (*b*) plano-convex; (*c*) concavo-convex or positive meniscus; (*d*) double concave or biconcave; (*e*) plano-concave; (*f*) convexo-concave or negative meniscus. In the cases of (*a*) and (*d*) the curvatures of the two surfaces may be equal or unequal.

The *Optical Centre*, or simply the centre, of a lens is that point on the principal axis through which pass all rays having their directions parallel before and after refraction. Such a ray actually undergoes a small lateral displacement, but, unless the lens is very thick, it is convenient to assume that the emergent ray is in the same straight line as the incident ray. The optical centre is so situated that its distances from the surfaces are directly proportional to their radii of curvature. In the cases of double convex and double concave lenses the centre lies within the thickness: in plano-convex and plano-concave lenses it is situated on the curved surface, while in meniscus lenses it is outside the lens and on the same side as the surface of smaller radius.

A *Secondary Axis* is any straight line, other than the principal axis, passing through the centre of a lens.

11. Refraction through Lenses. The nature of the refraction of rays traversing a convex lens is such that a beam of light is on emergence more convergent, or less divergent, than at incidence. Such lenses may be referred to as converging lenses in contra-distinction to concave or diverging lenses, which have the opposite effect. The influence of the curvatures of the surfaces of a lens is to produce different angles of deviation on the individual rays of a beam according to the positions of their points of incidence, and a result of this is that if all the incident rays are concurrent through a single point then the refracted rays, produced if necessary, will pass more or less exactly through some other point on the same axis through the centre of the lens.

In Fig. 6 (a), a parallel beam of light is shown incident upon a convex lens, in this case the source of the light being an infinitely

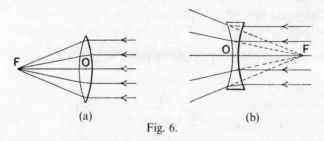

(a)   (b)

Fig. 6.

distant point on the principal axis. The rays, after refraction, converge to a point F, also on the principal axis. This point is termed the *Principal Focus* of the lens, and, in the case of a thin lens, its distance from O, the optical centre, is called the *Focal Length*, a quantity which will be designated by the letter f. In the case of the concave lens (Fig. 6 (b)), the parallel rays are, after refraction, made to diverge from the principal focus, which lies on the same side of the lens as the incident beam.

The principal focus may be regarded as the point at which there is formed an image of the distant source of light. In the case of the convex lens, the refracted rays actually pass through F, and the image is real, so that it could be received on a screen. With a concave lens, however, the diverging rays have to be produced backwards to locate the point of divergence, F: the image in this case has no real existence, and it is said to be a virtual image.

The *Power* of a lens is the reciprocal of its focal length, and is considered to be positive for a convex lens and negative for a concave one. The unit of power is the *Diopter*, which is the power of a lens having a focal length of one metre. Thus, a convex lens of 20 cm. focal length has a power of 5 diopters.

If two or more thin lenses are in contact, the power of the combination is the algebraic sum of the powers of the individual lenses. If the focal lengths of the several lenses in contact are f', f'', etc., the relationship

$$\frac{1}{f} = \frac{1}{f'} + \frac{1}{f''} + \cdots$$

enables the focal length f of the combination to be obtained. Thus, if a convex lens of $7\frac{1}{2}$ in. focal length is cemented to a concave lens of 30 in. focal length, the focal length of the combination is given by

$$\frac{1}{f} = \frac{1}{7\frac{1}{2}} - \frac{1}{30},$$

whence $f = +10$ in., the resulting compound lens being converging.

12. Conjugate Foci. If rays of light proceeding from a point P on the principal axis are refracted through a lens, the emergent rays pass through, or rather very close to, another point P′, also on the principal axis (Fig. 7). The object point P and the image P′ are

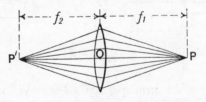

FIG. 7.

termed *conjugate foci*. If the distances from O to P and P′ are denoted by f_1 and f_2 respectively, it may be shown that

$$\frac{1}{f_1} + \frac{1}{f_2} = \frac{1}{-},$$

and this formula can be applied to any type of lens and any positions of conjugate foci provided that a suitable convention of signs is adhered to. The focal length of a convex lens is positive, and that of a concave lens is negative: the distance f_1 from the lens to the source of light is positive if measured against the direction of travel of the light, while f_2 is positive if measured with the travel of the light. For instance, f_1 is negative if the lens receives an already convergent beam which, in the absence of the lens, would come to a focus at a point beyond the position of the lens.

Source and image are related reciprocally, in the sense that the direction of travel of the light is reversible through an optical system.

The lens equation given above is applicable when the thickness of the lens is small compared with the conjugate distances: quantities computed from the formula will be accurate enough for considering the geometrical optics of simple telescopes, etc. The lens formula does not apply to conjugate distances along secondary axes; it applies to distances measured along the principal axis.

The reader may accustom himself to the sign convention by verifying the following positions of the image for the given positions of an object point in the cases of lenses of 10 in. focal length. When the object point is behind the lens, it is to be understood that the rays incident upon the lens converge towards that point, but do not actually pass through it.

Object Point at	Convex Lens Image at	Concave Lens Image at
Infinity in front	10 in. behind	10 in. in front
20 in. in front	20 in. behind	6·7 in. in front
10 in. in front	infinity	5 in. in front
10 in. behind	5 in. behind	infinity
20 in. behind	6·7 in. behind	20 in. in front
Infinity behind	10 in. behind	10 in. in front

13. The Image of a Body. Considered optically, a body is simply a collection of points. Cones of rays proceeding from these points, if incident upon a perfect lens, are each brought to a focus, thereby forming images of the points, and in the aggregate an image of the body. To determine the position of the image of any point on the body, it is evidently sufficient to trace the paths of any two rays from it to their intersection after refraction. Three rays are very easily located: these are the ray that passes through the centre of the lens, and the two rays each of which passes through one of the principal foci. These are shown in Fig. 8, which illustrates the formation, by a convex lens, of a real inverted image of an object situated at a distance from O greater than f. The distances f_1 and f_2 along the principal axis are related by the lens equation for conjugate focal distances, and it is evident, from the similarity of the triangles Oab and OAB, that the ratio of the size of the image to that of the object is f_2/f_1: this ratio is the *optical magnification*

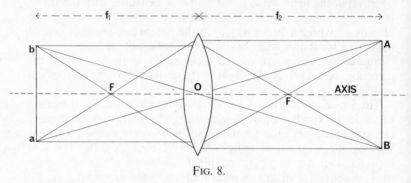

FIG. 8.

and if written $-f_2/f_1$ it indicates that the image is inverted in relation to the object.

14. Defects of a Single Lens. A simple lens is found to have several inherent imperfections affecting the formation of the image, and it is necessary to consider the means adopted to reduce these defects, in order to understand the construction of the telescope. Some of the errors to which lenses are subject need not be emphasised as, although their correction is of considerable importance in the photographic lens, they are of little consequence in the surveying telescope, the aim of which is to afford a good image of the central portion only of the field of view.

The principal defects are: (*a*) Chromatic Aberration; (*b*) Spherical Aberration; (*c*) Coma; (*d*) Astigmatism; (*e*) Curvature of field; (*f*) Distortion. The first two are by far the most important, since they affect the sharpness of the image at the principal axis. The others do not influence the image at the centre of the field, but are included on account of their possible effects in stadia tachymetry and in certain astronomical observations in which the image of a star is observed outside the principal axis of the lens system of the telescope.

15. Chromatic Aberration. White light may be regarded as composed of a series of undulations, of different wave-lengths. Difference of wave-length corresponds to difference of colour and also of refrangibility, and, as a consequence, a ray of white light after refraction at a single surface has its components separated from each other. The visible rays which have the shortest wave-length and greatest refrangibility are those of a violet colour, and wave-length increases and refrangibility decreases, through the range of the colours of the spectrum–violet, blue, green, yellow, orange and red.

Chromatic Aberration is that defect of a lens whereby rays of white light proceeding from a point are each dispersed into their components and conveyed to various foci, forming a blurred and coloured image. In Fig. 9, which illustrates the dispersion produced

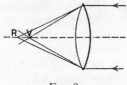

FIG. 9.

by a simple convex lens, it will be seen that a sharp image of the radiant point is nowhere formed. A screen placed at V receives an image surrounded by a halo of all the colours of the spectrum and bounded by a red fringe, and on removal of the screen to R the halo has a violet margin. If the rays of white light, instead of proceeding from a point, emanate from a white body, the middle portion of the image of the body is white, because the individual colours are recombined there, and the chromatic effects are seen only at the boundary of the image.

The circumstance that in different varieties of glass dispersive power bears no relation to refractive power renders the use of compound lenses available as a means of correcting the defect. Two kinds of glass are employed, crown and flint, both of which are manufactured in a considerable range of optical qualities. Flint glass has slightly the greater refracting power, but its dispersive power is roughly double that of crown. By combining a crown glass convex lens with a concave lens of flint glass, as in telescope objectives (Fig. 15), the proportions are arranged not only to yield the required focal length, but also so that the dispersion produced by the convex lens is largely neutralised at emergence from the concave.

Practically, however, complete achromatism cannot be achieved by the use of only two kinds of glass, since the ratio of their dispersive powers varies at different parts of the spectrum. A certain amount of residual colour, or secondary spectrum, is unavoidable, but does not have a serious effect on the sharpness of the image.

A lens which is corrected for two colours of the spectrum is said to be achromatic, although it is only incompletely so.

In eyepieces the correction for chromatic aberration is commonly made without compound lenses by the use of two plano-convex glasses of such proportions, and at such a distance apart, that the dispersion produced at the first is eliminated by the second.

It can be shown that, if a combination of two thin lenses of focal lengths f and f' and separated by a distance d, is to be achromatic, d must satisfy the relation: $d = \frac{1}{2}(f + f')$. As d is necessarily a positive quantity, one or both of the lenses must be convergent.

16. Spherical Aberration. Spherical aberration is a defect whereby the component rays of a beam proceeding from a point on the principal axis are not refracted to pass through a single point, but are differently focused according to their positions of incidence on the lens. It arises from the use of spherical surfaces; curved surfaces of any other shape are much more difficult to produce.

Fig. 10 illustrates spherical aberration in a convex lens, and

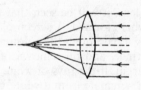

FIG. 10.

shows that rays passing through the margin are brought to a focus nearer the lens than are the central rays, so that the image formed is not sharp. The relative amount of error present in lenses of a given power is largely dependent upon the relationship between the curvatures of the two surfaces and upon the direction of the incident light. The surface of smaller radius should face the more nearly parallel rays. Thus, if parallel light is incident upon the curved surface of a plano-convex lens, the aberration produced is only about a quarter of what it would be if the plane surface faced the source of light.

Spherical aberration may be reduced to any required extent by diminishing the aperture so as to cut out the marginal rays but this has the serious disadvantage of reducing the brightness of the image. The correction is actually effected by replacing the simple lens by a combination of two lenses such that the positive aberration of one is neutralised by the negative aberration of the other. The combination may consist of a convex lens in contact with a concave, as in telescope objectives, or of two plano-convex lenses placed at a certain distance apart, as in eyepieces. A lens or combination of lenses in which the defect is eliminated is said to be *aplanatic*.

The theoretical condition to be satisfied in order that a combination of two thin lenses, of focal lengths f and f' and separated by distance d, should be aplanatic is given by: $d = f' - f$. Hence, if the combination is to be both achromatic and aplanatic, we must have: $d = \frac{2}{3}f' = 2f$. These are the proportions used in the Huygens' telescope eyepiece, but this kind of eyepiece is unsuitable for use in telescopes of measuring instruments that need diaphragms to determine line of sight.

17. Coma. The spherical aberration described above refers to an image produced on the principal axis, and may be distinguished as axial spherical aberration. The remaining defects to be noticed result from the spherical aberration of oblique pencils of light, the axes of which are secondary axes of the lens.

Rays emanating from a point on a secondary axis do not fall upon the lens in the symmetrical manner of an axial pencil. In consequence, the resulting image is neither sharp nor with its confused outline of circular form as in axial aberration. The image of a point source of light takes various forms, such as pear- or comet-shaped, with an axis of symmetry directed towards the principal axis of the lens. The effect of oblique spherical aberration is known as *coma*.

This defect impairs the sharpness of an image away from the principal axis. It may be reduced considerably by a moderate decrease in the aperture of the lens, but in the surveying telescope the resulting loss of brightness would be more objectionable than imperfection of the image towards the margin of the field.

18. Astigmatism. This effect of oblique spherical aberration is caused by the lens refracting the rays from an extra-axial point so that, instead of passing through a focal point, they pass through a focal line. The directions of the refracted rays are such that, on travelling onwards, they pass through a second focal line at right-angles to the first and radial to the principal axis. Between these two lines, the best image of the point is obtained as a circular disc.

If the object consists of a sheet of paper on which are ruled two lines, one radial and the other tangential, a satisfactory image of one of the lines will be received on a screen placed at one of the focal lines, but the image of the other object line will be confused. On moving the screen to the other focal line, the lack of definition applies to the image of the first object line. Astigmatism therefore prevents sharp definition in all directions simultaneously. Its correction in a system corrected for axial aberration necessitates the use of at least three component lenses. A lens combination corrected in this respect is said to be *anastigmatic*.

19. Curvature of Field. Curvature is a further effect produced by the spherical aberration of oblique rays whereby the image of a plane surface normal to the principal axis is formed as a curved surface. When received upon a plane screen, such an image is not equally sharp all over but, according to the position of the screen, will be distinct at the centre or the margin or along an intermediate circle. The two sets of line-foci, as well as the discs of least confusion, formed by a lens uncorrected for astigmatism really lie upon curved surfaces, but curvature may also be present in the anastigmatic lens.

If the object glass or eyepiece of a telescope possessed the defect in a marked degree, it would become necessary to adjust the focus to obtain a clear view of different parts of the field. Curvature may be reduced by the use of a diaphragm or by means of two lenses placed at a suitable distance apart, as in the eyepiece of the surveying telescope.

20. Distortion. Distortion is the defect, arising from the same cause as the last, whereby straight lines on an object are not reproduced as straight lines in the image. It is always present in single lenses and in achromatic combinations of two lenses.

The error can be reduced by covering the margin of the lens by a diaphragm placed on the surface of the glass, but it is accentuated if the stop is at some distance from the lens. The action of a stop set at a distance from the lens is illustrated in Fig. 11, which shows

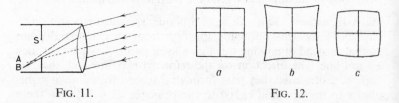

Fig. 11.   Fig. 12.

that rays passing through or near the centre of the lens may be intercepted by the stop S and prevented from contributing to the formation of the image at its true position A, while the marginal rays pass through the aperture of S, and, because of their greater refraction, cause the image of the point to be formed at B. If the object is rectilinear in outline (Fig 12 (a)), the sides of the image will in these circumstances be convex inwards, giving rise to what is termed pin-cushion distortion (b). The opposite effect is obtained when the stop is placed in front of the lens, and barrel distortion (c) is then produced. Since the defect does not influence the centre of the field, and is therefore of minor importance in the measuring telescope, the use of two compound lenses with a stop between is not warranted in the telescope objective, as it is in the photographic lens.

TELESCOPES

21. Types of Telescopes. In its simplest form, a refracting telescope consists optically of two lenses, the principal axes of which coincide as the optical axis of the telescope. The lens nearer the object to be viewed is convex, and is termed the object glass or objective, the function of which is to collect a portion of the light emanating from the object. The rays transmitted by it, passing onwards, suffer a second refraction at the other lens, called the eyepiece or ocular, and are then suitably presented to the eye. According to the form of the eyepiece, and its position relative to the object glass, a telescope may belong to one or other of two main types.

In *Kepler's* or the *Astronomical* telescope (Fig. 13), rays from the

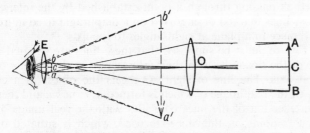

Fig. 13.

object AB are, after refraction at the objective O, brought to a focus before they enter the eyepiece E, and in consequence a real inverted image *ba* is formed in front of the eyepiece. If this lens is so placed that *ba* is situated within its focal length, the rays after refraction at E appear to the eye to proceed from *b'a'*, a virtual image conjugate to *ba*. The object AB thus appears magnified, inverted, and placed at *b'a'*.

In *Galileo's* telescope (Fig. 14), the rays refracted by the objective O are intercepted by a concave eyepiece E before a real image is formed. On entering the eye, they therefore appear to diverge from the virtual image *ab*, which is magnified and erect.

Both these arrangements meet one of the requirements of a telescope for use in surveying, since they afford distinct vision of distant objects, and from this point of view Galileo's telescope has the advantage that the image is erect. But the telescope, as an adjunct to a measuring instrument, must also be capable of furnishing a definite line of sight, and it will be shown that only the Kepler telescope fulfils this condition.

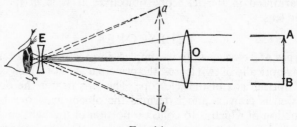

FIG. 14.

22. Line of Sight and Line of Collimation. By a line of sight of a telescope is meant any line passing through the optical centre of the objective, traversing the eyepiece, and entering the eye. *The* line of sight, or line of collimation, is one such line further and visibly defined as passing through a point established by the intersection of cross-lines marked or engraved on a diaphragm fixed in front of the eyepiece in a plane at right-angles to the axis.

In Fig. 13, let it be supposed that these lines are situated in the plane of the image ba with their intersection at the point c on the optical axis. The line of sight cO meets the object in a certain point C, the real image of which is formed at c on ba, and therefore the intersection of the lines coincides with the real image of that particular point on the object focused, which is situated on the line of collimation. Now, on viewing the image ba through the eyepiece, the crosslines are simultaneously seen focused in $b'a'$, and the coincidence of the plane of the lines with that of the real image ensures that the observed position of the lines relative to the points on the virtual image $b'a'$ is the same as on the real image. The lines and the image ba are equally magnified, and distortion or other defect produced in the passage of the rays through the eyepiece affects both to the same degree. The observer consequently sees the intersection of the lines apparently coinciding with that point of the object which lies on the line of sight.

The establishment of a telescopic line of sight therefore involves two essential conditions:

(*a*) A real image must be formed in front of the eyepiece,

(*b*) The plane of the image must coincide with that of the cross-lines (see Parallax, section 28).

In Galileo's telescope no real image is produced, and it is therefore useless for quantitative observations. Cross-lines are sometimes fitted in such telescopes, but they serve merely to indicate the centre of the field. Galileo's telescope may therefore be regarded as a viewing, as distinct from a measuring, instrument. It is adapted for field-glasses, and is used in sextants to facilitate the sighting of distant points.

23. The Surveying Telescope. The telescope of a surveying instrument is therefore of the Kepler type. It must be provided with some mechanism whereby an object at any distance beyond, say, 10 or 15 ft. may be clearly focused on the diaphragm.

The simplest form of telescope consists of an object glass, a diaphragm with cross-lines, and an eyepiece. The objective is a compound lens, of which the outer component is a double-convex lens of crown glass, and the inner is of flint glass, convexo-concave. As illustrated in Fig. 15, the objective is at the front end of a tube

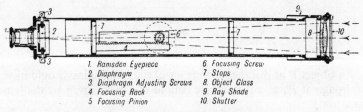

1. Ramsden Eyepiece
2. Diaphragm
3. Diaphragm Adjusting Screws
4. Focusing Rack
5. Focusing Pinion

6. Focusing Screw
7. Stops
8. Object Glass
9. Ray Shade
10. Shutter

FIG. 15—EXTERNAL FOCUSING TELESCOPE.

which is a close sliding fit inside the fixed body-tube of the instrument, and the focusing movement required for placing the real image in the plane of the cross-lines is effected by movement of the sliding tube by a rack and pinion.

This design of telescope is however almost obsolete, having been superseded by the internal-focusing telescope as illustrated in Fig. 16. In this telescope, the objective is at a fixed distance from the

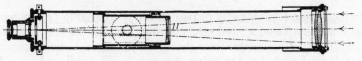

FIG. 16.—INTERNAL FOCUSING TELESCOPE.

cross-lines, and focusing is done by the sliding of a third, usually divergent, lens which is situated at about the middle of the tube. With this arrangement, the telescope tube can be almost air-tight, keeping away dust and moisture from moving parts.

Referring to Fig. 17, let O and I be the optical centres of the objective and internal lens respectively, and let f and f' be their focal lengths. The distance x between the lens centres is variable, and for any observation the internal lens must be adjusted into such a position that the image of the distant object is focused in the plane of the cross-lines at P_1, which is at a fixed distance l from O.

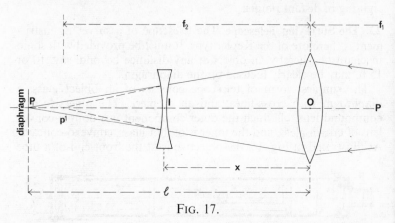

FIG. 17.

An object P at distance f_1 in front of the object glass would have an image at P', at a distance f_2 behind this glass given by the lens equation:

$$\frac{1}{f_2} + \frac{1}{f_1} = \frac{1}{f}$$

However, the divergence of the focusing lens causes the beam to converge further on, at P_1 on the diaphragm. In relation to the divergent lens, therefore, P' is the 'virtual object' and P_1 is the real image: the relationship derived from the lens formula is therefore:

$$-\frac{1}{f_2 - x} + \frac{1}{l - x} = -\frac{1}{f'}$$

In focusing on an infinitely distant object, f_2 is equal to f and x has its minimum value x_o, and a relationship between the four quantities $f f' l x_o$ is obtainable from the above equation. As an example, it will be found that the values $f = 8$, $f' = 20$, $l = 9$ and $x_o = 4$ satisfy the relation. If the focusing lens is moved so that $x = 5$, f_2 is found from the above equation to be $8\frac{1}{3}$ and f_1, from the previous equation, is 200. Thus, taking the numbers above to be inches, it is seen that a range of 1 inch for the sliding of the focusing lens is sufficient for focusing at all distances beyond $16\frac{2}{3}$ ft.

24. With the object of improving the distinctness of the image in a telescope, the interior of the body is painted dull black to prevent reflections from internal surfaces, and annular stops may be fitted to intercept rays not required for the formation of the image. In inferior instruments, the front stop may be found placed near the objective for the purpose of diminishing spherical aberration, but the reduction of aperture thereby occasioned is a serious defect.

A cap is usually provided to protect the objective when the instrument is not in use: a tubular shade may also be provided so that direct rays from the Sun can be kept off the glass and thus prevented from causing glare and optical confusion in the telescope.

The end of the body tube remote from the objective is often of smaller diameter than the objective end. It carries fittings to hold the diaphragm and the eyepiece: the diaphragm must be capable of lateral adjustment, and the eyepiece must have a short range of axial movement for focusing the cross-lines.

A Sun-glass may be provided for solar observations. This is a piece of very dark glass fixed in a ring which may be fitted over the eyepiece, or in some instruments the dark glass is permanently attached to the eyepiece and can be turned into position when needed, an excellent arrangement because a separate glass is easily mislaid.

Immediately in front of the eyepiece is the diaphragm. In older instruments this had spider webs stretched across it but usually nowadays it is a thin plate of glass with lines engraved on it and filled with dark material.

25. Diaphragm and Reticule. This is usually mounted on a ring which is held in place in the telescope tube by four screws, as shown

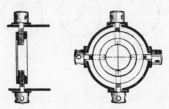

FIG. 18.—DIAPHRAGM.

in Fig. 18. The screws pass through slots, and some rotational adjustment of the diaphragm is possible as well as the necessary lateral movements to get the cross-lines properly positioned. In many modern instruments, the diaphragm is not mounted directly on the ring, but is on a separate cylindrical cell which fits frictionally into the ring, and can be easily removed and replaced by screwing an extractor device on to it. See Fig. 19. The advantage of these interchangeable cells is that if one has to be replaced, little or no re-adjustment of the holding ring will be required, because the manu-facturer should have taken care that the cross-lines are centrally placed on the cell.

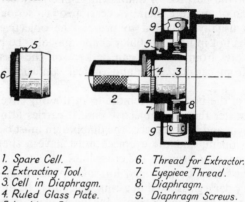

1. Spare Cell.
2. Extracting Tool.
3. Cell in Diaphragm.
4. Ruled Glass Plate.
5. Locking Pin.

6. Thread for Extractor.
7. Eyepiece Thread.
8. Diaphragm.
9. Diaphragm Screws.
10. Cover Ring.

FIG. 19.—INTERCHANGEABLE DIAPHRAGM.

Many different geometrical arrangements of the reticule of lines on glass diaphragms are to be found, and a selection of designs is illustrated in Fig. 20. Design A is commonly used in Levels; the observer can set the staff between the two vertical lines and can readily see if it is not vertical: B and C are not common nowadays,

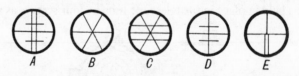

FIG. 20.

but are preferred by some users who like to set signals in the V corners: in designs A and D the stadia lines are distinguished from the centre line by being short and thus helping to avoid confusion; this is probably the design most commonly found in smaller instru-ments: the two close parallel lines in design E are useful for setting on long fine signals, like plummet strings or distant ranging poles which would be completely covered by a single line; some obser-vers prefer, in any case, to set between two lines rather than on one.

26. Eyepieces. The type of eyepiece used almost exclusively in this country on surveying telescopes is the Ramsden Eyepiece. As indicated in Fig. 21 (a), it consists of two equal plano-convex lenses with the curved surfaces facing each other and separated by a distance equal to $\frac{2}{3}$ of the focal length of either lens. If the distance between the diaphragm and the front lens of the eyepiece is $\frac{1}{4}$ of the focal length of a lens, the rays from a point on the diaphragm

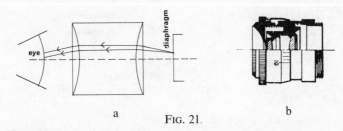

a   FIG. 21.   b

will enter the eye as a parallel beam. This eyepiece is well corrected for spherical aberration and curvature of field but is not strictly achromatic. In most modern instruments, the focusing of the cross-lines is done by rotating the eyepiece in a high-pitched thread (Fig. 21 (b)), and there are graduation marks that move against a fixed mark so that the observer may note and remember the setting that best suits his eyesight.

It is possible to have an eyepiece that gives an erect image, but this needs more lenses: all surveyors soon become accustomed to seeing everything upside-down!

27. Diagonal Eyepiece. When the telescope is at high inclination, it is inconvenient or impossible to sight directly into the ordinary eyepiece, and some instruments are provided with special eyepieces that turn the line of sight through a right-angle, as indicated in Fig. 22. Some of these attachments make it possible to set and read elevations up to 90°. However, the modern instrument with its additional eyepiece for the optical reading system has introduced difficulties in this respect, and some of the arrangements for taking high elevations are far from successful. Probably the best solution, if frequent observations of high elevations are necessary, is to use a theodolite with the setting eyepiece at the end of the transit axis and the micrometer reading eyepiece independent of the telescope.

28. Parallax and Focusing. It is important for the surveyor to note that the focusing of the observed object and the focusing of the cross-lines are completely independent operations. If the image of the object does not fall on the plane of the cross-lines, parallax

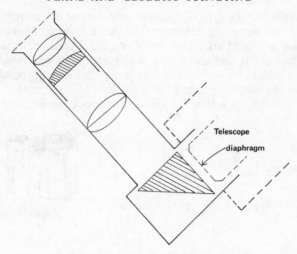

FIG. 22.—DIAGONAL EYEPIECE.

exists and can only be removed by correcting the focusing: parallax has nothing to do with the eyepiece. To bring a telescope into proper adjustment for sighting, the observer should first point it to the sky, or hold a piece of white paper in front of the object glass, and adjust the eyepiece for clear vision of the cross-lines. Then, the telescope may be turned in the desired direction and the object or signal brought to clear vision by use of the telescope focusing screw. If the observer moves his eye slightly from side to side, or up and down, and finds that the observed signal apparently moves in relation to the cross-lines, then parallax is present. This can *only* be eliminated by re-focusing until there is no apparent relative movement. After removing parallax, the observer may find that a slight re-adjustment of the eyepiece will give more comfortable vision. The best setting of the eyepiece is such that the observer feels no strain of muscular accommodation if he momentarily shuts his eye, or turns aside, and then looks again into the telescope.

29. Telescope Functions. The important functions of a telescope on a surveying instrument are to define a line of sight, and to provide resolving power, that is, ability to distinguish between directions to objects very close together at considerable distances. The un-assisted human eye cannot distinguish between marks unless their directions differ by at least one or two minutes of angle: the 0·01 ft. markings on a levelling staff placed at a distance of 20 ft. are separated by 1/2000 radian, which is nearly 2 minutes.

Resolving power depends primarily on the diameter of the aperture that receives the light: the pupil aperture in the case of the eye, or the diameter of the object glass of the telescope, provided that the whole area of the lens is effective.

There is no point in designing a theodolite to read to the accuracy of a few seconds unless it has a telescope of comparable resolving power. Four seconds is about 1/50,000 radian, so, at this accuracy, an object glass some 25 times the diameter of the pupil of the eye should be required. Telescopes of ordinary theodolites and levels have objectives about $1\frac{1}{2}$ inches wide, giving a theoretical resolving power of some 3 or 4 seconds.

30. Telescope Quality. Besides adequate resolving power, a telescope should give a clear undistorted picture of an object, free from haziness and coloured edges. A telescope will not give accurate results unless these conditions obtain at least near the centre of the field of view. The lenses must be correctly designed and accurately centred along a common axis.

Some lenses have specially treated surfaces to minimise loss of light by reflection. About 20% of the light may be lost in passage through a telescope, by reflection at surfaces, scattering by impurities and defects in the glass, etc. Brightness of the image depends on getting as much light as possible through the eyepiece, and the best results are obtained when the diameter of the beam emerging from the eyepiece–the exit pupil–is the same as that of the pupil of the eye.

31. Field of View. This is the angular diameter of the extent of view seen at one time through the telescope. Its value does not depend on the diameter of the object glass but on the diameter of the area of diaphragm 'covered' by the eyepiece. Thus, if the eyepiece field lens–the lens immediately behind the diaphragm–has a diameter of $\frac{1}{4}$ inch, and the length of the telescope is 8 inches, a rough estimate of the field of view would be 1/32 radian, nearly 2°. The exact calculation is rather complicated and of no particular interest.

There are several obvious ways for finding the field of view if it is required. The telescope can be set so that a fixed distant mark is at the apparent 'highest' point of the field, and the vertical angle read, then the telescope is moved to bring the same mark across to the opposite edge, and another reading of the vertical arc will provide the required answer. If there are stadia lines on the diaphragm, they will have a known angular separation, normally 1/100 radian, and an estimate of the angle of the whole field is easily made from this.

Most telescopes have external open sighting fittings for preliminary setting on the signal.

32. Magnification. The apparent, or visual, magnification is naturally to be defined as the ratio of the apparent diameter of a distant object seen through the telescope, to the diameter as seen directly from the same position. This ratio depends slightly on the distance of the object, but a good estimate of it is easily obtained by setting up a boldly-graduated staff about 100 ft. from the instrument and sighting it directly with one eye, and through the telescope with the other. The two pictures can be brought apparently alongside and the sizes of the graduations compared.

More precisely, if a beam of light from a distant point makes a small angle α with the optical axis of the telescope, and emerges from the eyepiece making an angle β with the axis, as indicated in Fig. 23, the observer will see the object apparently at angle β to the axis, so the apparent magnification is the ratio β/α.

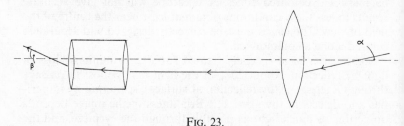

FIG. 23.

Given the optical and mechanical dimensions of the telescope, this ratio can be calculated. For the simplest kind of telescope with an objective of focal length F and single eye-glass of focal length f, the magnification ratio is just F/f. If a Ramsden eyepiece is used, the ratio is $4F/3f$, and with an internal-focusing telescope the magnification is

$$\frac{4F}{3f} \left(\frac{g+c}{g} \right)$$

where g is the focal length of the divergent focusing lens and c is the distance from the focusing lens to the diaphragm.

If the resolving power of the objective is 4 seconds and we assume that the human eye needs 100 seconds angular separation to distinguish adjacent marks, then a magnification of 25 is in order, and this is about the magnification found in ordinary surveying telescopes. There would be little point in trying to go much beyond this figure by using a stronger eyepiece, as that would reduce the illumination of the field without much improvement in the setting precision of the instrument.

33. Care of Telescope. The modern, almost airtight, telescope should require little attention beyond cleaning of the object glass and eye lens, and occasional cleaning of the glass diaphragm. Lens surfaces should be cleaned by gentle rubbing with soft cloth or chamois leather, moistened with spirit if necessary: they should not be touched by the fingers, because a trace of grease will soon collect dust.

The 'moving parts', which are the focusing mechanism and the eyepiece, should work smoothly and should stay in any position whatever the inclination of the telescope.

In tropical climates, the surfaces of soft glass may be attacked by a micro-organism, or 'fungus' as it is called. This is difficult to deal with, once it has started, but manufacturers now seem to have evolved a treatment that has much reduced the risk of this kind of damage.

BUBBLE TUBES

34. Since nearly all field measurements, angular or linear, are made in either a horizontal or a vertical plane, it is essential to have some ready method of determining the positions of these planes through any point, so that particular lines or planes in an instrument may be made to lie in them. The most convenient and sensitive device yet invented for doing this is the Bubble Tube, the action of which depends on the fact that the free surface of a still liquid, being at every point normal to the direction of gravity, is a level surface.

35. Construction. The device consists of a glass tube, shaped and carefully ground on the inner surface so that a longitudinal section of this surface is a circular arc. The radius of curvature of the arc may be as much as 100 yards. Some bubble tubes have the curvature formed all round the tube, that is they are barrel-shaped, and then they will function if turned upside-down.

The tube is nearly filled with a liquid such as alcohol, chloroform or ether: these liquids, besides being very mobile, have the advantage of low freezing-points. However, they have rather high coefficients of thermal expansion, so the bubble space is shortened in hot weather.

To indicate a central position of the bubble, some equally-spaced lines are etched on the tube on either side of the centre, or sometimes on a separate but attached scale. The lengths of the divisions are usually either 2 millimetres or 0·1 inch.

The bubble tube is usually mounted in a metal casing that leaves only the graduated part of the glass unprotected: see Fig. 24. The metal case is attached to some part of the instrument in such a way as to permit some adjustment of the position. A bubble for use independently of an instrument, as for levelling a plane-table, for instance, should be fitted into a casing with a polished plane bottom surface.

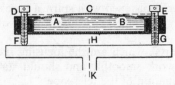

FIG. 24.

Some bubbles found on modern instruments are in non-graduated tubes. These are viewed by means of a system of prisms that cause the bubble to appear longitudinally split. Images of opposite halves of the ends of the bubble are brought to lie in juxtaposition, as indicated in Fig. 25. If such a bubble is disturbed, the two portions appear to move in opposite directions, and the observer is

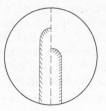

FIG. 25.

required to make some setting adjustment that brings the two images into coincidence. This method of setting a bubble into a specific position is extremely sensitive, and a tilt of one second can easily be detected.

36. Principle of the Bubble Tube. It is convenient to consider that a bubble tube has an axis, which may be taken as a straight line parallel to the free surface of the liquid when the bubble is in the central position determined by the graduation marks.

Suppose that a bubble in its casing AB is laid on a plane surface CD, and the bubble takes up the central position. If the bottom of the casing is parallel to the bubble axis, the plane surface must in fact be horizontal, as indicated in Fig. 26 (a). But if the bottom of the casing is not parallel to the bubble tube axis, the bubble may still be central although the surface CD is not horizontal, as shown in Fig. 26 (b). However, if the bubble tube is now reversed, as shown in Fig. 26 (c), the bubble will go off centre, and it is obvious that the tilt angle indicated by the bubble graduations will be twice the tilt of the surface CD. This surface can now be made horizontal by moving it so that the bubble is brought half-way back, as indicated

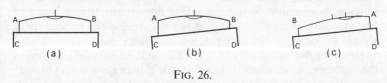

(a)   (b)   (c)

FIG. 26.

in Fig. 27 (a). If the bubble is then reversed, to the position shown in Fig. 27 (b), it will still be in the same, non-central, position in relation to the graduations.

Thus, the basic principle of the bubble tube is: *if the bubble, placed in any orientation on a plane surface, comes to rest in the same position in its tube, the surface must be horizontal, regardless of the direction of the bubble axis as defined above.*

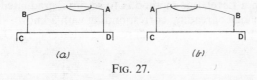

(a)   (b)

FIG. 27.

It is convenient of course, though not strictly necessary, to re-set the bubble in its mounting so that it is central when on a horizontal surface: then the 'correct' position of the bubble does not have to be marked or remembered by the observer. In any case, it is very difficult to set a sensitive bubble in exactly the right position in its casing, because the adjustment device, usually a simple screw, does not permit precision of movement.

The principle of reversal as indicated above should always be used, and it should not be assumed that the 'correct' position of a levelling bubble is necessarily exactly central.

The fact is that a bubble does not, of itself, indicate horizontality of anything it is attached to; but if a bubble is moved it will show any change of inclination. This principle is used in levelling an instrument such as a theodolite. In this operation, 'levelling' really means 'making the axis of rotation vertical', and from what has been stated above it should be clear that, if a bubble, attached anywhere on the instrument, takes up the same position in its tube at all orientations, then the rotation axis must be vertical. As before, it is convenient to fix the bubble holder so that the proper position is

central or nearly so, otherwise the levelling process is made unnecessarily troublesome.

37. Sensitivity. This refers to the capability of indicating small
changes of tilt, and it depends on the radius of curvature of the
longitudinal section of the tube. For high sensitivity, a liquid of
low viscosity and surface tension is needed, also the bubble space
should be long and the tube not too narrow.

However, the sensitivity of a bubble must be matched to that of
the instrument on which it is mounted, otherwise a surveyor may
waste time in levelling to an unnecessary precision. Thus, a bubble
on a theodolite reading to one minute need not be very sensitive.
A precise level, on the other hand, needs to be set to an accuracy of
1/1000 ft. at a distance of 200 ft., and this represents an angle of just
about 1 second. If the bubble is to run 0·02 inch for a tilt of 1 second,
its radius of curvature must be about 4,000 inches–over 100 yards.

To find the sensitivity of a bubble attached to an instrument
several methods are available, and one suitable to the type of
instrument must be chosen. If the bubble is fixed directly to a telescope, as in a Level, one method is to set up a graduated staff and
note the change of reading corresponding with a known shift of the
bubble. In Fig. 28, AB is the staff reading intercept; AEO and BFO

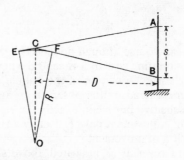

FIG. 28.

can be regarded as two positions of the line of sight and the bubble
radius perpendicular to it. It is obvious, then, that the radius of
curvature R bears the same ratio to the staff distance D as the bubble
shift EF bears to the staff intercept S. If the bubble shift is n divisions
of length d, then

$$R/D = nd/S \text{ or } R = Dnd/S.$$

Alternatively, if the staff is held at distance of $206\frac{1}{4}$ ft. from the
instrument, a staff intercept of 1/10 ft. corresponds with a tilt of

100 seconds in the line of sight, so the equivalent in seconds of a bubble division is thus easily determined.

38. Circular Bubble. This holds liquid in a circular glass container of 'pill-box' shape. The underside of the top of the container is formed to a concave spherical shape, and on the upper surface there is a central circular engraved line. The small vapour bubble left inside the container provides indication of tilt in any direction. These box bubbles are usually of low sensitivity; they are commonly fitted to tripods or instruments in order to facilitate rapid preliminary levelling, and they are also used as attachments to levelling staffs, plane-tables, etc.

THE VERNIER

39. This simple and ingenious device was invented as long ago as 1631, and it is still used widely on many types of measuring instrument. The principle of the vernier is illustrated in Fig. 29.

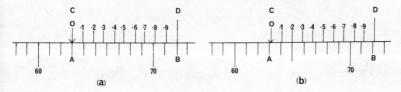

(a)   (b)

FIG. 29.

AB is a portion of the main scale, and the section marked by the letters AB covers 9 divisions. The vernier CD covers the same length but it is divided into 10 equal parts. The reading point is marked by the arrow, and the reading in Fig. 29 (a) is exactly 63. It is obvious that if the vernier is moved 1/10 of a main division to the right, the vernier graduation numbered 0·1 will now coincide with a division mark of the main scale: similarly if the vernier moves another 1/10 the graduation numbered 0·2 will make coincidence, and so on. The reading 63·2 is illustrated in Fig. 29 (b). The coincidence of graduation marks is very easily picked out by eye: direct reading to 1/10 would require the main scale to be marked at every tenth, and even if this presented no manufacturing complications the reading of such fine divisions would cause severe eye-strain. Of course, subdivision to 1/10 interval by eye estimation is not difficult, but the vernier makes it possible to provide much finer subdivision, beyond the capability of estimation. Thus, the scales of a small vernier theodolite are usually divided to $\frac{1}{2}$ degree, and reading by vernier to $\frac{1}{2}$ minute is quite straightforward. Such a vernier is illustrated in Fig. 30, where the reading, right to left as usual on a theodolite, is 345° 14½′.

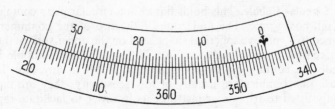

FIG. 30.—VERNIER READING TO 30 SECONDS.

The vernier principle also works if the space occupied by $n+1$ divisions of the main scale is divided into n parts on the vernier. This is shown in Fig. 31, where the reading 63.2 is again illustrated. Such a vernier must be numbered in the direction opposite to that of the main scale, and this may cause some confusion.

FIG. 31.

An *extended vernier* has graduations at more open intervals than the main scale, and this may make it easier to read. On the vernier illustrated in Fig. 32, the 60 divisions of the vernier cover 119 of the main scale, and the reading shown is 29° 54′ 10″.

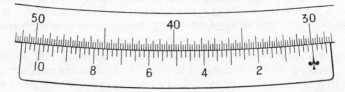

FIG. 32.—EXTENDED VERNIER.

The interval corresponding with a vernier division is sometimes called the *least count* of the instrument.

It is common practice to fit a small magnifier lens to assist the reading of the verniers on surveying instruments.

Verniers reading both ways will be found with the vertical circles of vernier theodolites.

40. Reading a Vernier. The observer who is not familiar with a particular vernier instrument should examine it carefully, because it is easy to make mistakes in reading these devices.

The observer should:
 (i) note the value of a division of the main scale, and of a division of the vernier,
 (ii) note whether the reading of the vernier is direct or retrograde: most verniers are direct-reading.
(iii) find the fiducial mark against which the reading is to be taken,
(iv) estimate the reading on the main scale, being careful not to omit a whole division and not to read the scale in the wrong direction,
 (v) examine the vernier scale to find the place of coincidence of graduations, and note the reading.

There may be doubt as to which graduation is making coincidence. If two adjacent marks seem to be displaced in opposite directions, a mean reading may be noted. More likely, two consecutive graduations may seem to coincide with main scale graduations, or the coincidence may seem to occur at different places depending on the direction from which the observer is looking. This may mean that the vernier is rather too finely divided, or the instrument is worn so that the two graduated edges are not in close relationship. Some guidance may be obtained by looking at graduations on either side of the coincidence position.

THE MAGNETIC NEEDLE

41. A magnetic needle is the essential component of the various forms of compass used by surveyors, and it is also a convenient accessory with several other instruments, notably the plane-table and the theodolite. It consists of a thin strip of magnetised hard steel supported on a hard sharp steel pivot, usually with a jewel or agate bearing on the magnet itself.

In the Earth's magnetic field, such a magnetic needle tends to move into a position lying in the magnetic plane, thus indicating the direction of magnetic north-south, which does not coincide with the true or geographical meridian. The difference is the *magnetic variation*, also called *declination*, referred to below (Sec. 199).

If a magnetic needle is supported at its centre of gravity, it will not float horizontally, but one end will tend to dip down towards the nearer magnetic Pole of the Earth. At present (1966) the dip in England is about 66° below horizontal. To counteract this, the usual method is to pivot the magnet at its centre before it is magnetised and then balance it by means of a suitable riding weight, held on by friction, which can be moved along the needle to the position required to make it float horizontally.

A magnet designed to be observed directly against a scale or zero mark is usually in the form of a narrow strip with its broader faces vertical: a widened portion in the middle carries a brass cell containing the hollowed bearing-stone: each end of the magnet is formed to a fine point or edge.

The rotating graduated disc, found in small prismatic compasses, can have two or more magnets of simple shape fixed to its underside, as they need not be centrally placed. A magnet used for directing an instrument into the magnetic meridian, as fitted to some theodolites, can be pivoted in a long narrow box.

In any case, the magnet must be raised from its support when not actually in use, and mechanisms for doing this should always be provided. In some magnets, the raising mechanism is automatically actuated when the cover of the box is replaced. It is helpful if the raising mechanism can also be used as a brake so that oscillations can be damped down and a reading or setting obtained as soon as possible. If a magnetic needle comes to rest very readily, its pivot is probably broken or dirty! Some compasses are filled with liquid, which serves the double purpose of damping oscillations and taking most of the weight off the pivot, so that a raising mechanism may not be needed.

The observer using a magnet to give direction must of course be careful to avoid the vicinity of iron and steel structures, and to keep his penknife at a reasonable distance from the instrument. On some kinds of engineering work, the use of magnetic instruments is quite out of the question.

THE TRIPOD STAND

42. The most important requirement of a tripod is rigidity or steadiness, after which lightness and portability are desirable. The legs may be solid pieces of wood or light metal, or they may be built as open frameworks. Rigidity of a tripod depends largely on its having wide hinges at the tops of the legs, so the open frame legs are generally to be preferred for precise work. Telescopic legs make for portability, but must be very well designed if they are not to be shaky when opened out to full length.

The lower ends of the legs should have pointed steel shoes for thrusting into the ground, and the shoes should be provided with tongues of metal so that the surveyor can use his foot to press them down. Rubber covered shoes may be safer on concrete and smoother surfaces.

At the top of the tripod is a frame or casting on which the legs are hinged. The hinges may be tightened with wing-nuts, or by a screw which tightens all three legs simultaneously. The connection of the legs to the head piece is the weak point of many tripods, rapid development of looseness in the screws necessitating their frequent tightening.

Various forms of tripod head are found with modern instruments. The flat top with a central screw is suitable for plane-tables and is used also for theodolites, because it is a simple matter to provide an inch or so of lateral adjustment to facilitate the precise centering of the instrument over a ground mark. Some instruments are screwed to their tripods. Manufacturers nowadays are tending to make interchangeable sets of instruments–theodolites, levels, sighting signals–which all fit on the same design of tripod head.

It is usual also to fit a small circular bubble into a tripod head so that it can be approximately levelled before the instrument is mounted.

Some tripods have an additional central telescopic 'plumbing rod': in this rather convenient arrangement, the plumbing rod is rigidly attached to the head on which the instrument is to be clamped, and is fitted into the tripod head on a ball-and-socket joint which can be tightened. The tripod is set up with the bottom of the plumbing rod on the centre-mark and is moved about until the circular bubble on the centre-piece indicates the verticality of the rod. The plumbing rod may be graduated so that the height of the instrument may be directly read off. See Fig. 263.

CHAIN SURVEYING INSTRUMENTS

43. Chain. The chain (Fig. 33) is formed of 100 pieces of straight

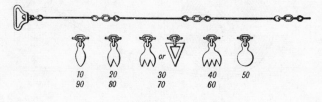

FIG. 33.—CHAIN.

wire, the ends of which are bent over and connected by small links, preferably three at each joint. Stiff steel wire, 8 to 12 gauge (0·16 to 0·10 inch) should be used. At the ends of the chain there are brass handles which are adjustable and are included in the length of the chain: the handles are on swivel joints to prevent twisting of the chain.

Chains used in English-speaking countries are of two lengths; the 100 ft. chain with 100 long links is usually called the Engineer's Chain and the 66 ft. chain with 100 links of 7·92 inches spacing is called the Gunter's Chain. When the words 'chain' and 'link' are used as indicating units of length, the Gunter's chain quantities are to be understood.

At every tenth foot or link of a chain, there is attached a distinctive tag or tally of brass, of patterns shown in Fig. 33. The arrangement of the tags is the same in relation to either end of the chain, so care must be taken in reading, particularly to avoid confusing the tags 40 and 60. Taking readings between tags, one must count from the previous tag, and estimate tenths if required. Some chains have small tags at the intermediate fives, and this is quite helpful.

In countries using the metric system, chains are mostly of 20 metres length, and are made up with 2-decimetre links with tags at 2-metre intervals. (See Appendix.)

44. Steel Band. The long steel band or tape is a ribbon of steel with brass handles attached at the ends by swivel joints. These bands can be obtained in various lengths, widths, and methods of marking divisions. For details, reference may be made to section 142. As indicated in Fig. 34, the band may be wound in a leather, plastic,

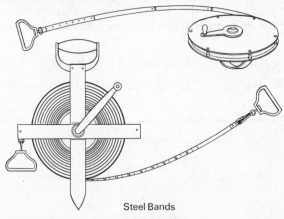

Steel Bands

Fig. 34.

or metal case, or it may be on an open-framed reel, the open reel being the most usual method nowadays. The handles may be included in the length, but for high accuracy it is best to have the terminal marks on the band itself.

45. The Foot and the Link as Units. Most engineering and constructional work is measured in feet, so any surveys required in this kind of work will be done with the 100 ft. chain or tape.

For land surveys, the Gunter's chain is sometimes preferred because it facilitates the calculation of areas, since 10 sq. chains make an acre.

Further, the Gunter's chain is a subdivision of a mile, 1/80, and for this reason it is sometimes used in setting out road and railway centre-lines.

46. Relative Merits of Chain and Steel Band. The chain is robust, and is easily repaired if broken; a broken link can be temporarily repaired, even with string. Disadvantages of the chain are that it catches easily in shrubbery and other obstructions, and its length increases rapidly due to wear at the 800 or so contact points and to stretching of the small links; it is also rather heavy and tends to collect mud, etc.

The steel band keeps its correct length very much better than the chain, and it is light in weight: it can be obtained with graduation in feet on one side and links on the other. It is easily pulled along the ground, provided the rear handle is removed. However, it is easily broken by excessive pulling or by getting twisted into loops or being run over by vehicles, and it needs more maintenance in cleaning and greasing: it is less easily read than a chain, and etched graduation marks may eventually become illegible.

All setting-out work should be done with a steel tape, as a chain cannot be regarded as accurate enough for such purpose.

47. Accessories. In high-class measurement with a steel tape, the air temperature is taken with a thermometer, the correct tension is applied with a spring balance, and special clips are used for gripping the tape at any point. These accessories and their use are described in Chapter II.

48. Testing Chains and Tapes. A chain should be tested frequently against a chain or tape of known accuracy, or against any available standard of length. Chains can be shortened by expanding some of the small links, and some chains and tapes have adjustable handles. For further details on testing, see Chapter II.

49. Arrows. These are pieces of thick steel wire about a foot long, pointed at one end, with a loop at the other, and they are used for marking the ends of whole chain lengths. They should be painted bright yellow, or have strips of coloured cloth tied to them. Longer ones may be useful for chaining in high grass.

50. Short Tapes. In ordinary surveying, short measurements are often made with a flexible tape, which is attached to a spindle in a leather, metal or plastic case into which the tape is wound when not in use. See Fig. 35. These tapes are usually $\frac{5}{8}$ inch wide, and of

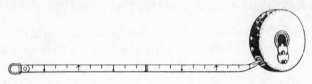

FIG. 35.—LINEN TAPE.

lengths 33, 50, 66 or 100 ft., the 66 ft. tape being the most serviceable on the whole. They are made of woven linen, painted and varnished or covered with flexible plastic, and they may have metal wire woven in, to help prevent stretching.

These tapes should not be used where precise results are required, as they are subject to serious variations in length. They stretch when pulled, and may easily become permanently elongated, since a considerable pull is needed to straighten them in wind. Exposure to wet may cause them to shrink. They soon become worn and illegible, so frequent replacement must be allowed for.

51. Ranging Poles. For the conspicuous marking of points and the ranging of lines, poles of lengths such as 6 ft., 8 ft. or 10 links are generally used. They can be obtained made of wood, metal tube, fibre-glass, etc., of circular, octagonal or square cross-section. They taper towards the top and have long steel points fitted to the lower ends. Ranging poles are usually painted in alternate bands of red and white, or red, black and white, and if these bands are made one foot, or one link, wide, the pole can be used for short measurements such as offsets.

For some kinds of surveying, short poles two or three feet long are convenient, as they need not be removed when a tripod is to be set up at the same place.

Collapsible metal tripods, constructed on the umbrella-frame principle, are obtainable for holding ranging poles vertically over ground marks on pavements or other hard surfaces.

52. Line Ranger. The line ranger is a small instrument whereby intermediate points can be established in line with two distant signals without the necessity of sighting from one of them. It consists essentially of two reflecting surfaces, either small plane mirrors or square prisms, as in Fig. 36, one above the other, and with their reflecting surfaces normal to each other.

In locating an intermediate point in line with the poles A and B, the observer stands approximately in line and places the instrument on a rod, or holds it at the level of the eye, turning it until the image of one of the poles, A, is seen in the field of view. He then moves backwards or forwards at right angles to the line until the

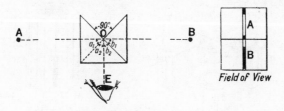

FIG. 36.—LINE RANGER.

image of B appears. A point in the line AB is reached when the images of A and B lie in the same vertical line. The reflected rays from both A and B are then situated in the same vertical plane OE. The direction of OE depends upon the position of the eye, and is not necessarily normal to AB, but, by the laws of reflection, $a_1 = a_2$ and $b_1 = b_2$, and, since $(a_2 + b_2) = 90°$, we have $(a_1 + a_2 + b_1 + b_2) = 180°$. AOB is therefore one straight line, and O can be transferred to the ground with the pole.

One of the mirrors or prisms is commonly made adjustable for securing the necessary perpendicularity between the reflecting surfaces. To perform the adjustment, three poles are ranged by theodolite, and the line ranger is held over the middle one. If the images of the others do not lie in the same vertical line, they are made to do so by turning the adjusting screw.

53. Cross Staff. The cross-staff is the simplest instrument used for setting out right angles. It is mounted on a pole shod for fixing in the ground, and is arranged in the form of a frame or box furnished with two pairs of vertical slits yielding two lines of sight mutually at right angles. Figs. 37 and 38 illustrate two common types. The open form has usually a longer base between the slits than the octagonal box pattern. The latter has additional openings, so that an angle of 45° may also be set out. In another form of the instrument, the head consists of two cylinders of equal diameter placed one on top of the other. Both are provided with sighting slits. The upper cylinder carries an index or vernier, and can be rotated relatively to the lower, to which is attached a graduated circle. This arrangement enables any angle to be set out or measured roughly.

To set out a perpendicular to a line from a point on it, it is only necessary to fix the staff in the ground at the given point and bring one pair of slits in range with the poles marking the line. The slits at right angles then define a perpendicular line of sight, in which poles or marks may be set. To drop a perpendicular to a line from a point outside it, the surveyor, holding the staff on the line, must

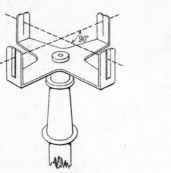
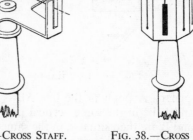

FIG. 37.—CROSS STAFF.   FIG. 38.—CROSS STAFF.

proceed by trial and error until a point on the line is obtained at which the poles in the line and the point outside it are simultaneously in the two lines of sight of the instrument.

The cross staff is non-adjustable, and is not susceptible of high accuracy, but is useful for setting out long offsets.

54. The Optical Square. This is a compact hand instrument for setting out right angles, and is capable of greater accuracy than the cross staff. It is based on the principle that a ray of light reflected successively from two surfaces undergoes a deviation of twice the angle between the reflecting surfaces. In the instrument these surfaces are placed exactly at 45° to each other, and belong to small mirrors mounted in a circular box or on an open frame, or, alternatively, they form two sides of a prism, the instrument in this case being sometimes distinguished as a prism square. The latter is the more modern and better form of the instrument.

Fig. 39 shows a plan of the essential features of the box form, the top cover being removed. The periphery is formed of two cylinders, of about 2 in. diameter and about $\frac{1}{2}$ in. deep, the one capable of sliding on the other, so that, when not in use, the eye and object openings can be closed to protect the mirrors from dust. Mirror A is silvered over half the depth only, and the other half is of plain glass. Mirror B is completely silvered. To an eye placed at E objects such as C are visible through the transparent half of A, and at the same time objects in the direction of D are seen in the silvered part after re-flection at B. If the object D is so placed that its image appears directly above (or below) C, so that the eye receives the rays from D in the same vertical plane AE as those from C, then C and D subtend a right angle at the instrument. The angle G formed by

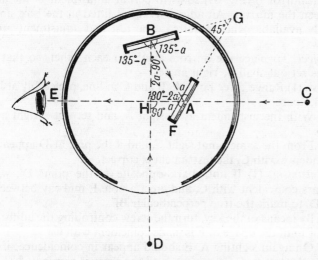

FIG. 39.—OPTICAL SQUARE.

the reflecting surfaces is made 45°, and if EAF be denoted by a, then BAG = a by the laws of reflection of light, and the values of the remaining marked angles are readily deduced, giving CHD a constant value of 90°.

55. Use of Optical Square. To set out a perpendicular to a line EC from a point on it, the observer holds the instrument over the point, preferably resting it against a ranging pole, and turns it until the pole C, marking the line, is seen through the clear glass. A chain-man, having been sent out with a pole in the direction of D, is then directed to right or left until the reflected image of his pole appears coincident with the line pole.

To find where a perpendicular from a given point D would meet the line EC, the surveyor, holding the instrument to his eye, walks along the line until he obtains the coincidence as before, when the instrument will be over the required point.

It is helpful to have two forward poles marking the survey line. By keeping them in range, the surveyor can maintain the instrument in the line without trouble.

When the perpendicular lies to the other side of the survey line, the instrument is held upside down.

56. Testing and Adjusting an Optical Square. The instrument is often made with both mirrors permanently fixed, but some makers mount mirror B (Fig. 39) so as to permit adjustment of the angle between the mirrors. An adjusting key is fitted in the box, and is readily available when required. The test and adjustment are as follows:

Object. To place the mirrors at 45° to each other, so that the angles set out shall be right angles.

Test. (1) Range three poles A, B and C in line, preferably at least 300 ft. apart.

(2) With the instrument at B, sight A, and set out a right angle ABD.

(3) From the same point sight C, and if the pole at D appears in coincidence with C, the instrument is correct.

Adjustment. (1) If not, mark opposite D the point D′, which appears coincident with C, and erect a pole E midway between D and D′ to mark the true perpendicular BE.

(2) By means of the key, turn the screw controlling the adjustable mirror until the image of E is made coincident with C.

(3) On again sighting A, E should appear in coincidence: if not, repeat the test and adjustment until the error is eliminated.

The above test is a straightforward case of reversal, and corresponds exactly to the testing of a set-square (section 3).

57. The Prism Square. The same principle applies to the prismatic form of the instrument, shown diagrammatically in Fig. 40, and it is used in the same manner. The prism square has the merit that no

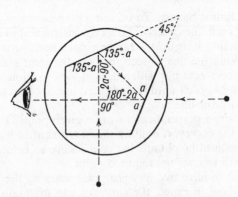

FIG. 40.—PRISM SQUARE.

adjustment is required, since the angle between the reflecting surfaces cannot vary. The deviation of this angle from 45°, due to errors of workmanship, may be regarded as negligible for the purposes of the instrument.

58. The Box Sextant. The sextant, of which the box sextant is the most compact form, is a reflecting instrument capable of measuring angles up to about 120°. Although the box sextant is not strictly a chain surveying instrument, it can be used in place of an optical square, and may appropriately be dealt with here. The principle that the deviation of a ray of light reflected successively from two mirrors is twice the angle between them is applied as in the optical square, except that in this case one of the mirrors, called the index glass, is mounted on an axis about which it can be rotated. The variable angle between the mirrors is read on a scale.

The box has a diameter of about 3 in. and a depth of about $1\frac{1}{2}$ in., and is provided with a cover, which is removed and screwed on underneath to form a handle when the instrument is in use. Fig. 41 represents an interior plan of the instrument. The fixed mirror, or horizon glass, is silvered on the top half only, and the index glass is silvered all over. Attached to the latter is a toothed segment gearing with a small pinion which is actuated by the milled head *3* on the top of the box, shown in the exterior plan (Fig. 42). The axis of the index glass carries an index arm with a vernier, which on rotation of the mirror is carried over the graduated arc. The graduation is marked to half degrees, subdivided to single minutes by the vernier, which is viewed through the hinged reading glass.

The observations are usually made by sighting through an eyehole. A small telescope may be provided for long-distance sighting, and can be attached to the instrument when required.

The theory of the sextant is as follows (Fig. 41).

Let a = the angle EAH between the horizon glass and the line of
sight EC,

  b = the angle AHB between the mirrors when the image of a
pole D appears in line with pole C.

By the laws of reflection, BAF = a, and ∴ EAB = $(180° - 2a)$.
But BAF is an exterior angle of triangle ABH.
∴ ABH = $(a - b)$ = GBD, so that ABD = $180° - 2(a - b)$.
But ABD is an exterior angle of triangle ABJ, whence
AJB, the angle between the signals, is $180° - 2(a - b) - (180° - 2a)$
  $= 2b$ = twice that between the mirrors.

The point J is not fixed in position, and, in general, the angle read on the instrument is not the angle subtended at the centre of the box by the observed objects. The difference, however, is negligibly

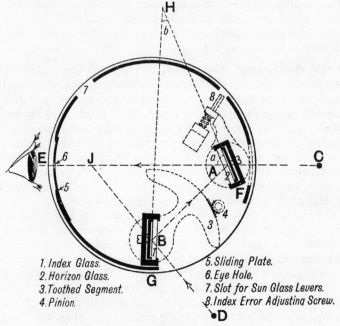

1. Index Glass.   5. Sliding Plate.
2. Horizon Glass.   6. Eye Hole.
3. Toothed Segment.   7. Slot for Sun Glass Levers.
4. Pinion.   8. Index Error Adjusting Screw.

FIG. 41.—INTERIOR PLAN OF BOX SEXTANT.

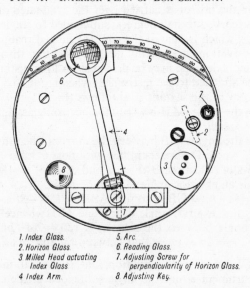

1. Index Glass.   5. Arc.
2. Horizon Glass.   6. Reading Glass.
3. Milled Head actuating   7. Adjusting Screw for
   Index Glass   perpendicularity of Horizon Glass.
4. Index Arm.   8. Adjusting Key.

FIG. 42.—EXTERIOR PLAN OF BOX SEXTANT.

small, unless the objects sighted are very near the instrument.

To avoid the necessity of doubling the angle between the mirrors at each observation, the graduated arc has the angles figured twice their real values.

The sextant differs from other angular instruments in that it measures the actual angle between the directions to two objects, not the horizontal projection of the angle. Thus, in Fig. 43, the

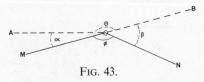

FIG. 43.

directions OA and OB make angle θ, measured by the sextant. OM and ON are horizontal lines and the angle between them is ϕ. If this angle is required, the elevations α and β of the two objects must be found, and then

$$\cos \phi = (\cos \theta - \sin \alpha . \sin \beta)/(\cos \alpha . \cos \beta)$$

59. Observing with the Box Sextant. The instrument is rested against the station pole, and one hand is left free to turn the milled head *3* (Fig. 42). To measure the angle subtended by two distant signals, the instrument is turned in the hand until one of them is visible through the transparent part of the horizon glass, and screw *3* is then slowly rotated until the image of the second signal is .brought in line with the first. Unless the signals are at the same level as the instrument, the latter must be tilted to lie in their plane.

In setting out an angle, the index must first be set to read its given value. Keeping the line of sight, through the clear glass, on the given signal, the surveyor directs a chainman into the position at which a pole held by him appears coincident with the signal. As a rule, the left hand object should be sighted through the horizon glass.

60. Testing and Adjusting a Box Sextant. The requirements of the astronomical sextant (Vol. II, Chapter II) are applicable to the box sextant, but in the latter instrument both the index glass and the telescope are mounted without provision for their adjustment. The horizon glass has the two movements required for setting it perpendicular to the plane of the arc and for the elimination of index error.

61. First Adjustment. *Object.* To set the horizon glass perpendicular to the plane of the instrument, so that both mirrors may be perpendicular to the same plane.

Test. Set the vernier to zero and sight on a distant object, preferably one with a conspicuous horizontal feature. If there is lack of continuity between reflected and direct images, as indicated in Fig. 44 (a), adjustment of the horizon mirror is necessary.

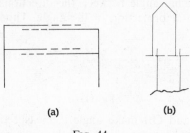

(a)   (b)

FIG. 44.

Adjustment. By means of the key, kept at *8* (Fig. 42) when not in use, turn screw *7* until the vertical displacement is eliminated.

62. Second Adjustment. *Object.* To eliminate index error, so that, when the mirrors are parallel, the vernier will read zero.

Test. Set the vernier accurately to zero, sight any distant object furnishing a definite vertical line, and observe whether the reflected image appears displaced laterally with respect to the object (*b*, Fig. 44).

Adjustment. By means of the adjusting key turn the screw *8* (Fig. 41) until the lateral displacement is eliminated.

THE THEODOLITE

63. The theodolite is an instrument designed for the measurement of horizontal and vertical angles. It is the most precise instrument available for such observations, and is of wide applicability in surveying.

The line of sight of a theodolite is provided by a telescope, the optical principles of which have already been explained (sections 21–33). For setting on points at different elevations, and for measurement of vertical angles, the telescope must be capable of rotation about a horizontal axis; for measurement of horizontal angles, the instrument must be rotated about a vertical axis.

A theodolite may be regarded as a mechanical realisation of the geometry of three concurrent straight lines–the telescope axis, the so-called transit axis to which the telescope axis should be perpendicular, and the main rotation axis to which the transit axis should be perpendicular.

If the geometry of the instrument is perfect, then when the rotation axis is vertical the transit axis will be horizontal and the line of sight will sweep out a vertical plane when the telescope is elevated or depressed.

64. Measurement of Horizontal Angles. For this purpose it is evident that the requirements are:

(1) a horizontal graduated circle,

(2) the upper part of the instrument carrying the telescope must rotate about the vertical line through the centre of the graduated circle, and the telescope must be capable of being pointed in any direction,

(3) there must be an index mark on the rotating part, placed so that readings can be taken against it on the graduated circle,

(4) a horizontal angle is measured by setting the telescope on each of two signals and taking the difference of the two readings on the graduated circle.

(The alternative possible arrangement, with the graduated circle on the rotating part of the instrument and the index mark fixed, is not used.)

65. Measurement of Vertical Angles. Unlike the horizontal angle, a vertical angle (elevation or depression) is not a difference between two settings, but is referred to the horizontal plane as zero. The requirements in this case are therefore:

(1) a graduated circle attached to the telescope and perpendicular to the transit axis,

(2) an index mark suitably placed on the main body of the instrument,

(3) adjustment of the relevant parts of the instrument in such relationship that when the vertical circle reading is zero the line of sight is in fact horizontal.

(The possible alternative arrangement, with the index mark on the telescope and the vertical graduated circle attached to the main body of the instrument, is inconvenient and is not used.)

66. Construction Principles. The basic mechanical design of a theodolite, so as to realise the requirements mentioned above, is illustrated by very diagrammatic drawings in Fig. 45. There are four main sections of a theodolite, denoted by letters A, B, C and D on the diagrams.

Standing on the tripod is the *levelling head* or *tribrach*, consisting essentially of a strong frame A_1 with a central vertical hole, of conical or cylindrical section, and three levelling screws A_2.

Next, fitting into the central bearing hole, is the *lower plate* B,

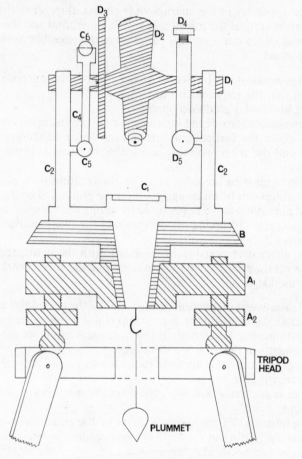

FIG. 45.

which has the graduations around its periphery. This also has a concentric hole to take the upper main body of the instrument.

The main body C, sometimes called the upper plate, rotates in the central bearing of the lower plate. An essential fitment on the upper plate is the levelling bubble C_1. Standing on the *upper plate* are the two main supports or frames C_2, at the tops of which are the coaxial cylindrical bearings that hold the axis of the telescope assembly D.

The essential parts of D are the axle D_1, the telescope D_2 which should be perpendicular to the axle, and the vertical graduated circle D_3.

A practical theodolite has various other attachments which are necessary for its successful functioning as an accurate angle-measuring instrument.

67. Clamp and Slow Motion. For accurate setting of the telescope on a signal mark, it is necessary to be able to clamp the rotating parts and to give them small movements under conditions of smooth and positive control. A simple clamp and slow-motion arrangement for the lower plate of a theodolite is illustrated in Fig. 46. The clamping screw presses a metal shoe against the axle

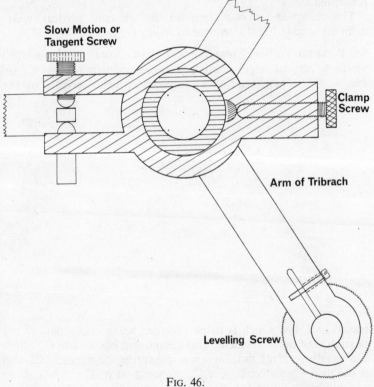

Slow Motion or Tangent Screw

Clamp Screw

Arm of Tribrach

Levelling Screw

FIG. 46.

of the plate, and the tangent screw operates against a tongue fixed to the tribrach: the tangent screw is held positively against the tongue by a spring-loaded plunger on the opposite side to the tangent screw.

In an instrument of the design illustrated in Fig. 45, a clamp will also be required to hold the portions B and C together, because, for the measurement of a horizontal angle, the plate B must remain fixed while the setting of the telescope on the signal mark is done with this upper slow-motion arrangement.

Some instruments are provided with double-threaded tangent screws giving coarse and fine rates of movement.

A good feature of some modern instruments is the so-called impersonal clamp. In this device, the operation of a lever allows a spring to apply to the clamp a pre-determined force which is independent of the force exerted by the operator. Thus, the application of excessive force which might cause mechanical damage is prevented.

The clamp and slow-motion for the telescope vertical movement are seen in Fig. 45 denoted by letters D_4 and D_5.

68. Parts for Vertical Measurements. The 'fixed' marks against which a vertical angle reading is taken are denoted by C_3 on Fig. 47. These marks are on a component which is pivoted on the

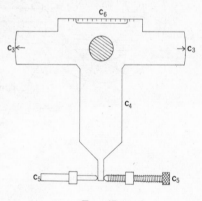

FIG. 47.

transit axis, but which is to be regarded as, in effect, part of the portion C of the instrument. This component has an arm C_4 which can be adjusted by a tangent screw and spring arrangement C_5, and it has also a sensitive bubble C_6 mounted on it. The slow-motion and bubble are used for setting the index marks in a desired position.

69. Fixing on Tripod. It is of course very unwise to place an instrument on a tripod without some arrangement for preventing its being tipped off. All theodolites are therefore held on the tripod by some device which allows the necessary movements for levelling up, and the arrangement for doing this takes various forms. In Fig. 48

the thin springy plate A is fitted over the necks of the levelling screws, and is held by loose-fitting screws to the base-plate B; both these parts are permanently attached to the theodolite. The holding screw C passes through a slot in the arm D which is hinged at one end to the underside of the tripod head; parts C and D remain

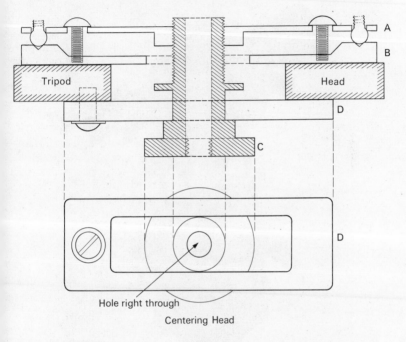

Hole right through

Centering Head

FIG. 48.

with the tripod. The holding screw is hollow if the instrument has an optical centering fitment, and has an internal screw thread for attaching a plummet.

The springy plates referred to above are clearly seen in Figs. 49 and 50.

70. Levelling Screws. The function of these screws is for tilting the instrument so as to get its rotation axis in the truly vertical position. Most modern instruments have three screws, though the four-screw arrangement is still to be seen in the United States.

The threads of levelling screws should be of fine pitch, $\frac{1}{32}$ inch being suitable. It is essential that there should be no looseness of the fitting of these screws, so the holes in which they rotate should be of good length, and there should be some means of adjusting the

FIG. 49.—THE 'ELTHAM' VERNIER THEODOLITE.
(By courtesy of W. F. Stanley & Co. Ltd.)

tightness: for example, by splitting the tribrach arms and providing tightening screws as shown in Fig. 46.

The levelling screws rest in sockets or grooves on a base-plate, and, as mentioned above, it is now usual for the base-plate to be permanently fitted to the theodolite, not a separate piece or a fixture on the tripod head.

FIG. 50.—MICROPTIC THEODOLITE No. 1.
(By courtesy of Hilger Watts Ltd.)

Levelling adjustment by supporting the instrument on three eccentric cams is a mechanical possibility, and at least one manufacturer provides this arrangement. Cams have less range of movement than screws, so it is necessary to have a reliable circular bubble on the instrument for preliminary approximate levelling.

71. Levelling Bubbles. Some theodolites are provided with two tubular bubbles set at right-angles on the upper plate; this helps to speed up levelling, but most instruments nowadays have a circular bubble for preliminary levelling and one sensitive tubular bubble for the final adjustment.

72. Magnetic Compasses. Theodolites of simpler types may be obtained fitted with a circular compass-box in the centre of the upper plate. The pivot of the compass needle is on the theodolite axis and the direction determined by the 0° and 180° graduations

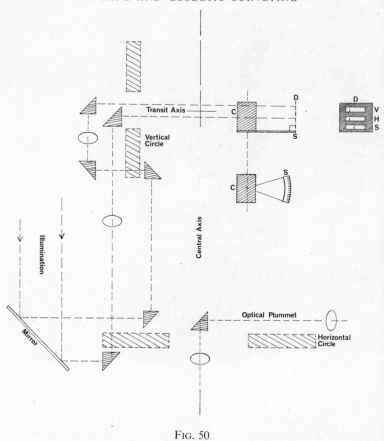

FIG. 50

on the compass ring is intended to be perpendicular to the transit axis. With such an instrument, the telescope may be used merely as a sighting device for finding magnetic bearings.

Another kind of fitting is a long trough-compass inserted into a slide on the underside of the lower plate: when the compass is swinging freely with its ends against reference marks, the zero of the graduated plate should be in such a position that the horizontal vernier reads 0° when the telescope is directed into the magnetic meridian. Then, with the lower plate clamped, a reading of the upper plate is a magnetic bearing, but of course the accuracy is no more than that obtainable in magnetic surveying generally.

Theodolites are also made with magnets attached to the lower plate: this plate is independently supported so that it can rotate quite freely when unclamped and will set itself in the meridian. If it is then clamped, horizontal readings of the theodolite are magnetic bearings.

73. Centering Head. A very useful feature of most theodolites is illustrated in Fig. 48. The hole in the tripod head is much larger than the fitting which goes through it, and this allows some lateral movement, over a range of one or two inches, before the instrument is finally clamped down. This is a great help in centering the theodolite over a mark on the ground. The designs of these centering heads are various, some being much more convenient in use than others. It is essential to have a centering head when optical centering (described below) is being employed.

74. Centering over Mark. All theodolites are supplied with a plummet to be hung from a hook or other suitable fitting at the lower end of the vertical axis. Many instruments also have *optical centering*. This is done by fitting lenses into the hollow central axis of the instrument so as to form a small telescope pointing vertically downwards; the line of sight is brought, by means of a reflecting prism, to an eyepiece at the side of the instrument, and the observer can view the ground mark in relation to a diaphragm mark in the optical system. A centering eyepiece can be seen in the illustration, Fig. 50.

Optical centering is very useful in some circumstances. It is of course not effective until the sight line of the centering telescope is vertical, so the levelling and centering operations have to be done in conjunction: this can be time-consuming, and it is generally found that the use of the hanging plummet is more expeditious, except in a high wind, or when the ground mark is at the bottom of a hole, or other special circumstance. The optical centering device should be on a rotating part of the instrument, otherwise there is no simple means of checking that it is in line with the vertical axis.

75. Types of Theodolite. Almost all theodolites in use nowadays are *transits*; that is, the telescopes can be rotated through the vertical position so that the graduated circle attached is either on the left or on the right side of the telescope as viewed from the eyepiece end. These positions are called *face left* and *face right* respectively.

The advantage of the non-transiting telescope was that it did not require high supports, so the instrument could be made comparatively compact: as the telescope of an ordinary theodolite nowadays is only 5 or 6 inches long, this consideration has lost its importance.

Some time ago, theodolites could be divided into two classes, those reading by verniers and those reading by microscopes.

Vernier instruments are simple and comparatively cheap, and are quite commonly used. The precision of a vernier instrument is of course limited by the least count which can conveniently be accommodated. It is found that a graduated circle 5 inches in diameter can be divided to 20-minute intervals and read by vernier to 20 seconds; this is about the limit, a smaller circle should be divided to half-degrees and read to half-minutes. A typical vernier theodolite is illustrated in Fig. 49.

In the other class of theodolite, the divided circle is viewed by microscopes. With the aid of this magnification, precise readings of minutes and seconds are obtainable by movement of a cross-wire in the image plane of the microscope. In fact, the function of the microscope is similar to that of the vernier, in giving an accurate measure of the small interval between the reference point of the rotating part and the adjacent graduation of the divided circle.

However, 'micrometer-microscope' instruments have been largely superseded by theodolites in which the circle graduations are on glass plates: these are viewed through a more or less elaborate optical system by means of which the readings of the horizontal and vertical circles are taken at the position in which the observer stands at the sighting telescope, and he does not have to walk round the instrument in order to read it. This feature represents a very substantial improvement of speed of observation in comparison with the microscope instruments.

The optical reading instruments are of two main classes:

(i) precise instruments with divisions to single seconds on the micrometer setting device, or even finer division in first-order instruments. In these theodolites the optical reading system is arranged so that diametrically opposite graduations of the circle are viewed simultaneously, and the observer operates a setting device (Fig. 51) to obtain a mean reading free from eccentricity error (see section 91).

(ii) theodolites of lower precision. In some of these the readings are obtained after a micrometer setting; in others, called *scale-reading* theodolites, the optical reading system provides a magnified picture of the main scale, and the subdivision is done by estimation of the position of a main scale division against a short finely divided scale fixed at a suitable position (see Fig. 52). These less precise instruments usually have circles of about three inches in diameter, reading is done at only one point, and the precision is of the order of $\frac{1}{4}$ minute. The theodolite illustrated in Fig. 50 is read by a micrometer on which the graduation interval is 20″, and readings can be taken to 10″.

The optical reading systems of this instrument are shown diagrammatically in Fig. 50. Part of the light directed into the instrument by the mirror is sent through the graduations of the horizontal circle, and part through the vertical circle. By suitably placed lenses and prisms these two beams are sent along the hollow

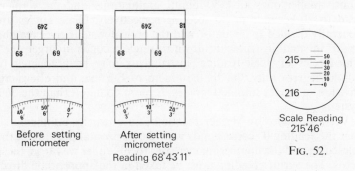

Before setting
micrometer

After setting
micrometer

Reading 68°43′11″

FIG. 51.

Scale Reading
215°46′

FIG. 52.

transit axis and brought to focus at the plate D. Both beams pass through a parallel-sided glass block C which can rotate about a vertical axis. The block is rotated by the micrometer setting knob and its position is indicated by the graduated sector scale S.

Looking through the reading eyepiece at the plate D, the observer sees a portion of the vertical circle graduations in the window V, a portion of the horizontal circle graduations in window H, and a portion of the sector scale in window S. Each window has a fixed reference line or marker and, as the block C is rotated, the pictures of the scales move across the windows.

The thickness of the block C and the length of the scale S must be calculated so that a movement of the scale from reading 00′ 00″ to 20′ 00″ causes the pictures in the windows V and H to move a distance of exactly one division, 20′, of the circle scales.

To make a reading, the micrometer knob is operated so as to place a graduation line of either the vertical or the horizontal scale in coincidence with the fixed reference mark, the reading of the circle scale is taken to 20′, and the reading of the micrometer scale S is taken to 20″ (perhaps estimated to 10″) and added to the reading of the main scale.

76. Graduated Circles. The diameter of the circle of graduations on the horizontal graduated plate of a theodolite is usually quoted to designate the size of the instrument. Vernier instruments are commonly in the range 4-inch to 5-inch, and their vertical circles are usually of the same size as their horizontal circles. Optical reading theodolites for ordinary surveying purposes, having glass circles, are nowadays generally in the range 3-inch to 4-inch, and in many cases the vertical circle is smaller.

Horizontal circles are graduated clockwise as seen from above, 0° to 359°. The larger vernier theodolites have 20-minute graduations, smaller ones have 30-minute graduations, and the verniers read to 20 seconds or 30 seconds respectively. Glass circles are generally divided to 20′ or 10′ and the optical system gives reading to 20″ in small instruments, down to single seconds in larger ones. The circles of 'scale-reading' instruments may be graduated to whole degrees only. On the continent of Europe, and elsewhere, division of circles into 400 'grades' is more common than the 360° system.

Vertical circles on vernier theodolites are usually numbered so that the reading 0° corresponds with the horizontal position of the telescope, and the numbering increases to 90° either way from these zeros. The vernier reading must be taken in the appropriate direction, according to whether the angle is elevation or depression; in some instruments the same vernier is numbered both ways, and in others the vernier is of double length and numbered each way outwards from the central reference point: the latter arrangement is perhaps the less confusing. Figs. 53 and 54 show a reading of depression angle 17° 38′.

In optical reading instruments the numbering of the vertical circle is usually continuous from 0° to 359°, and it is usually so arranged that readings increase as the telescope is elevated in the 'face-left' position. This means that the readings will decrease when the telescope is elevated in the 'face-right' position. The position of the numbering varies from instrument to instrument; in some the reading is 0° with telescope horizontal face left, and 180° with telescope horizontal face right, in others the numberings may be 90° and 270° respectively.

For example, suppose the telescope is set on a signal, face left, and the reading is 8° 17′ 50″: this is an elevation of 8° 17′ 50″. On changing face and setting on the same signal, the reading will be 171° 42′ 10″ (assuming no vertical index error in the instrument, see section 98). If the signal was at depression 6° 55′ 20″ the readings would be FL 353° 04′ 40″, FR 186° 55′ 20″.

Another system of numbering goes from 0° to 180° twice and the 90° marks correspond with horizontal telescope. In this case, an elevation of (say) 5° 28′ 10″ will read: FL 95° 28′ 10″, FR 84° 31′ 50″. A convenient feature of this system is that the vertical angle can be calculated by subtracting the FR reading from the FL reading and dividing by 2. In the same system, a depression of 5° 28′ 10″ would read : FL 84° 31′ 50″, FR 95° 28′ 10″, and the same procedure would give the angle as −5° 28′ 10″, the minus sign indicating a depression angle. A surveyor who has to use an instrument with which he is not familiar should examine the graduation and reading systems carefully as these vary greatly from instrument to instrument.

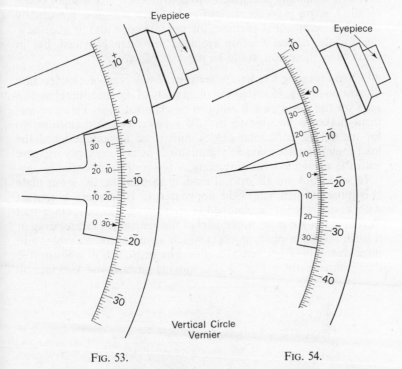

Fig. 53.   Fig. 54.

77. Telescopes. The telescope of a modern theodolite for ordinary surveying purposes is only 5 or 6 inches long, has internal focusing lens and stadia lines, and transits at the objective end or at either end. It should be fitted with external sights lined up parallel to the optical axis, preferably two pairs of sights so that in either position there is a pair above the telescope tube.

78. Illumination. Many theodolites are provided with electric illumination for use in tunnels or at night. This is powered from a box of dry cells with a switch-rheostat control and a connecting flex to be plugged into the body of the instrument. Illumination is required for the optical reading system, the vertical arc bubble and

the inside of the telescope. The batteries usually run down rather
quickly, and a spare set should always be available if extensive use
is to be made of the internal illumination. The illumination systems
provided on some theodolites are not all that could be desired.

79. Accessories.   Screwdriver, spanners and tommy bars may be
required in making adjustments to instruments; also a small bottle
of oil. A spare glass diaphragm and a dark sun-glass are usually
provided. A diagonal eyepiece for the telescope and, if necessary,
one for the optical reading eyepiece, are often provided, but in
some cases these, too, could be more satisfactory.

80. Non-repeating Circles.   A feature of the type of construction
illustrated in Fig. 45 is that the main body of the theodolite, part C,
rests on the lower part B and can be clamped to it. This arrangement makes it possible to use the *repetition method* (section 86)
for measuring horizontal angles, and is an advantage when the
instrument is rather coarsely graduated, as is necessarily the case
with the vernier system of reading.

In some, but not all, optical reading theodolites, the lower plate
B is quite independent of the upper part C, is nowhere in contact
with it, and cannot be clamped to it. This has the advantage that
movement of the main upper part of the instrument cannot drag or
disturb the lower plate, an effect which would introduce errors into
measurements of horizontal angles. The construction is illustrated
diagrammatically in Fig. 55; the moving parts B and C rotate on

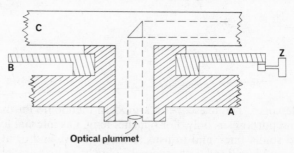

Optical plummet

FIG. 55.

separate machined bearings on the fixed tribrach A. The part Z
represents an adjusting wheel and axle engaging with cogs on the
lower plate, so that the horizontal circle can be placed in any
position and will stay there steadily because of the amount of
friction in the bearing. The operation of moving the lower plate
into different positions is called *changing zero*. The setting screw
must be protected by a hinged cover, or located in a well-guarded

position, as a precaution against accidental disturbance during a round of observations. Instruments with independent lower plate cannot be used for the repetition method of measurement (section 86), or for the direct bearing method in traversing (section 225).

TEMPORARY ADJUSTMENTS
OF THE THEODOLITE

81. The temporary adjustments are:
   (1) Setting over the Station,
   (2) Levelling up,
   (3) Focusing and Elimination of Parallax.

82. Setting Up. This includes both centering the axis of the instrument vertically over the station mark and approximate levelling by manipulation of the tripod. The required accuracy of centering depends on the circumstances, and on the nature of the survey work. Where angle errors are likely to accumulate, centering must be carefully done, and this is particularly the case when the work is a theodolite traverse with short lines. A movement of an inch at one end of a line 100 yards long can alter its bearing by one minute. With a modern theodolite having a shifting attachment to the tripod, there should be no difficulty in centering to a small fraction of an inch.

In the setting up of an instrument, the legs of the tripod can be moved radially and sideways: radial movement serves for centering, and sideways movement for approximate levelling after centering is nearly right. In setting up an instrument on steeply sloping ground, place one leg uphill.

If ground chaining is to be done from the station mark, avoid setting a leg of the tripod on the chain line.

83. Levelling Up. Levelling up a three-screw instrument is a simple and rapid operation if carried out systematically.

Assuming that there is one bubble fixed to the upper plate, rotate the theodolite until this bubble is parallel to the line joining two of the levelling screws; rotate these screws in opposite directions so as to bring the bubble to centre position in its tube (the bubble moves in the same direction as the operator's left thumb). Then turn the instrument round 90° and centre the bubble, using the third levelling screw *only*. Carry out the above process rapidly, without precise setting of the position of the bubble, so as to get the levelling nearly right as quickly as possible.

Now return the instrument to the first position and centre the bubble accurately. Rotate the instrument 180°, and if the bubble does not rest in the central position bring it back *half-way* to centre

by equal movements of the two levelling screws. Now rotate 90°
and bring the bubble to the *same*, possibly off-centre, position by
the third levelling screw *only*.

The rotation axis of the instrument will now be vertical.

As mentioned above (section 36), the 'proper' position of the
bubble may not be exactly central, and much time can be wasted
in trying to centralise it with the levelling screws. If the surveyor
already knows the 'proper' position, he can put the bubble in this
position from the beginning and considerably reduce the time
needed for levelling up.

Some instruments have two plate levels, and if both are in good
adjustment the levelling can be done without rotating the theodolite, but, as a rule, one of the bubbles will be much more sensitive
than the other, and this one should be used for the final levelling.

Many theodolites have a graduated bubble tube attached to the
vertical angle reading fitment (part C_4 in Fig. 45), and this can be
used for the levelling-up. Often it is the most sensitive bubble on the
instrument. The general rule applies—the axis is vertical if the
bubble rests in the same position in its tube for all positions of
rotation.

84. Focusing and Elimination of Parallax. See section 28.

ANGLE MEASUREMENT BY THEODOLITE

85. To Measure a Horizontal Angle. The temporary adjustments at
a station having been made, several procedures are possible, depending on the type of instrument, for measuring the angle subtended at the instrument by two signals.

(1) With a theodolite in which the upper and lower plates can be
clamped together, it is possible to set the upper plate so that the
reading is exactly 0° on the horizontal circle. With both clamps
loose, the reading can be brought by hand to be nearly zero, the
plates are then clamped together and the upper tangent screw used
to make the reading exactly 0°. Then the instrument is turned
towards the left hand signal, set on it by hand as nearly as possible,
the lower clamp is tightened, and the setting made exact by use of
the *lower* tangent screw. Thus, the telescope is now sighted on the
left hand signal and the reading is exactly 0°. Now the *upper* clamp
is loosened and the telescope is set on to the right hand signal by
use of the *upper* clamp and slow-motion only. The reading of the
horizontal circle is now the required angle.

(2) It is of course possible to measure an angle by simply setting
the telescope on one of the signals with both clamps tight and
using either tangent screw, reading the circle, unclamping the

upper plate, setting on the other signal by use of the *upper* clamp and slow-motion *only*, and taking the second reading. The difference of the two readings is the required angle.

This is the procedure which must normally be used with an instrument having independent lower plate, because the positioning of the horizontal circle on such an instrument is not sensitive enough for setting it to an exact reading.

If the 360° reading happens to be between the two signals, the reading on the right hand signal must be increased by 360° before subtraction of the other reading.

(3) In measuring horizontal angles it is necessary to take great care not to move the wrong tangent screw. A check can be obtained if the telescope is set back on the first point after the readings have been made.

It is very helpful to the user of a theodolite if the screw heads for the upper and lower clamps and slow-motions are so different as to be immediately recognisable to the touch.

An advantage of the instrument with independent lower plate is the absence of the lower tangent screw.

(4) When setting on a ranging pole or similar signal, it is best to intersect the pole as low down as possible, so as to minimise the effect of any non-verticality.

(5) A single measurement of an angle, as described above, will contain the full effects of any mechanical mal-adjustments in the instrument. If the theodolite is in good condition and has been carefully adjusted, such effects are likely to be too small to influence appreciably the results of ordinary simple surveying operations.

For high quality work, precise instruments and elaborate methods of observation should in any case be employed. However, one way to improve accuracy while using a simple instrument is the *repetition method*, as described below.

86. The Repetition Method. We suppose that an ordinary vernier theodolite reading to 30″ is in use. Let the actual value of the angle to be measured be 17° 06′ 37″. If a single measurement is made as described above, the reading will be booked as 17° 06′ 30″. Now, the *lower* clamp is loosened and the telescope set again on the left hand signal using the *lower* clamp and slow motion *only*: the exact reading of the instrument is still 17° 06′ 37″, and if the upper plate is now loosened and the telescope is set again on the other signal, the exact reading of the instrument will be 34° 13′ 14″, and this will be booked as 34° 13′ 00″. The process is repeated, and after the third setting on the right hand signal the exact reading will be 51° 19′ 51″, which will be booked as 51° 20′ 00″. This is obviously three times the required angle, and division by 3 gives 17° 06′ 40″, which is only 3″ in error.

It is clear that with more repetitions a closer and closer approach to the true value could be obtained, in theory. In practice, 6 to 10 repetitions are considered to be enough, and it is usual to change 'face' half-way through the group, so as to get rid of mechanical errors as well.

In general, if *n* repetitions are made with an instrument graduated to 30″, the resulting angle should be correct to 15″/*n*: however, random errors will not divide down like this, but they should tend to cancel out in the course of the numerous settings that are made in the use of this method. The repetition method is no more than a means of partially overcoming the limitation set by the least count of an instrument.

87. To Set Out a Horizontal Angle. This operation is the converse of a simple measurement. Being given one point, it is required to locate the direction in which a second point should lie from the instrument, so that a given angle is subtended between them. We suppose that the given angle is to be measured clockwise.

The reading of the instrument is set to zero, as described before, and the telescope directed on to the left hand signal, using the lower tangent screw.

The upper clamp is released and the theodolite turned to read the given angle approximately, the upper clamp is tightened, and the reading made exact using the *upper* tangent screw.

Now, the telescope is pointing in the required direction, and a signal or mark may be established on the line of sight. Since most telescopes invert, the assistant who sets up the mark must be asked to move in the direction opposite to that in which it appears he should move when viewed through the telescope.

88. To Measure a Vertical Angle. Most theodolites nowadays have separate slow motion controls for setting the telescope on the signal and for positioning the vertical circle reading device (vernier, or optical type).

The procedure for measuring a vertical angle is then:

(1) After levelling up the instrument in the normal way, set the telescope with the central horizontal cross-line on the object or signal,

(2) Set the reading device so that the bubble attached to it is central, or, if an optical split-bubble is fitted, set this to have an apparently continuous edge,

(3) Read the vertical circle. According to the position and graduation system of the instrument, the reading will be the vertical angle
θ, or it may be $180° - \theta$, $90° + \theta$, or some obvious similar function of
θ. With a vernier theodolite, the angle is read directly, but the
functions mentioned above will be obtained with an optical reading
system. Care must be taken to distinguish between elevation and
depression if this is important, but when the vertical angle is only
required for calculating slope correction the distinction is not
necessary.

(4) A single reading of a vertical angle will be affected by any
vertical zero error in the instrument, see section 98. If the vertical
reading device has been adjusted, the error may perhaps be neglected in ordinary survey work, but it is a simple matter to repeat the
measurement with the instrument on the other 'face', and the mean
of the two measures will be free from zero error. Even if the zero
error is large, it will cancel out in this way: it is a check on the
performance of the instrument, that the zero error, whatever its
amount, should remain more or less constant.

TESTING AND ADJUSTMENT
OF THE THEODOLITE

89. Requirements of the Theodolite. Through inevitable imperfections of workmanship and the development of defects by continued
use, the mechanical ideals of a theodolite are imperfectly fulfilled,
and its successful operation is very largely dependent upon a
knowledge of the nature and relative importance of the effects
produced by instrumental errors. Most theodolites are provided
with means of adjusting some of the relationships between parts of
the instrument, while the influences of defects in non-adjustable
parts can be reduced or completely eliminated by the adoption of a
proper routine in observing.

The more important requirements as regards non-adjustable
relationships are:

(1) The whole instrument should be stable, that is free from
looseness or slackness between components that should fit,

(2) The axes of rotation of the upper and lower plates should
coincide and should not deviate in direction when the parts are
rotated,

(3) The geometrical centres of the graduations of the circles should
be on their respective axes of rotation,

(4) There should be a reasonable correspondence between the
resolving power of the telescope, the sensitivity of the bubbles, and

the smallest change of reading that can be definitely detected on the graduated circles.

(5) Graduation of the circles should be accurate.

90. The following conditions can be established by adjustments:

(1) When the rotation axis of the instrument is truly vertical, the plate level bubble or bubbles should be central in their tubes,

(2) The line of sight of the telescope should be perpendicular to the horizontal, or transit, axis,

(3) The horizontal axis should be perpendicular to the vertical axis,

(4) When the instrument is levelled up, and the bubble attached to the vertical circle reading device is central, and the line of sight is in fact horizontal, then the reading of the vertical circle should be zero, or some multiple of 90°, see Sections 76 and 88.

91. Testing Non-adjustable Parts.

(1) Stability of tripod can be tested by inspection and manipulation. Most tripod heads have arrangements for taking up any slackness in the hinges. Telescopic legs can be a source of unwanted movement if the design is not mechanically sound.

To test the stability of an instrument, it should be set up on a firm tripod, directed on a clearly defined mark and clamped. If gentle but firm pressure is applied to the eyepiece or any other point, the cross-wire will probably be deflected visibly from the mark, but it should return accurately once the force is removed.

A little looseness of the eyepiece itself is not a serious matter, as the eyepiece plays no part in defining the line of sight.

(2) If the instrument is of the type in which the upper plate rests on the lower plate and can be clamped to it, non-parallelism of the two vertical axes is easily detected by carrying out the levelling-up process round one of the axes and then rotating the instrument about the other. If the axes are not parallel, the bubble will behave as if the instrument is not levelled.

(3) The axes may be parallel but not coincident: this condition is called *eccentricity*. If the theodolite has two verniers placed at diametrically opposite points, the difference of the readings of the two will not remain constant but will vary periodically round the circle, if there is eccentricity.

Thus, in Fig. 56, let A be the centre of graduation of the circle, and B that of the verniers, the distance AB being greatly exaggerated. When the line of sight occupies the position CBD, the vernier readings at E and F differ by 180°. On turning the line of sight through *a* into the position GBH, the verniers are actually moved through a to K and L, but they do not record a. The reading at K is EAK, say $(a-e)$, and that at L is $(180° + FAL) = (180° + a + e)$, since AKB = ALB = e. The readings therefore now differ by $(180° + 2e)$.

The mean of the readings of the two verniers will be free from eccentricity effect. In taking the mean, it is usual to record the number of degrees indicated by one particular vernier and assume the reading of the opposite vernier to be altered by 180°.

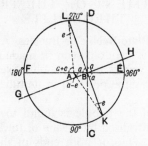

FIG. 56.

If the readings of the two verniers differ by a constant amount not exactly 180°, only one vernier need be used but it should always be the same one.

With an instrument having independent lower plate, levelling must necessarily be done about the upper plate axis, and no simple test of the lower plate axis is possible.

(4) If the telescope is set on a mark and the vertical and horizontal readings are noted, and the telescope is then moved just perceptibly off the mark by the vertical or the upper horizontal tangent screw, the appropriate circle reading should show the change. But if the displacement is done in the vertical direction by means of a levelling screw, there should be a perceptible movement of the vertical index bubble. These operations test the compatibility between telescope resolution, bubble sensitivity, and the smallest reading interval.

(5) Errors of graduation of a circle can be classified as (*i*) periodic errors that recur at regular intervals round the circle according to some law, and (*ii*) random errors which are quite irregular. Accuracy of division cannot easily be tested, but the refinement of modern dividing engines is such that any graduation errors are likely to be much too small to affect the surveyor's results in any ordinary survey work; indeed, many modern small instruments graduated to read to 10″ or 20″ have no provision for reading at two opposite points.

In practice, the only way to reduce the effects of graduation errors is to measure an angle a number of times on different settings of the circle, a procedure which is essential when highly accurate work is being done with a precise theodolite.

In any case, change of zero cannot be done on the vertical circle, and there would be little point in doing so because vertical angles are influenced by atmospheric refraction, the effects of which are somewhat irregular and cannot be accurately determined.

ADJUSTMENT OF THE TRANSIT THEODOLITE

92. Adjustments of Plate Bubble or Bubbles. As stated previously, the rotation axis of a theodolite can be made truly vertical by adjusting the levelling screws until a bubble fixed to the instrument comes to rest in the same position in its tube for all positions of rotation.

If this position of the bubble is found to be considerably off centre, it may be centralised by raising or lowering one end of the bubble tube holder. One or both ends of the holder should be adjustable by turning nuts on a threaded support, or by turning a capstan screw to move the bubble holder against the pressure of a spring. As a rule, this adjustment mechanism is rather coarse, but it is not necessary to spend a lot of time trying to get the bubble precisely centered.

After the bubble has been adjusted as well as possible, the levelling-up should be checked by the usual procedure.

93. Line of Sight. A modern internal focusing telescope has a fixed object glass, and a focusing glass which has to be moved longitudinally by some mechanical device to bring into focus objects at distances from 'infinity' down to some 10 or 15 feet. Ideally, the optical axes of the two lenses should be coincident at all focusing positions, in other words the optical axes and mechanical movement should be 'collimated'; moreover, this line of collimation should be perpendicular to the transit axis of the theodolite.

The ideal condition may not be exactly fulfilled, but any deviation should be very small. If such deviation exists to an extent that appreciably influences the accuracy of the instrument, there is little or nothing that the surveyor can do about it, and the instrument should be regarded as unserviceable.

In practice, the surveyor is concerned with what may be called the 'line of sight', which depends on the position of the diaphragm cross-lines in relation to the optical axis of the lens system. If the telescope is sharply focused and accurately set on a distant mark, the line from the axis of the telescope to the distant mark is the *de facto* line of sight, and this is probably as good a definition as any that could be stated.

94. Adjustment of Horizontal Collimation. The placing of the line of sight so that it is perpendicular to the transit axis must obviously be the primary object of one of the adjustments of the instrument, and this adjustment is in practice made by lateral movement of the diaphragm.

In the examination and elimination of errors in the adjustable parts of the theodolite, considerable use is made of the principle of reversal (Section 3). An important reversal is that produced by transiting the telescope and then turning it horizontally (termed 'wheeling' or 'swinging') through 180°, so that, if the vertical circle is initially on the right-hand side of the telescope, it will lie to the left after reversal. These positions are distinguished as 'face right' and 'face left', the operation being termed 'changing face'.

As stated in section 25, the diaphragm is held on a ring which, by loosening its supporting screws, can be given a certain amount of lateral or rotational movement. Perpendicularity of the line of sight and the transit axis can best be tested by a double reversal procedure as follows:

  (*i*) set up the instrument on fairly level ground with clear view up to 100 yards or so in two opposite directions,

 (*ii*) establish a mark A (see Fig. 57) at about 100 yds. from the instrument, by putting a chaining arrow in the ground, or otherwise, and set the telescope on it accurately after clamping all horizontal movement,

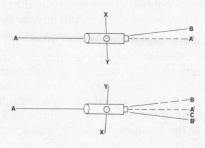

Fig. 57.

(*iii*) transit the telescope; put in a mark B on the line of sight at about the same distance as A from the instrument,

 (*iv*) unclamp, swing through 180°; again clamp and set on A,

  (*v*) again transit: the line of sight may now appear in a different direction and if so, put in a mark B′.

On the diagram, the transit axis is indicated as XY, and the angles XOA and XOB may be written as $90° - e$, where e is the amount of departure from perpendicularity between the line of sight and the transit axis. AOA' being a straight line, it is obvious that the angle BOA' will be $2e$. Then, after the second transiting, the angle BOB' will be $4e$.

The procedure just described is *double* reversal, so it produces an effect of *four times* the error being tested (section 3). It is obvious that if C is half-way between B' and A', that is, a quarter of the distance from B' to B, then the direction OC will be exactly perpendicular to the transit axis in its second position.

Thus, a mark can be put in at C and the diaphragm can be moved by means of its positioning screws, so that the cross line is set on C. The test should be repeated as a check.

95. A simpler test for horizontal collimation is as follows:
  (*i*) set the telescope on a mark which is at about the same level as the instrument, and read the horizontal circle,
  (*ii*) release upper clamp, transit telescope, and set on the same mark, using the upper tangent screw.
(*iii*) if the reading is not exactly 180° different from the first reading, the discrepancy is $2e$. In fact, this procedure, in effect, measures the angle BOA of Fig. 57.

This method is a simple reversal. Since it depends on readings of the circle it will be rather insensitive on a coarsely divided instrument.

The reason for using marks on the same level as the theodolite is that the above tests could otherwise be confused by the existence of any transit axis error, tests for which will now be described.

96. Adjustment of Horizontal Axis. After adjustment No. 2 (Section 90) has been done, the line of sight will sweep out a plane perpendicular to the transit axis when the telescope is transited. However, if the transit axis is not perpendicular to the vertical rotation axis of the theodolite, it will not be horizontal when the instrument is levelled up, and so the plane of the line of sight will not be a vertical plane. A consequence of the situation would be, that if two signals are in fact on a vertical line, one exactly above the other, the horizontal circle readings when the telescope is set on the two signals will not be the same.

To test for transit axis error, the instrument should be set up near to a high building or a tower so that it can be directed on to a well-defined mark at a considerable altitude, say 40°. With the horizontal plates clamped, the telescope is set on the elevated mark, A in Fig. 58, and then brought down to approximately horizontal position, and a mark is placed at B′ on the line of sight. The telescope is then transited and the instrument is swung round, and the procedure is repeated. If there is transit axis error, the line of sight will not intersect B′, and another mark B″ may be placed. Obviously, if there were no transit axis error, the telescope should come down to point C which is midway between B′ and B″.

Many instruments are provided with means of raising or lowering one end of the transit axis: in such case, the telescope can be

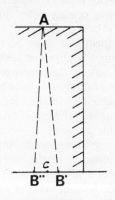

FIG. 58.

directed on to point C and the axis adjusted until, when the telescope is raised from C it exactly intersects the mark A.

If there is no provision for transit axis adjustment, and the error is found to be considerable, correct angles could be obtained by measuring on both faces and taking the mean value, but an instrument in such condition should really be regarded as unserviceable.

It may be convenient, instead of putting in marks at B′ and B″, to fix a graduated scale or a levelling staff in a horizontal position and take readings on it.

Some high-precision instruments are provided with striding levels which make the transit axis test a very simple operation: see Vol. II Chapter III.

97. Adjustment of the Diaphragm. The two cross lines on the diaphragm of a theodolite are placed perpendicular to each other accurately by the manufacturer. When any adjustment of the diaphragm is made, a rotational displacement may be introduced so that the cross lines are not truly vertical and horizontal when the instrument is levelled up.

This can be tested by setting the telescope on a well-defined mark and rotating the instrument to see if the mark appears to run accurately along the horizontal cross-line. Alternatively, a plummet can be hung up quite near the instrument and the vertical cross-line compared against it.

98. Adjustment of the Vertical Index. The procedure for measuring a vertical angle with a modern theodolite is to set the horizontal cross-line on the signal or object, set the vertical reading attachment (vernier or micrometer) to a position indicated by a bubble, and read the vertical circle. If this is done on both right- and left-face positions, the two angles may differ but the mean of the two values will be the true vertical angle.

The object of any adjustment must be to arrange the reading device so that the true angle is obtained with only one observation. This is a simple matter; leaving the telescope set on the signal, alter the position of the reading attachment until the true angle is read on the circle; this will displace the vertical index bubble, which must then be re-centered by moving the bubble holder itself by whatever means is provided for this purpose.

As a rule, the adjusting arrangement for the bubble is rather coarse, and precise elimination of the vertical index error is not easy. However, this is not really important: in simple surveys an error of half a minute or so in a vertical angle is not likely to be significant, and in situations where accurate vertical angles are needed the measurement should always be done on both faces, so that any vertical zero error, however large, is completely eliminated.

EFFECTS OF INSTRUMENTAL ERRORS OF A THEODOLITE

99. The two principal mechanical errors that can exist in a theodolite are non-perpendicularity of axes that should meet at right angles. The effects of these errors on measured horizontal angles will now be considered.

100. Transit Axis Error. This error exists if the transit axis of the telescope is not perpendicular to the vertical axis of rotation. Assuming that the vertical axis is made truly vertical by proper application of the levelling routine, the result of transit axis error is that the transit axis will not be horizontal, and the line of sight of the telescope will sweep out a plane which is perpendicular to the transit axis and therefore not vertical.

The effects of this condition can perhaps best be visualised by considering the telescope set to zero elevation: then the line of sight will be horizontal and in fact in the same position as if there were no transit axis tilt. In Fig. 59, let P be an elevated point and

P_o the point vertically below it at the same level as the instrument.

Suppose the telescope is directed to P_o and the transit axis has a small tilt α from the horizontal with the right hand end higher than the left hand end. Then, when the telescope is raised to the elevation of P it will come up to a point P′ which will be at a distance H . α from P, where H is the height of P above P_o. To set the telescope on P, the horizontal reading of the instrument will have to be increased; if P″ is vertically below P′, the length of P″P_o will also be H . α, and the change of horizontal angle reading will be H . α/D, where D is the *horizontal* distance of P from the theodolite. But H/D is *tanE*, where *E* is the angular elevation of P. Thus the effect of the transit axis error α on the reading of the horizontal circle is α . *tanE*, and the effect on an angle measured once will be

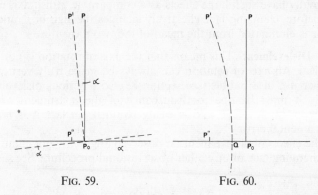

FIG. 59.   FIG. 60.

the difference $\alpha(tanE_2 - tanE_1)$, where E_1 and E_2 are the elevations of the two points concerned. Since depressions are to be regarded as negative elevations, the total effect on an angle can be considerable.

Note that if the face of the instrument is changed the error α will effectively become one of $-\alpha$, and the mean of the two readings will be the true reading on P_o.

(N.B. In formulae like H . α, where an angle is used in calculating a linear quantity, the angle must be expressed in radian units, see section 271).

101. Horizontal Collimation Error. This means that the line of sight is not perpendicular to the transit axis. Assuming no transit axis tilt and the instrument correctly levelled, the transit axis will be horizontal but the line of sight will trace out a cone centred about the transit axis and having a semi-angle not exactly 90°: let this semi-angle be $90° - \beta$.

First suppose there is no collimation error and the telescope is sighted on P_o, Fig. 60. Now introduce the collimation error β deflecting the line of sight to the observer's left, and the telescope will be sighted on Q, where $P_o Q = D \cdot \beta$. But, as the telescope is raised, the deflection from the vertical line $P_o P$ will increase because it is proportional to the actual distance from the centre of the theodolite, and the actual distance of P from the instrument is $D \cdot sec\ E$, so $PP' = \beta \cdot D\ sec\ E$, and the effect on the horizontal setting of the theodolite will be $(\beta\ D \cdot sec\ E)/D$, that is $\beta \cdot sec\ E$.

The effect of the collimation error on a single measure of an angle will thus be $\beta (sec\ E_2 - sec\ E_1)$. Since secants do not change sign with the angle, the expression inside the brackets will always be actually a difference, so it is seen that the collimation error will not normally have such large effects as a comparable transit axis error. As before, changing face effectively changes the sign of β, and the error is eliminated from the mean of the two measures.

102. Dislevelment. This means that the axis of rotation is not truly vertical. An axis of rotation can always be made truly vertical by proper use of a bubble (see section 83), so if serious dislevelment exists it must be due to disturbance of the instrument or just observer carelessness, and is easily rectified: in fact it is not an instrumental error.

The effects of dislevelment on observed angles are not eliminated by changing face or any other observational procedure. See section 235.

THE COMPASS

103. Surveying Compass, or Circumferentor. This is an instrument of the type illustrated in Fig. 61. The needle itself is of edge-bar

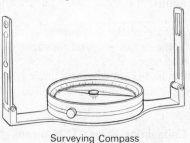

Surveying Compass

Fig. 61.

pattern (see section 41), 5 or 6 inches long, pivoted in a shallow circular box with a glass cover. Two sighting vanes fold down when not in use, and there is a device for raising the needle off the pivot.

The instrument is provided with a fitting for screwing it on to a single rod, called a Jacob Staff, or on to a light tripod.

In sighting the vanes on to an object, the whole instrument is rotated, the needle remains in the magnetic meridian, and therefore the numbering of the degrees must go anti-clockwise round the circle.

For general surveying purposes, these instruments are not much used nowadays.

104. Prismatic Compass. Small and more portable compasses, particularly those called *prismatic compasses*, are, on the other hand, extremely useful for rapid traversing and reconnaissance surveys.

The construction of a prismatic compass is indicated in Fig. 62.

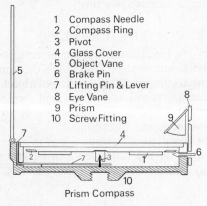

1   Compass Needle
2   Compass Ring
3   Pivot
4   Glass Cover
5   Object Vane
6   Brake Pin
7   Lifting Pin & Lever
8   Eye Vane
9   Prism
10 Screw Fitting

Prism Compass

FIG. 62.

The graduations are on a ring or disc pivoted at its centre: the magnet attached to the underside of the disc may be in the form of a flat strip.

The 45° prism, from which the instrument gets its name, is opposite the sighting vane and has a sighting notch immediately above it. The prism acts as a mirror and its horizontal and vertical faces are convex so that it also acts as a magnifier by which the graduated scale below the prism is seen clearly by the observer and a reading against a fixed mark can be taken while the object sighted is seen correctly lined up with the vane. The prism can be shifted vertically over a small range and thus placed to suit the user's eyesight.

The graduations go clockwise round the scale but, because the reading is taken at a point opposite the sighting vane, the zero of the scale is at the south-seeking end of the magnet, and the reading increases from right to left as seen in the prism.

These compasses are normally 3 or 4 inches in diameter; some are provided with a screw fitting for a Jacob Staff. The smaller compasses of Service pattern, only about $2\frac{1}{2}$ inches diameter, and very robust, are easily carried in the pocket and are excellent instruments for rapid traversing.

The convenience of these instruments is much increased if they are filled with liquid, and then they are usually known as *liquid compasses*. The liquid takes most of the weight off the pivot and minimises frictional effects; it also damps movements so that the scale comes to rest very quickly after setting on a mark. A compass not filled with liquid must have a device which lifts the needle off the pivot when the vane is closed down, and it should also have a braking device so that the operator can stop excessive oscillations and get a reading in a reasonable time.

THE LEVEL

105. The Surveyor's Level is an instrument designed primarily to furnish a horizontal line of sight. The line is determined by a tele-scope with the usual components consisting of object glass, focusing arrangement, diaphragm with cross-lines, and eyepiece. In practice, the telescope must be capable of rotation about a vertical axis so that it can be pointed in any direction.

Neglecting for the present various forms of rough levelling in-struments, the levels in use nowadays can be grouped into three main classes: (*a*) Dumpy Levels, (*b*) Tilting Levels, (*c*) Automatic Levels.

106. Dumpy Level. This type of instrument has been in use for a long time, but now seems likely to become superseded by the other types mentioned above.

A Dumpy Level consists of two principal components, the tribrach lettered A in Fig. 63, and the telescope assembly marked C. The tribrach is of normal design with three levelling screws and a central bearing, and is usually provided with a circular bubble A_1 for preliminary levelling. Methods of fixing the instrument to the tripod head are similar to those used for modern theodolites.

The telescope and the vertical column which rotates in the tribrach bearing form one rigid component, and there is a sensitive bubble C_1 fixed on the top of the telescope.

It will be seen that an essential permanent adjustment of this type of level is the requirement that the line of sight shall be accu-rately perpendicular to the axis of rotation, so that when the instru-ment is levelled (axis of rotation truly vertical) the line of sight will be truly horizontal when set in any direction. Moreover, any adjustment necessary to achieve this condition must be made by moving the diaphragm. This is the distinctive feature of Dumpy Levels.

The bubble C_1 is used in carrying out the levelling process, and it should be adjusted to be, as nearly as possible, central when the rotation axis is vertical.

A small Dumpy Level is illustrated in Fig. 65.

107. Tilting Level. This type of Level is slightly more complicated than the Dumpy, but it is preferred by many surveyors because it is quicker in use. It has three principal components. See Fig. 64.

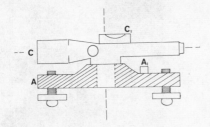

FIG. 63.

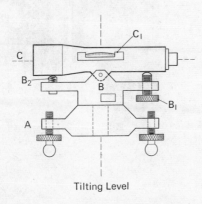

Tilting Level

FIG. 64.

In this type, the telescope C is pivoted near its centre on a short horizontal axis supported by the component B. When the setting, or tilting, screw B_1 is turned, the eyepiece end of the telescope is raised or lowered, and this movement is resisted by the compression spring B_2 so that the movement is free from looseness or backlash.

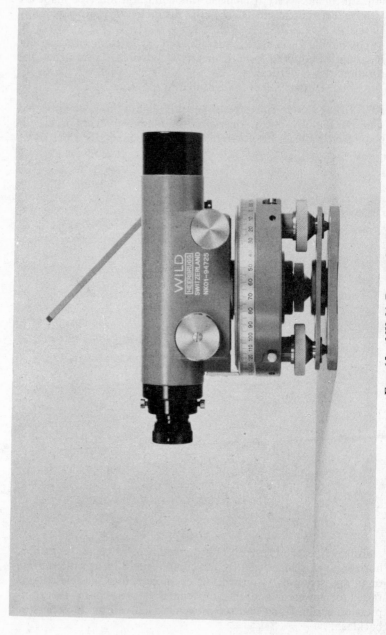

FIG. 65.—NK 01 DUMPY LEVEL

Parts B and C rotate together about the vertical bearing in the tribrach A.

The advantage of the tilting system of construction is that the line of sight can be adjusted by the tilting screw with reference to the sensitive bubble C_1, and it is therefore not necessary to make the axis of rotation exactly vertical. After approximate levelling with the circular bubble, the line of sight is re-set precisely by small adjustments of the screw B_1 whenever the telescope is rotated into another direction: for this purpose, there must be a mirror or prism device arranged so that the observer can see the bubble clearly from his observing position at the eyepiece. In some levels, the bubble can be seen in the telescope eyepiece itself, so the observer can check the setting at the moment when he takes the staff reading.

The adjustment required for a tilting level is therefore that the line of sight shall be horizontal when the sensitive bubble is central or in some specified standard position. Any necessary adjustment can be done by moving the diaphragm or, more easily as a rule, by resetting the bubble on its supports after the horizontality of the line of sight has been established by a suitable test.

On some of the larger tilting levels, the tilting screw is fitted with a graduated drum by means of which the line of sight can be set to a pre-determined gradient. In this case, the line of sight must be horizontal when the gradient drum is at zero.

108. Automatic Levels. These instruments require only approximate levelling by reference to a good circular bubble, and they have no sensitive bubble. The path of the line of sight through the telescope includes reflection at mirrors or prisms which are suspended on fine threads or on a component hanging as a pendulum and the mechanical design of the suspended portion is such that if the body of the instrument is slightly disturbed the consequent motion of the suspended component re-sets the line of sight into the former direction. That is, the horizontality of the line of sight is restored, provided it was horizontal in the first place.

The Level illustrated in Fig. 66 has prisms hanging on fine metal threads.

Most automatic levels are comparatively expensive, but they speed up the work because the observer is not constantly watching and readjusting the setting of a bubble, nor is he anxious that he may forget to level.

The adjustment of an automatic level will usually be done by moving the diaphragm; adjustment of the setting of the automatic device is a job for instrument experts.

109. Other Components of Levels. Most levels of the types described above will be provided with clamp and slow motion arrangement for the rotation movement.

Almost all levels have stadia lines on the diaphragm, the pattern shown in Fig. 20 (a) being most commonly found.

Some levels have horizontal graduated circles, to be read by a

FIG. 66.—S77 SELF-ALIGNING LEVEL
(By courtesy of Messrs. Vickers, Ltd.)

simple pointer to perhaps 1/10 degree, or by a vernier, or in some cases by means of an optical scale reading arrangement estimating to 1 minute (See Fig. 65).

A level fitted with horizontal circle and stadia lines can be used to make a complete three-dimensional survey of a limited area round the instrument. See Chapter XI.

A level is not ordinarily required to be accurately centered over a ground mark: in Levelling, the staff positions are the significant points. As with modern theodolites, many levels have permanently attached base-plates and these are directly screwed to the tripod head. Some of the smaller tilting levels are provided with clampable ball-and-socket for preliminary levelling adjustment.

LEVELLING STAFFS

110. A levelling staff (or 'rod' in America) is essentially a graduated scale which is held vertically and viewed through the telescope of the Level. Thus the staff reading at the place where the central horizontal cross-line of the telescope apparently cuts the scale is the vertical distance between the line of sight of the instrument and the support on which the staff is held.

The length of a levelling staff is generally in the range 10–15 feet, and most staffs can be hinged, telescoped, or separated into 3 or 4 pieces for convenience of transport.

A staff should have a strong metal plate at its lower end, to resist wear. A small circular bubble may be fitted to the back of a staff to indicate when it is vertical.

The graduations of a staff should if possible be on a recessed surface so that they do not get rubbed by contact with the ground or with other parts of the staff.

111. Graduation of Staffs. The reading of the staff is noted by the observer at the Level, hence the design and marking of the graduations must be such that the reading can be recognised at all distances, from the minimum focusing distance of the telescope to the maximum at which the numbers can be read: in practice, sights are not normally taken at distances much greater than 200 ft.

In British practice, division into feet with decimal subdivision is normal. The commonest forms of subdivision is in alternating bands of black and white each 1/100 ft. deep, with the numerals indicating feet in red and the numerals indicating 0·1, 0·3, 0·5, 0·7, and 0·9 in each foot in black. To avoid confusion, the 0·5 is often printed as V and the 0·9 as N. These figures are exactly 1/10 ft. high, and the usual convention is that the *top* of each numeral is at exactly the value indicated. Additional small red numerals are provided so that the surveyor can read the number of feet in case the staff is so near that a whole foot numeral does not appear in the field of view. This type of graduation is illustrated in Fig. 67.

At short distances, readings can be estimated to 1/1000 ft. if such accuracy is necessary. Some staffs are graduated to 1/20 or 1/10 ft. and readings are estimated to 0·01 ft.; this accuracy is quite sufficient for ordinary survey purposes.

On a staff graduated in metric units the small subdivisions are generally 1 centimetre deep; a common design of graduation and numbering is shown in Fig. 70 (c).

112. Types of Staff. The well-known Sopwith Staff is illustrated in Fig. 67. It is made of wood with brass fittings at points likely to wear, and has two hollow sections. Each upper section, when pulled out, is located in correct position by a springy clip which engages with the metal rim of the section below. The Sopwith staff is usually 14 ft. long fully extended. The graduations are printed on a strip of paper which is stuck to the wood.

The Scotch Staff, Fig. 68, and similar types are of metal girder

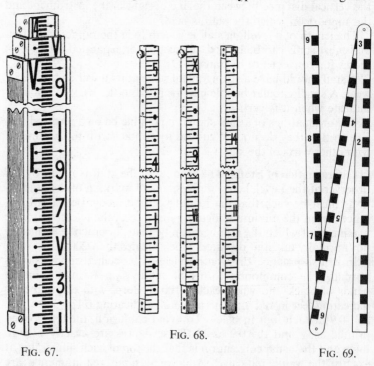

FIG. 67.   FIG. 68.   FIG. 69.

sections fitting together in sockets; they are usually of the same cross-section throughout, but some are tapered towards the top.

Graduations on a metal staff may be positioned by machining the surface, and such graduations are likely to be more accurate than those printed on paper strips.

Staffs are also made of light alloy metals in tubular cross-sections about 3 inches in diameter.

Another type of lightweight staff, illustrated in Fig. 69, is made to fold in 3 or 4 sections with a tightening nut at each joint. These staffs are generally made of wood in simple rectangular cross-section; they are inexpensive, but they will not stand up to use in rough conditions.

113. Staff Reading. The surveyor must of course get used to reading the staff apparently from top to bottom. As mentioned above, the actual tops of the normal-size numerals indicate the values; this means that if the cross-line does not cut one of the black numerals the first number after the decimal point must be odd. On some staffs the numerals are painted upside down but this tends to be confusing. Some readings are illustrated in Fig. 70.

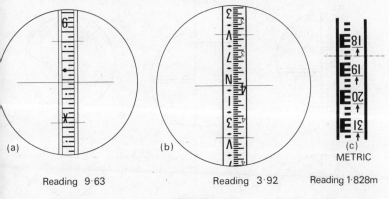

(a)   (b)   (c)
                                                          METRIC

Reading 9·63   Reading 3·92   Reading 1·828m

FIG. 70.

114. Target Staff. A staff may be provided with a sliding target of conspicuous design, that is raised or lowered by the staffman in obedience to signals from the surveyor: when the target is on the line of sight the reading is noted by the staffman. With such a staff, the graduations need not be so boldly marked, as the surveyor does not have to read them; also, very long sights may be taken if necessary.

However, it is the surveyor who should take the readings, and in any case the staffman is fully occupied in moving from station to station and holding the staff upright.

A target staff is necessary in conjunction with the Cowley Level as described below.

115. The Cowley Automatic Level. This instrument belongs to the 'automatic' class and operates by means of a pendulum. In appearance it does not look like an ordinary level but resembles a small cine-camera. It is particularly suitable in building and engineering work for setting out foundations, gradients, etc., and, although it can be used for running short lines of levels from point to point, it is not so suitable as an ordinary level for carrying forward long lines of levels. It has the advantage of being very simple and quick to use and it is very cheap considering the degree of accuracy obtainable with it. It also has the advantage of not requiring a skilled surveyor to manipulate it.

The level is used in conjunction with a graduated staff to which is attached an adjustable target at right angles to it, which can be slid up and down the staff by the staff holder and clamped at any position in obedience to signals from the observer. With the staff

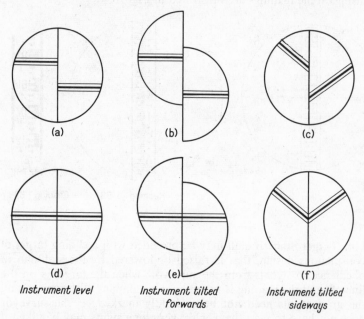

(a)   (b)   (c)

(d)   (e)   (f)

Instrument level   *Instrument tilted*   *Instrument tilted*
                     forwards   *sideways*

FIG. 71.—IMAGES OF THE TARGET IN THE FIELD OF VIEW OF THE
COWLEY AUTOMATIC LEVEL.

set vertically on any point, the observer looks through the aperture on top of the case and sees two images of the target separated by a vertical line. If the black horizontal line marking the centre of the target is not at the same elevation as the optical centre of the instrument, the two images of it will be separated as shown in Figs. 71 (a), (b) and (c). If the centre of the line is at the same elevation as the optical centre of the instrument, the two images will coincide as shown in Figs. 71 (d), (e) and (f). Hence, the observation consists in sighting on the staff and getting the staff holder to move the target up or down until the two images of the horizontal line at its centre appear to coincide at their edges. The reading of the height of the target above the bottom of the staff gives the amount by which the point on which the staff is held is below a horizontal plane through the optical centre of the level.

Fig. 72 shows sectional side and front views of the instrument.

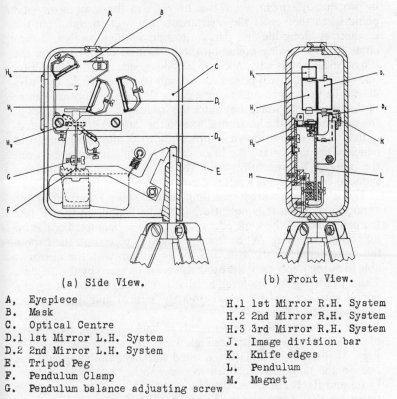

(a) Side View.   (b) Front View.

A, Eyepiece   H.1 1st Mirror R.H. System
B. Mask   H.2 2nd Mirror R.H. System
C. Optical Centre   H.3 3rd Mirror R.H. System
D.1 1st Mirror L.H. System   J.   Image division bar
D.2 2nd Mirror L.H. System   K.   Knife edges
E. Tripod Peg   L.   Pendulum
F. Pendulum Clamp   M.   Magnet
G. Pendulum balance adjusting screw

FIG. 72.—SECTIONAL VIEWS OF THE COWLEY AUTOMATIC LEVEL.
(By courtesy of Messrs. Hilger & Watts Ltd.)

There are two mirror systems, one on each side of a vertical plane passing longitudinally through the centre of the front of the instrument. The left hand system consists of the mirrors D_1 and D_2 and the right hand system of the mirrors H_1, H_2 and H_3. The mirrors D_1, D_2, H_1 and H_2 are fixed relative to the sides of the casing but the mirror H_3 is attached rigidly to a pendulum in such a way that it is always horizontal when the pendulum is vertical. A ray of light passing through the centre of the front aperture to meet D_1 is reflected to the mirror D_2 and thence to the eyepiece A on top of the case. Another ray passing through the centre of the front aperture and meeting the mirror H_1 is reflected to the mirror H_2 and thence to the horizontal mirror H_3, from which it is reflected upwards to pass through the eyepiece A. If, when the instrument is in proper adjustment, these two rays are in the same horizontal plane when they enter the instrument, they will emerge from the eyepiece A alongside and parallel to one another, so that, if they come from adjoining points on the horizontal central line on the target, these points will appear to lie side by side and the horizontal line on the target will be seen as an unbroken line. If the two rays do not lie in a horizontal plane passing through the optical centre of the instrument when they enter the front aperture, the images of the two halves of the horizontal line on the target will be separated.

In order to damp down the oscillations of the pendulum, and hence of the mirror H_3, the blade of the pendulum swings between the poles of a powerful magnet M and, as the pendulum swings, induced eddy currents quickly damp down the oscillations.

The level is supported on a long vertical spike E carried by the tripod, and, when it is not attached to the tripod, an arm F lifts the pendulum off the knife edges K so that it is held fixed and is no longer free to swing. When the instrument is placed on the tripod, the spike depresses the arm, and the pendulum with the mirror is able to swing freely until the oscillations are damped by the magnet. Thus, in setting up the level, all that is necessary is to set up the tripod with the spike E approximately vertical and then slip the instrument on to the spike.

If the instrument is tilted forwards or backwards, the two halves of the field of view are displaced relative to one another as shown in Figs. 71 (b) and (e), and, if it is tilted sideways, the images of the line on the target are inclined to one another as shown in Figs. 71 (c) and (f). In all cases, the observation consists in moving the target up and down the staff until the left hand and right hand images coincide as shown in Figs. 71 (d), (e) and (f).

When the level is to be set up on a brick course or other place where it is not possible to use the ordinary tripod, a special 'brick-layer's stand', provided with a spike similar to that on the tripod, can be employed. This stand is obtainable as an extra.

The adjustment of the instrument can be tested by the two-peg test described in section 118· but any lack of adjustment is best corrected by the makers.

The Cowley level weighs only $2\frac{3}{4}$ lbs., the tripod under 2 lbs., and the staff and target $2\frac{1}{2}$ lbs. The accuracy is stated to be about $\frac{1}{4}$ in. at 100 ft., and better at shorter distances.

TEMPORARY ADJUSTMENTS OF THE LEVEL

116. After a Level is fixed on its tripod, a preliminary levelling is done with the aid of the circular bubble. Then:

(i) with a dumpy level, the procedure of levelling is carried out as for a theodolite, by use of the footscrews and reference to the sensitive bubble attached to the telescope; the object is to make the axis of rotation truly vertical. See section 83.

(ii) with a tilting level, the sensitive bubble is set precisely into its standard position, by use of the tilting screw, whenever the direction of pointing of the telescope is changed.

(iii) with an automatic level, no further levelling is necessary. It is desirable, however, to make occasional tests to ensure that the automatic device is in operation, by slightly disturbing a foot-screw and noting if the crossline returns quickly to the reading previously noted. The automatic device is heavily damped and has a small range of movement. Some automatic levels are provided with a fitting by which the observer can deliberately disturb the device, to check that it is actually operative.

As for a theodolite, the diaphragm must be focused by adjustment of the eyepiece. In levelling, focusing of staff and removal of parallax must frequently be done because the sights taken in levelling are short and of various lengths.

TESTING AND ADJUSTMENT OF THE LEVEL

117. In the examination of the Level with reference to the quality of the non-adjustable parts, the important requirements are stability and a proper relationship between the sensitiveness of the bubble and the resolving power of the telescope. These tests are the same as for the theodolite (section 89).

There is only one fundamental requirement for the permanent adjustment of a level, and it is that the line of sight shall be horizontal when:

(i) in a dumpy level, the axis of rotation has been made truly vertical,

(ii) in a tilting level, when the sensitive bubble has been set in its standard position,

(iii) in an automatic level, when the automatic device is operative.

118. To test a Level, it is first necessary to establish a line which is known to be horizontal. The simplest way to do this is the well-known 'two peg' method, indicated in Fig. 73. Staffs are held at two points A and B some 150 ft. apart, and the instrument is set up equidistant from them (Fig. 73 (a)). If the temporary adjustments are made and readings are taken of the staffs, the difference of the readings is the true difference of height between A and B, because

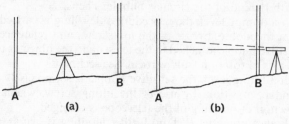

FIG. 73.

even if the line of sight is not horizontal it will be equally inclined at both pointings, and the equidistance of the sights causes any error to cancel out.

Let us suppose that the readings at A and B are 5·878 and 4·014 respectively; then the true difference of height is 1·864. Therefore, a line joining a point on staff A to the point on staff B where the reading is 1·864 less will be horizontal.

Now, the instrument is removed and set up at a place much nearer to staff B than to staff A, as indicated in Fig. 73 (b); it is convenient to put the level nearly but not quite on the line AB produced. The instrument is levelled and readings are taken. If the same difference of reading is not obtained, the line of sight cannot be horizontal. Let us suppose the new readings are 6·224 and 4·282: the difference of these is 1·942, which is too great. Now it is obvious that if the tilt of the line of sight is altered, both readings will increase or both will decrease, but it is the reading on A that will alter most.

In the above case, it can be seen therefore that the line of sight is directed upwards, because if it is now lowered both staff readings will decrease but the reading on A will be reduced much more than that on B, and so the difference will be reduced, which is what is required.

The necessary adjustment of the instrument can be made by trial and re-testing, until the true difference of level is given by the readings. With a tilting level, the adjustment is easy; the level is tilted until the correct difference of staff readings is obtained and then the bubble attached to the telescope is centralised by re-setting its supporting screws. With other types of Level, it will probably be necessary to move the diaphragm up or down—up in the case described above.

If the distances from instrument to staffs are measured, it is possible to work out the correct readings by a simple bit of arithmetic. Suppose the staffs are 140 ft. apart, and the Level in the second position is 20 ft. from staff B and 160 ft. from staff A. Then any alteration of inclination of the line of sight will result in an alteration of the A reading 8 times as great as the alteration of the B reading: the alteration of the difference of the readings will therefore be 7 times the alteration of the B reading; hence the alteration of the B reading must be $\frac{1}{7}(1\cdot942 - 1\cdot864) = 0\cdot011$, and the alteration of the A reading will then be $\frac{8}{7}(1\cdot942 - 1\cdot864) = 0\cdot089$. The level must be set to new readings $6\cdot135$ and $4\cdot271$, which have the correct difference.

HAND LEVELS AND CLINOMETERS

119. These levelling instruments are designed with a view to lightness and compactness rather than a high degree of precision, since the absence of a steady support does not allow of the employment of a sensitive bubble tube. They are convenient for rapid work, especially where serious error cannot be accumulated, as in cross sectioning. The clinometer is simply a hand level adapted for measuring vertical angles, and, as it can be employed as a hand level, is the more useful instrument.

120. The Hand Level. The hand level (Fig. 74) consists of a tube, about 4 in. long, circular or square in section, and having a small

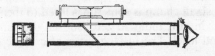

Fig. 74.—Hand Level.

bubble tube mounted on top. A line of sight parallel to the axis of the bubble tube is defined by a pin hole at the eye end and by a cross bar, or one edge of a rectangular opening, at the other. To afford the observer a view of the bubble to enable him to hold the instrument level, the metal body is cut away under the level tube, and a reflecting surface is inserted in the tube at 45° to its axis. This reflector extends half-way across the tube, and distant objects can be viewed through the other half. When the centre of the bubble appears opposite the cross-bar sight, or lies on a line ruled on the reflector, the instrument is being held level, and all points inter-sected by the line of sight are at the same level as the eye. Telescopic forms of the instrument are also available.

If provision is made for adjustment of the bubble tube relatively to the line of sight, the test is carried out by the two-peg method as for the dumpy level.

121. The Abney Clinometer. This instrument (Fig. 75), usually known as the Abney level, is deservedly the most popular instru

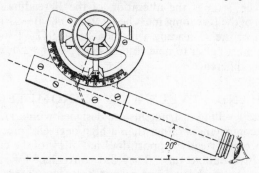

FIG. 75.—ABNEY CLINOMETER.

ment of its class, and is simply the hand level of Fig. 74 adapted for measuring vertical angles. The bubble tube is pivoted, and carries an index arm perpendicular to its axis. Rotation of the milled wheel moves the index over a graduated arc attached to the sighting tube, and, when the bubble is centered, the reading on the arc represents the inclination of the line of sight above or below the horizontal. A small clamp is sometimes fitted to the index arm, and is useful for fixing the index to zero, so that the clinometer can be used as a hand level. The instrument is also made in a telescopic form.

In using the Abney, the sighting tube is directed towards the object, and the milled wheel is turned until the reflected image of the bubble appears on the index line. The required angle is then read on the arc. To test and adjust the instrument, the index is set to zero, and the two-peg method is applied. Alternatively, two points at different elevations are selected, and the vertical angle between them is observed from both. The angle of elevation observed from the lower station should equal the angle of depression taken at the upper. If these differ, their mean is the correct value of the inclina-tion, and the instrument is made to record this by means of the adjusting screws controlling the level tube.

122. The Water Level. This simple piece of apparatus consists of two glass, or hard transparent plastic, tubes connected by a length, say 100 ft., of rubber, or flexible plastic tubing; transparent plastic is best because the observer can see if there are air bubbles in it. The two tubes are attached to graduated strips of wood as shown in Fig. 76.

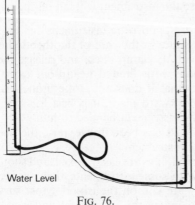

Water Level

FIG. 76.

A suitable quantity of water, preferably slightly coloured with a dye, is run into the apparatus so that the water surface is about half way up each scale when they are placed at the same height. There must be no air bubbles in the water.

When the graduated scales are held vertical at separate places, the difference of the readings of the water surfaces is the difference of height of the places.

This simple instrument can be used by almost unskilled personnel. It is advantageous to fit a tap, or a spring-loaded valve, at the top end of each vertical tube, and to close them during movement from one position to another, otherwise it is very easy to lose water from open tubes.

The Water Level is best when used in rather flat country because its vertical range is limited to about 5 feet per set-up.

The accuracy of a simple water level is comparable with that of a small telescopic level. However, the method of hydrostatic levelling, as it is called, has been used to transfer levelling values across wide estuaries, etc., in connection with levelling operations of high accuracy.

PRECAUTIONS IN USING INSTRUMENTS

123. Setting Up. (*a*) In lifting the instrument from the box, handle it in such a way as to minimise the possibility of strain. It must not be lifted by the telescope, the standards, or axes, but should be supported by the hand placed under the levelling head.

(b) Screw the instrument firmly home on the tripod head, so that the connection may not slacken during observations.

(c) If the instrument has three levelling screws held by a locking plate, see that the plate is gripping the screws and is properly locked, otherwise the instrument will fall off the tripod when lifted.

124. Carrying. (a) Support the instrument against the shoulder as uprightly as possible. In the case of the theodolite, place the telescope vertical with its clamp slack, and release the lower clamp.

(b) In situations with limited headroom carry the instrument in front, and see that it does not strike branches, etc. In crossing a fence or any awkward place, it is best for one man to go over first and have the instrument handed to him.

(c) If the compass has been in use, raise the needle off its pivot before lifting the instrument.

(d) Modern theodolites are easily boxed, and should not be carried on the tripod. With some instruments, the cover can be fitted over them while they are still on the tripod. Most surveying organisations require boxing of the instrument and removal from tripod during transport.

125. In Wind. (a) Spread the tripod legs well apart, and thrust them firmly into the ground.

(b) On a pavement or other smooth surface, if sufficient spread is given to prevent overturning by wind, the instrument may collapse by the feet sliding outwards: they should therefore be inserted in joints or cracks, or tied together with strong cord. See Section 42.

(c) Do not leave the instrument unattended in high wind, even though the above precautions have been taken.

126. In Sun or Rain. (a) Protect the objective in strong sunlight by fitting the sun shade. Use a brush for absorbing moisture from the eye lens. If a brush is not available, a corner of the handkerchief may be used, but the lens should not be rubbed.

(b) If work is stopped temporarily, put the cover on the objective. The telescope of a theodolite should be set vertically, with the eyepiece down. In wet climates a waterproof hood to slip over the instrument is useful.

(c) If water enters the telescope tube, it will cause dewing of the lenses, and observations are impossible. This cannot be remedied at once in the field; moisture will gradually evaporate if the instrument is left in the sun or in a moderately warm room, and drying is hastened by removing the eyepiece.

(*d*) Dry all exposed parts thoroughly before putting away the instrument.

127. General. (*a*) Do not force screws, or apply clamps too tightly.

(*b*) A camel-hair brush is best for dusting instruments. Levelling and tangent screws are cleaned with a nail brush, and a little oil should be applied, and finally wiped off. Oil should not be left on exposed parts as it collects dust. Vaseline or watch oil is used to lubricate axes, but this need seldom be renewed. Unless the instrument develops stiffness, it should not be taken apart.

(*c*) Avoid fingering silver graduated surfaces. If it is tarnished, apply a little Vaseline, and where necessary rub it and wipe clean with a soft rag. On no account use plate polish.

(*d*) For care of telescope, see section 33.

128. Packing. (*a*) There is only one position in which an instrument will fit into its box, and the difficulty of packing may be lessened if a note of the correct positions of the various parts is written in the box. The plates of the levelling head must be left approximately parallel, and clamps should be slackened.

Some instruments have coloured spots or other marks, and should be packed with all these marks showing on top.

(*b*) The instrument should pack without forcing, but it may happen that difficulty is experienced owing to shrinkage or warping of the wood of the box. The offending part can be discovered by chalking those portions of the instrument held by the bearing pieces inside the lid, and may be cut down a little. Most modern instruments are packed in metal boxes.

(*c*) The cap and ring of the tripod should be kept in the box while the instrument is in use: or there may be fittings for temporarily attaching them to the legs of the tripod.

REFERENCES

BANNISTER A. AND SCHOFIELD R. B. 'Modern Surveying Instruments'. *Journal of the Institution of Municipal Engineers*, Vol. 94, June 1967.

COOKE, TROUGHTON AND SIMMS LTD. 'Theodolite Design and Construction'. *Publication No.* 883, 1955.

HALLER R. 'Theodolite axis systems: their design, manufacture and precision'. *Surveying & Mapping*, Vol. XXIII, No. 4, 1963.

HARDY R. L. 'Pendulous cantilever principle applied to self-leveling instruments'. *Surveying & Mapping*, Vol. XVI, No. 3, 1956.

HIGGINS A. L. 'Defects of Surveying Telescopes'. *Empire Survey Review*, No. 55, Jan. 1945.

MUSSETTER W. 'Stadia characteristics of the internal focusing telescope'. *Surveying & Mapping*, Vol. XIII, No. 1, 1953.

PAGE B. L. 'The graduation of precise circles'. *Surveying & Mapping*, Vol. XIII, No. 2, 1953.

PECKMPAUGH C. L. (Jnr.). 'The Zeiss Opton Ni. 2 automatic Level'. *Surveying & Mapping*, Vol. XIV, No. 2, 1954.

SUTTON D. AND THOMAS T. L. 'New meridian indicator for the precise determination of underground azimuths'. *Trans. of the Inst. of Mining & Metallurgy*, Vol. 73, 1963.

TAYLOR E. W. 'The evolution of the Dividing Engine'. *Empire Survey Review*, No. 52, Apr. 1944.

— 'Testing a theodolite for accuracy'. *Empire Survey Review*, No. 40, Apr. 1941.

— 'The Surveyor's Telescope'. *Empire Survey Review*, No. 39, Jan. 1941.

— 'The effects of eccentricity and misplaced indices of divided circles'. *Empire Survey Review*, No. 38, Oct. 1940.

— 'The art of original circular dividing'. *Empire Survey Review*, No. 49, July 1943.

— 'The new Cowley automatic Level'. *Surveying & Mapping*, Vol. XIV, No. 2, 1954.

CHAPTER II

LINEAR MEASUREMENTS

129. There are many different ways of making the measurements of length or distance required in Surveying, but always the method chosen must be suited to the accuracy desired. For some classes of work an error of one or two feet in a hundred does not matter, but for others an error of one foot in a mile would be considered excessive, while for the very highest class of linear measurement done in survey work–that is in the measurement of a base for geodetic triangulation–the discrepancy between two separate measures of the whole line will rarely exceed one part in a million and the real accuracy may be almost as good.

In ordinary engineering work and simple land surveying, great accuracy in linear measurement is seldom necessary. Quite rough methods are permissible in surveying ill-defined detail such as the edge of a marsh, or in the measurement of lines of a rapid compass traverse. On the other hand, the setting-out for tunnels, large dams, and other extensive constructional works may demand measurement of the highest possible precision.

There are two main types of distance measurement, direct and indirect. The direct methods include the use of chains, wires or steel tapes stretched along the lines to be measured. Indirect methods make use of optical systems for relating a short known length to the angle it subtends at the distance to be determined: tachymetry, range-finding and the subtense method come into this class. On geometrical principles, the optical methods cannot have a very high accuracy, but in certain circumstances they can be extremely useful and efficient.

It is also necessary to mention, among the direct methods, the use of so-called 'electromagnetic distance measuring' (EDM) instruments, such as the Geodimeter, Tellurometer and Mekometer. These instruments send out beams of light or short radio waves which are modulated at perfectly regular intervals, and the operator is able to determine the exact number of modulation intervals in the distance from the sending instrument to the other end of the line and back. These instruments are expensive, and they are not likely to have wide application to large-scale surveying work in the immediate future.

In this chapter, only the direct methods with chain or tape will be considered. Optical methods will be described in Chapter XI, and EDM instruments will be referred to in Vol. II.

RANGING OR SETTING OUT CHAINAGE LINES

130. In measuring the lengths of lines it is important that the chain or band should follow, as far as possible, the straight line between the terminal points, and that deviations from that straight line should be as small as possible. If the line is short, or the distant end easily visible from every point along it, it is easy to maintain direction; but if the line is long, or the station at the distant end of it not visible from every point, it may be necessary to put some intermediate marks or ranging rods in positions where they will assist the rear chainman to control direction and prevent the path actually followed by the chain from deviating to any considerable extent from the line that it is supposed to follow. This 'ranging' as it is called, should generally be done before measurement commences; and, for most purposes, it can be done by eye without the use of any directing instrument, though, in the case of the more precise type of traverse, or in such work as laying out long 'straights' in railway surveys and construction, it will nearly always be advisable to set out the line by using a theodolite to control direction.

131. Ranging by Eye. The ranging of a line by eye is performed by the surveyor and an assistant. The former remains at one station, while the latter proceeds the required distance along the line and puts himself approximately in alignment. He faces the surveyor, and holds a pole vertically and nearly at arm's length, so that his body will not obstruct the surveyor's view. The latter directs him to right or left until the pole appears in line with the remote station pole. When signalled to mark, the assistant fixes his pole and examines it for verticality. He should then wait for a second signal indicating that the surveyor is satisfied with the fixing of the pole.

Accurate ranging can best be done if the surveyor stands a few yards behind the pole at his end and lines up the edges of the poles with one eye. The assistant should hold his pole lightly between finger and thumb, so that it can hang vertically. To minimise effects of any non-verticality the lining up should be done by sighting as low as possible on the poles.

In guiding the assistant into line, the surveyor should use a pre-arranged code of signals, such as:

Rapid sweeps with right (left) hand = Move considerably to right (left);

Right (left) arm extended = Continue to move to right (left);

Slow sweeps with right (left) hand = Move slowly to right (left);

Right (left) hand up and moved to left (right), left (right) hand
   down and moved to right (left) = Plumb pole in direction
   indicated;
Both hands above head = Correct.
Signals should be made clearly: when they have to be read at a
considerable distance a handkerchief should be held in the hand.

132. Ranging by Theodolite. In ranging a line by theodolite, the
instrument is set up and adjusted over one station and sighted so
that the vertical cross-line intersects the distant station. The vertical
axis is now firmly clamped, and, if necessary, the cross-line brought
on to the mark by the tangent screw. Intermediate points can then
be set out by signalling to an assistant holding the ranging rod until
the rod is seen behind, and apparently bisected by, the vertical cross-
line. The double vertical cross-line, Fig. 20 (e), can be useful here.

   If a straight line has to be prolonged beyond a marked end-point
the instrument is set up, centered and levelled over the forward
point and the telescope is directed to the back station. Much the
same procedure is now adopted as was done in adjusting the vertical
cross-line for collimation (section 94). Referring to Fig. 57, let A be
the back station. After setting on A, with plates clamped transit the
telescope and line out a wide-topped peg at B, the place where the
next forward station is required, and ask the assistant there to
put a tack on the peg to mark exactly a point where the vertical
cross-line intersects it. Swing the instrument through 180°, again
sight A, transit the telescope, and line in a tack on the peg at B'.
Unless the instrument is in perfect adjustment for collimation,
which it very seldom is, the tacks at B and B' will not coincide.
Take a point A' half-way between B and B', and put a tack or make
a mark there: then OA' is a prolongation of the line AO. If the
collimation error of the instrument is rather large, it may be necessary to use two separate pegs at B.

   If an appreciable collimation error exists, and steps are not taken
as indicated above to eliminate the effects of it, the line traced out
on the ground will consist of a series of chords to a flat curve, and
will thus not be a straight line.

   The prolongation can also be done by turning the theodolite
through exactly 180°, provided that the elevation of the telescope
does not have to be changed and that the theodolite is not so
coarsely divided that a precise setting of the 180° cannot be made.

LINEAR MEASUREMENTS WITH THE CHAIN

133. The wire chain, already described in section 43, cannot be
expected to give very precise measurements, nevertheless it is a

most useful article in general surveying and engineering. It will stand a lot of rough treatment, but it is liable to increase in length during use, so it should be frequently tested against a known standard length. Stretching due to distortion of the small links may be considerable, and is not readily noticeable.

Careful measurement with a reliable chain in good condition can give an accuracy of 1 in 2000, but it is best to reckon on 1 in 500 in normal practice, and perhaps a value as low as 1 in 200 if the measurement is done rapidly over rough ground.

134. Field Party. The party consists of:

The Surveyor, who determines what measurements are to be taken, directs their carrying out, and records the results in his field-book.

Two Assistants or Chainmen, who make the measurements.

Except in the case of a survey which may be completed in an hour or two, it is a mistake for the surveyor to work with only one assistant, as he must then take part in the chaining, etc., time is wasted, and it is difficult to keep the notes clean.

135. Equipment. Essential–One chain, 66 ft. or 100 ft. with ten arrows; one linen or plastic tape, 50 ft. or 66 ft.; six to twelve ranging poles, according to size of survey and character of ground; field-book, pencil and rubber.

Optional–Cross-staff, or Optical Square, or Box Sextant for laying out offset lines; clinometer; small compass; plumb line; offset rod; pegs; flags for attaching to poles; field glass.

136. Chaining the Line. The word 'chaining' denotes the measurement of a distance whether it is executed by the chain or the steel band. No explanation is required of the case where the distance is shorter than the length of the chain, but otherwise the chain must be stretched successively in the correct line, and it is important to realise the need for system in this apparently very simple operation if speed is to be acquired and serious error avoided.

Throwing out the chain. With both handles in his left hand, the chainman throws out the chain with his right, and, assisted by the second chainman, proceeds to free it from twists and knots. Otherwise, one man may retain the handles, while the other walks forward with the remainder of the chain, paying it out on to the ground. When the chain has been approximately straightened, and examined for bent links and badly opened joints, the work of laying it down successively may proceed.

Lining and Marking. The more experienced of the two chainmen remains at the zero end or rear of the chain, and is known as the *follower*; the other, called the *leader*, takes the forward handle. Before starting, the leader must provide himself with the ten arrows with which to mark the successive positions of his end of the chain, and he and the follower should each have a ranging pole.

While the follower holds the rear handle at the initial point, the leader stretches the chain as nearly on the line as he can judge, and, standing erect, holds his pole for alignment by the follower. It is best to range a point a little short of the chain length from the follower; the end of the chain being slack and held in the leader's left hand, the pole may therefore be placed about opposite the handle. When the pole has been lined, the position it occupied is marked by the hole made, or by a scratch, and the pole is removed, but in long grass or heath it should be left in. The follower now holds his handle exactly at the station point, while the leader proceeds to stretch the chain in a straight line over his mark or against the pole. This is accomplished by transmitting a series of gentle undulations along it by shaking the handle up and down, and at the same time bringing it gradually into line. Only sufficient tension to keep the chain straight is required. While straightening the chain, the leader should be stooping, and holding an arrow hard against the end of the handle and in the same hand, so that he can thrust the arrow vertically into the ground immediately the follower instructs him to mark.

To repeat the procedure for the next chain length, the leader, taking his pole and remaining arrows, walks forward and pulls the chain after him, having first swung it a little out of the line so that the arrow which has been placed will not be disturbed. He should count his paces so that he may know to stop and turn round when the follower, carrying the rear handle, has reached the arrow. The follower ranges the leader by planting his pole behind the arrow, and, having held his handle at the arrow (preferably flush with that side which was against the leader's handle) and got the second length marked by the leader, removes the arrow. The number of arrows in the possession of the follower at any time thus shows the number of completed chain lengths.

Transfer of Arrows. At the end of the tenth chain length the leader has inserted his last arrow. When the chain is next pulled forward, the follower removes the tenth arrow, and marks the place by pushing a nail or pencil into the ground. The two men then meet, and the ten arrows are handed over to the leader, who uses them over again. In a long line a careful record must be kept of the number of such transfers. Occasionally eleven arrows are carried, the additional one being used before the transfer to mark the forward end of the eleventh length.

End of Line. At the end of a line there will usually be a part chain length to be measure. The leader pulls his end of the chain beyond the station, and then straightens the chain up to the station pole, and reads.

The counting of his paces by the leader when pulling the chain forward should not be omitted, as it may save the follower a search for the arrow in long grass. The leader should not be stopped suddenly by the follower pulling the chain, as the jerk may lengthen the chain.

It requires a little practice to be able to line up the chain quickly. Beginners are apt to shake and tug violently, making it very difficult for the follower to hold his arrow. The tension should be only sufficient to make the chain lie straight.

When chaining on a hard surface, such as that of a road or street, the leader should mark each chain length by a scratch made with an arrow, or by a chalk mark, and should lay the arrow on the ground. It is advisable to write the chainage at each mark.

137. Making up the Chain. The chain must be folded up in a particular manner to form a compact bundle which can be tightly held by a strap. Having doubled up the chain by pulling it at the middle, the chainman holds it at the 50 tab in the left hand, and with the right hand grasps it at the joints corresponding to 48 and 52. He passes the four links to the left hand, doubling them as he does so to form a length of one link fourfold. He continues with the remainder, laying each four links obliquely across those in his hand. When finished, the bundle should have an hour-glass shape, all the links bearing on each other at the waist. The strap is passed round the narrowest part and through the handles.

138. Chaining on Sloping Ground. If the ground over which a measurement has to be made is not level, a procedure for getting the horizontal lengths of the lines should be adopted. This may be done directly by 'stepping' the measurement in horizontal lengths, or indirectly by deducing the horizontal equivalent of the distance along the slope.

Stepping. Chaining Downhill. In this case the follower's end of the chain is held on the ground, and his routine is as before. The leader must hold his end, either of the whole chain or of a suitable portion of it, above the ground so that the chain is stretched horizontally in the air (Fig. 77). It is ranged, and the end of the suspended length transferred to the ground by means of a plumb line held by the leader.

Chaining Uphill. It is more difficult to obtain good results in this case, as the follower's handle is now off the ground, and he must simultaneously plumb the end over the mark and range the leader. It is therefore a great convenience to have sufficient poles in the line so that the leader can align himself.

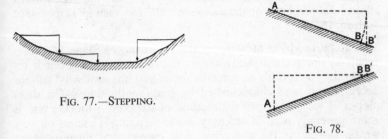

FIG. 77.—STEPPING.

FIG. 78.

The length that can be measured in one step is limited by the steepness of slope, as it is inconvenient to have the suspended end of the chain more than about 6 ft. above the ground. Sag of a heavy chain may also cause serious error, the effect of which is proportional to
$$(\text{weight of chain/pull})^2,$$
and for this reason it may be better to lay out the chain on the ground but use the linen tape for measuring the steps.

Transfer of the position of the free end of the tape or chain to the ground may be done by dropping an arrow, but it is better to use a plumb-bob.

It is difficult for the chainmen to judge the horizontality of the chain, so the surveyor should stand to one side of the line and direct them.

It is advisable to keep the lengths of the steps to some particular values such as 20 or 50 links, if possible. Arrows should not be used to mark steps, as this can confuse the counting of chain lengths; twigs or nails may be used.

Measuring along Slope. In this method, chaining is performed along the surface of the ground, and the various slopes encountered are measured by clinometer (section 121). If the measurement is composed of a series of lengths l_1, l_2, etc., inclined at angles θ_1, θ_2, etc., to the horizontal, the the required horizontal distance is:

$$l_1 \cos \theta_1 + l_2 \cos \theta_2 + \text{etc.}$$

If only the total distance is required, the calculation may be made on completion of the chaining, but, when numerous intermediate points have to be located, it is better to make a correction in the field at every chain length. The chain having been stretched in the position AB (Fig. 78), the leading arrow is shifted from B forward to B', where BB' = 100 (sec $\theta - 1$) feet or links, and the next chain length starts from B'. On moderate slopes the chainage of intermediate points may be read directly from the chain with sufficient accuracy for ordinary offsetting. In applying the method, a table should be prepared of the values of BB' for various slopes. (See section 164).

139. Relative Merits of Stepping and Measuring on Slope. Measurement on the ground yields better results than stepping but it requires the surveyor to carry a clinometer and it is somewhat tedious except on ground characterised by long gentle slopes. On short slopes of varying degree the method of stepping is quicker and is more generally used in ordinary work. Stepping is useless on very gentle slopes, as the sag error may exceed that introduced by assuming the inclined and horizontal lengths to be equal. At the same time, there is little point either in stepping or in measuring slopes when the chain is used and the slope of the ground is less than about 3°, or, say, 1 in 20, as the accuracy of the method will seldom justify such refinements; but, on gradients exceeding 1 in 20, it is as well to correct for slope, by one method or the other: the slope correction for this gradient, for a length of 100, is 0·13, or 1 in 800 approximately.

ERRORS IN ORDINARY CHAINING

140. An examination of the nature and effects of the various sources of error in ordinary chaining is necessary for the due appreciation of the relative importance of the precautions to be observed against them. The difference between cumulative and compensating, and between positive and negative, errors should be kept in mind. Errors and mistakes arise from:

    (1) Erroneous length of chain.
    (2) Bad Ranging.
    (3) Bad Straightening.
    (4) Non-horizontality.
    (5) Sag.
    (6) Careless Holding and Marking.
    (7) Variation of Temperature.
    (8) Variation of Pull.
    (9) Displacement of Arrows.
    (10) Miscounting of Chain Lengths.
    (11) Misreading.
    (12) Erroneous Booking.

    (1) *Erroneous Length of Chain.* Cumulative + or −. This is the most serious source of error in using the wire chain because of its

liability to stretch. It should be tested before commencing work, and on important surveys should be compared with a field standard every day or two. Knotting of the rings should be carefully avoided. Measurements made with a chain that is subsequently found to be in error can be corrected by adding the quantity:

(actual length minus nominal length) × number of chains measured.

If the chain is stretched the error is negative, and the correction is therefore to be added, because the chain does not go sufficiently often into the line and the recorded length will be too small.

(2) *Bad Ranging*. Cumulative +. In ordinary work this produces a relatively small error. If a whole chain length diverges a distance d from the true line, the error will be practically $d^2/200$ feet, or links, plus the effect on the rest of the line, part or all of which must be out of alignment. Refined ranging is unnecessary if only distance is required, but care is necessary if offsetting is to be done because the offsets will be put in error by the full amount of the divergence.

(3) *Bad Straightening*. Cumulative +. The effect produced by the chain lying in an irregular horizontal curve is similar to the last, but is more productive of error as the deviation is not so easily seen.

(4) *Non-horizontality*. Cumulative +. This error is common in stepping, but also arises from disregarding slight slopes. It is the second error above, reproduced in the vertical instead of the horizontal plane, but is much more important in practice.

(5) *Sag*. Cumulative +. When the chain is stretched in the air either in stepping or in measuring over small undulations or obstructions, it must necessarily sag, and the distance between the ends is less than that read. The error is reduced by suspending short lengths only and pulling firmly.

(6) *Careless Holding and Marking*. Compensating ±. The follower may hold his handle to one or the other side of the arrow, and the leader may not thrust his arrow vertically into the ground or exactly at the end of the chain. The possibility of inaccurate marking is much increased when plumbing. The error of marking developed by inexperienced chainmen is often of a cumulative character, but with ordinary care the distance, as marked, may be greater or less than a chain length, and such errors tend to compensate.

(7) *Variation of Temperature*. Cumulative + or −. The effect of temperature variation is negligible in ordinary chaining.

(8) *Variation of Pull*. Compensating ±. This also is unimportant in ordinary work with a chain. Inexperienced chainmen may consistently apply excessive tension and this will cause a small cumulative error.

The other four errors mentioned above are due to mistakes and should occur only very rarely. If an arrow is displaced and its proper position cannot be recovered, the line should be re-chained. Miscounting of chains can be avoided by adherence to a good routine: an error of a whole chain will certainly show up when the work is plotted, if there are sufficient tie lines. It may help, to write down each whole chain in the field-book, whether there is an offset or not. The commonest source of misreading is to confuse the 40 and 60 tabs on the chain: the surveyor should watch any offsetting done near the 50 tab. To avoid erroneous booking, the chainman should call out the measurements clearly and the surveyor should repeat them.

141. Relative Importance of Sources of Error. It is instructive to compare the conditions under which each source of error, taken by itself, may give rise to a definite amount of error. Supposing the 100 ft. chain to be used, an error of about 1 in 1,000 is produced by each of the following conditions:

(a) Length of chain 0·1 ft. or about $1\frac{1}{4}$ inches wrong,
(b) One end of chain $4\frac{1}{2}$ ft. off line,
(c) Middle of chain $2\frac{1}{4}$ ft. off line,
(d) One end of chain $4\frac{1}{2}$ ft. higher than the other end,
(e) Chain hanging with its mid-point 2 ft. below the chord joining its ends,
(f) Temperature 150°F different from the standard value,
(g) Excessive pull of 125 lbs. and upwards, depending on the cross-section of the chain.

These figures emphasise the futility of elaborate refinement in ranging and marking unless the chain is correct or its error is known. Speed in chaining, as in other survey operations, is best maintained by relating precautions against error to the seriousness of the errors likely to occur.

LINEAR MEASUREMENTS WITH THE LONG STEEL BAND OR TAPE

142. The use of the long steel band or tape, instead of the chain, gives greatly increased accuracy in Surveying, at little or no extra cost in time or labour. The bands can be obtained in many different lengths and widths, lengths from 50 ft. to 1000 ft., widths from $\frac{1}{20}$ inch to $\frac{3}{4}$ inch. For the measurement of long lines and traverses, and for laying out railways and roads, a tape 300 ft. long and $\frac{1}{8}$ in. wide, weighing 9 to 14 oz. per 100 ft. length, has been found to be a convenient size, though some surveyors say that they prefer to use a tape 500 ft. long. Where only comparatively short lengths have to be measured, and for much engineering work, a tape 100 ft. long and $\frac{1}{4}$ in. wide will probably be found as convenient as any.

The advantage of the long tape is, of course, that fewer applications of the tape are required to measure a given distance than would be the case if a short tape is used. Very long tapes, however, are inconvenient to handle, more especially when working over a line cut in bush or forest, and they are more liable to accidents than short tapes. Moreover, as these long tapes generally have to be of light section, they should not be dragged along the ground, but should be held clear of obstructions when being moved forward, and this means extra labourers to support them at intermediate points.

143. Tape Graduations. In some types of tapes the overall length is that between the outsides of the brass handles at the ends, but, for the better classes of work, it is far better to use a tape with the zero and end marks on the tape itself, either in the form of lines etched on the metal or on small sleeves or studs fastened to it. When ordering a tape, it is always necessary to specify very clearly which kind is required.

Graduations, marking intermediate lengths, are usually either etched on the metal or consist of small brass studs let into it at the required intervals, or brass sleeves wrapped round and brazed to the tape. Etched graduations sometimes tend to get rusted up, and, when the tape is dirty, may not be easy to see or read in bad light. Studs are easy to see, but tend to weaken the tape since the pins to which they are connected pass through the tape and thus reduce its cross-sectional area. Hence, if a tape with studs breaks, it usually does so at one of the studs. Brass sleeves are more satisfactory.

Tapes are graduated in various ways. Those intended to be used solely as standards of length may be marked at the 100 ft. points only. For ordinary purposes, the most generally useful system of graduation consists of numbered marks at every 10 ft., with small studs at every foot and the first and last foot subdivided into tenths by still smaller studs. Some people like to have also a one-foot length outside both the zero and end divisions graduated to tenths, so that the total length of the graduated part of the tape is 2 feet longer than the nominal length: this extra two feet is often useful when careful work is being done and pegs or marks have been put in beforehand at every tape length. In such a case, the peg may come just outside the zero or end mark, and the difference is then read off very easily on the extra graduated foot at the end. This system, however, is not usual. A convenient system of graduation of very long tapes used in traversing is to have marks at each 100 ft. and the first hundred feet finely graduated and numbered in the reverse direction to the rest of the tape: the zero mark is 100 ft. from the beginning of the tape. If a line of say 583 ft. is to be measured, the 500 mark is made to coincide with the forward station and the 83 is read by the surveyor at the back station.

144. Taping Methods. There are two ways in which a tape may be used. The first, and more common, method is to lay it flat on the ground like an ordinary wire chain, and to use it in a manner similar to that in which a chain is used. The second method is to work with it 'in catenary'—that is, suspended in the air in such a manner that it hangs, clear of the ground, in a natural curve, similar to the curve in which a telegraph wire hangs. On very flat and even ground the first method is quicker and more convenient than the second, but, unless very special precautions are taken, it is not so accurate. When, however, work is being done over rough ground and a reasonable degree of accuracy is desired, the second method is generally the more economical in the end. This is particularly the case when the survey is along lines cut through bush or forest. When cutting lines of this kind, it is easiest to cut trees and bushes about 18 to 24 inches above ground level, but, to cut lower and clear everything below that height is troublesome and adds considerably to the time, labour and cost of clearing. If a tape is to be used along the ground, all, or nearly all, of this low stuff has to be cleared away, as otherwise it is difficult to stretch the tape out straight without catching or kinking it on ground obstacles. Hence, by suspending the tape in such a way that the lowest point of it is about 2 ft. above ground level, all this additional cutting and clearing is avoided, and the time lost by using a somewhat slower method of taping is more than counterbalanced by the saving in money on cutting and clearing.

USING THE TAPE ON THE FLAT

145. When using the tape on the flat, varying degrees of accuracy may be obtained by slight variations in method, and, as time and costs can be reduced by reducing the standard of accuracy aimed at, the surveyor can exercise considerable discretion in adjusting his methods to what he considers to be the necessities of the case with which he is dealing. Thus, the tape can be used in exactly the same manner as a chain, without observing temperatures or taking special precautions to maintain constancy of pull, and, when so used, it should yield slightly better results than the chain. For a better class of work, precautions may be taken, by means of a small spring balance attached to the tape, to ensure that a constant known tension is applied while the tape is being used. Again, even when care is taken to keep a uniform pull, it may be decided that no special observations need be taken, or allowance made, for temperature. On the other hand, the thermometer may be read at each set-up of the tape, and a correction made for the mean temperature of each line, or for the mean temperature during the whole survey.

Hence, considerable latitude in the choice of details of method is available, but, when the standard of accuracy required has been decided, the particular means to be adopted to reach this standard can best be settled by a careful consideration of the different sources of error and the effect of each of them on the final result. For this, the detailed analysis given in sections 167 to 170 can usefully be consulted.

146. Unrolling and Rolling up the Tape. The end handle of the tape is usually attached to the drum or casing by a strap passing through a small slot in the latter. The strap is undone and the handle taken by the leader or front chainman, the follower holding the drum about waist level, with its circular sides in a vertical position. The follower now walks slowly backwards, letting the tape unroll from the drum. While doing this, he keeps his hand on the winding knob of the drum, if it has one, and sees that the tape does not run out too quickly or in a jerky manner. If he feels or sees any indication of a jerk, he should yield to it so as to prevent undue stresses being imposed on the tape. Some drums are furnished with a mechanical braking device. When the tape is very nearly fully drawn out the follower stops and then proceeds to detach the rear handle from the drum.

It is inadvisable to let the tape be drawn off the drum by the leader moving forward, and the follower remaining stationary with the drum, as this tends to cause sudden jerks to be applied.

If, at any time, the tape is dragged along the ground, the rear handle must be detached, otherwise it will soon catch in something and probably cause the tape to break.

When re-winding, the tape is laid out straight on the ground, and the handle of the rear end fastened to the axis of the drum. The latter is then wound slowly so as to bring the tape on to it fairly tightly, but not too tightly, and without jerks. During winding, the tape should not be drawn forward along the ground, but the drum should be carried along slowly, winding taking place all the time, until the other handle is reached. This is then wound on and fastened to the drum by the leather strap.

Tapes, when not in use, should be kept lightly covered with non-corrosive grease, and this should be cleaned off before use.

147. Attaching the Spring Balance. Most instrument makers supply special spring balances, of the barrel type, for use with steel tapes. Some tapes have small lugs fitted at the sides of the handle, for attachment of a fork as indicated in Fig. 79: small holes in the ends

Hook of Balance.

FIG. 79.

of the prongs slip over the lugs, and the hook of the spring balance is passed through the loop at the end of the fork. Otherwise, the spring balance can be attached by a strong string or leather thong passing through the hook of the balance and the ring on the handle of the tape: this is generally not so convenient as the use of the fork, because the handle hangs loose and is apt to catch on stones or twigs on the ground. Sometimes there is a small hole in the back of the tape handle, through which the hook of the balance can be passed, or, if not, a hole can be drilled if the handle is thick enough.

In any case, care must be taken to see that the pull is applied in the line of the tape.

The standard temperature and pull for a tape are usually to be found marked on it at the zero end. The standard pull will probably be 10 or 15 lbs.; if for any reason a different pull is used, a correction can be made, provided that the effect of a given tension on the tape is known. See section 163.

148. Applying the Tension. At the rear end, the tape can be held by an arrow, or two arrows, passed through the handle and stuck in the ground and used as a lever to make small adjustments to the position of the tape. But if there is a considerable distance between the handle and the zero mark, the chainman may have to use a ranging pole so that he can stand up to see clearly the position of the tape against the station mark, while making the adjustment.

At the forward end, the balance can be fastened, by strong string or a leather thong, to a point near the bottom of a ranging pole, which is pressed into the ground and used as a lever. After he has obtained line from the rear chainman, the leader chainman holds the pole at the top and presses on it until he sees the correct pull registered on the balance. During this process, the tape will have to be moved forwards or backwards by the rear chainman who must signal when the marks coincide at his end. It may be difficult, with some kinds of tape, for the forward man to maintain correct tension and at the same time read the tape or put in an arrow at the proper place: this should then be done by the surveyor himself or by another assistant. Indeed, if sufficient labour is available, it may be found more expeditious to have two men at each end.

149. Observing the Slope. After the measurement of one bay is complete and before the tape is moved forward, the slope should be observed. This may be done by Abney Level or Clinometer, which can be read by the rear chainman. For this, the leader chainman carries a mark on his ranging pole placed at the same height above the bottom of the pole as the rear chainman's eyes are above ground. While the leader holds the pole vertically, the rear chainman takes the sight and calls out the slope to the surveyor. Unless it is desired to carry heights forward through the survey, there is no need to distinguish between slopes of elevation and depression.

If there is a decided change of gradient at any point between the ends of the tape, both gradients should be measured and the length of each portion noted.

150. Observations for Temperature. Various types of thermometer are available for this purpose, but probably the best for field use is one contained in a metal tube or casing, in which there is a narrow slot running nearly the whole length of it, so that the mercury thread and the graduations can be seen clearly. It is desirable that the casing should be of similar material and have a surface of similar nature to that of the tape, because the temperature taken up by a piece of metal, especially when it is exposed to direct sunlight, is very dependent on the nature of its surface, and the metal may become very much hotter than the surrounding air. The thermometer should be placed alongside the tape if possible, and in similar conditions of temperature. To avoid breaking the mercury thread, the bulb of a thermometer should always be lower than the stem, during use as well as during transport and storage.

The thermometer can be read by the surveyor himself. He may read it at each set-up of the tape, or less frequently, according to the accuracy required: each line may be separately corrected for temperature, or an average temperature may be taken for a whole day's work.

151. Measurement of Odd-Length Bays. When the length of a line is not just a whole number of tape lengths, it is necessary to measure an odd section which is a fractional part of the tape length. The procedure in this case will depend on the type of graduation on the tape in use. If the tape is simply marked at every foot, a terminal graduation must be set against the last station mark and the fraction of a foot at the other end estimated, or measured by use of an ordinary foot-rule. A tape that has a finely divided section at one or both ends can be placed so that such sections can be used at one of the station marks. If many odd sections have to be measured, it is convenient to have a tape that is finely divided throughout: tapes divided throughout to tenths of a foot, or even to hundredths, are obtainable.

In stretching the tape when measuring odd length bays it is best to apply the tension at the ends of the tape in the usual way, but if this is inconvenient or impossible, the pull has to be applied at some intermediate point of the tape. This can be done by means of a special clip. Two forms of clip are illustrated in Fig. 80: one is a

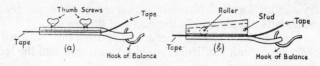

FIG. 80.

piece of metal with a longitudinal slot along one edge and a ring at one end; the tape can be placed in the slot and clamped by two wing-nuts: in the other kind of clip, the tape is held under a hard steel roller in a brass box of tapering shape, and a firm grip is obtained when the pull is applied.

152. Check Taping and Detail Survey. There are considerable possibilities of making mistakes, both large and small, in taping lines, so it is advisable that taping should be done twice. In ordinary work the object of the check taping is to detect gross errors rather than misreadings of a tenth or a hundredth which, if they occur only occasionally, do not matter very greatly. Hence, check taping may be done by less accurate methods than those used for the main taping, and the main taping is the accepted one for use in computations. Thus, in the check taping, it will not usually be necessary to take temperatures, use the spring balance, or observe slopes. Offsetting to details can be done from the check taping, so that the survey party can concentrate on the main taping without distraction or delay.

Check taping may be more effective if done in a different unit, say links if the main taping is in feet, or conversely: in this case, a handy conversion table should be available in the field. It is important for the surveyor to compare the two tapings as soon as possible: omission of this simple procedure has more than once led to the discovery of a gross error after the party has gone off elsewhere.

153. Booking the Results. Fig. 81 illustrates a page from a field-book, to show how the observations might be booked. The line is part of a theodolite traverse, and the figures on the extreme right

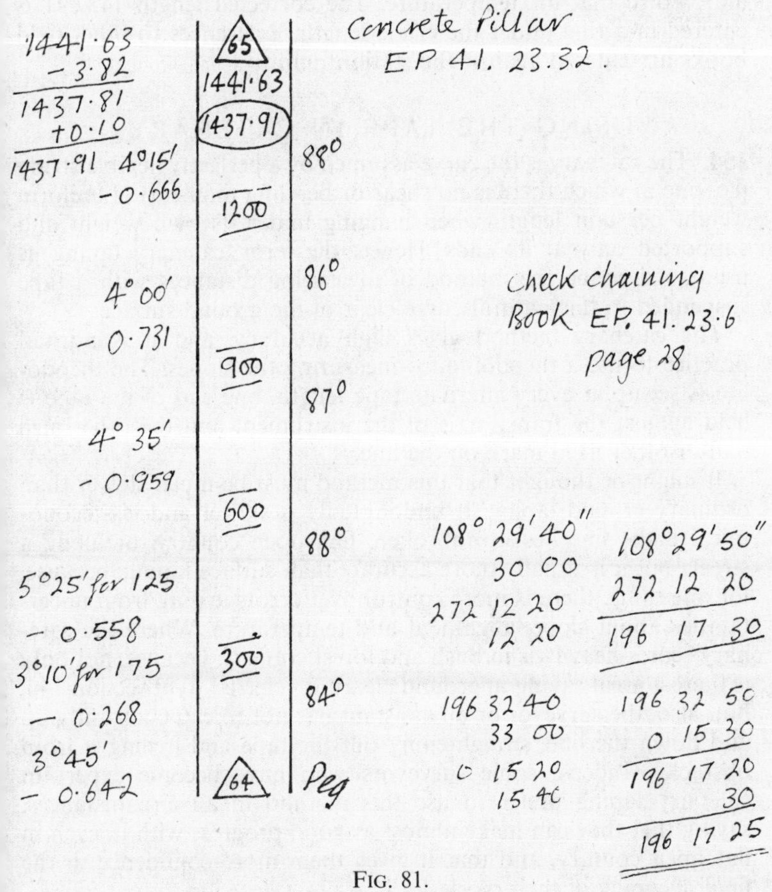

```
1441·63          /65\        Concrete Pillar
 - 3·82          1441·63       EP 41. 23.32 .
 1437·81        (1437·91)
 +0·10                    88°
1437·91  4°15'
  -0·666         1200

   4° 00'         86°        check chaining
  - 0·731                    Book EP. 41. 23·6
                 900             page 28
                 87°

   4° 35'
  - 0·959
                 600
                 88°      108° 29' 40"    108°29'50"
5°25' for 125                 30 00      272  12  20
  - 0·558                 272 12 20      196  17  30
3°10' for 175    300          12 20
  - 0·268        84°      196 32 40      196 32 50
3° 45'                        33 00        0 15 30
  - 0·642        /64\  Peg   0 15 20      196 17 20
                              15 40             30
                                         196 17 25
```

FIG. 81.

refer to the measurement of the included angle at station 64 (see section 217).

In this case, a 300 ft. tape, of correct length at 76·8°F and 15 lb. pull, was used. Bookings start from the bottom of the page and run upwards. Slopes are recorded on the left and temperatures on the right. The total recorded crude length was 1441·63 ft. Between the 300 and 600 ft. marks there was a decided change of slope, at 475, and the slopes of the two sections are separately measured. The correction for each length is worked out and entered, as shown, under the figures representing the angles of slope, and all corrections are summarised and entered under the crude length at the top left-hand side of the page, − 3·82 being the correction for slope and + 0·10 that for temperature. The corrected length 1437·91 is entered in a ring under the crude length. References to other field books are entered against the station numbers.

USING THE TAPE IN CATENARY

154. The catenary is the curve assumed by a perfectly flexible string (i.e. one in which there is no shear or bending moment) of uniform weight per unit length when hanging under its own weight and supported only at its ends. Hence, the term 'catenary taping' is used to describe the method of measuring distances with a tape suspended in this natural curve clear of the ground surface.

The catenary method gives high accuracy, and it is normal practice to use a theodolite for measuring the slopes. The theodolite is set up at every alternate tape length, one end of the tape is held against the transit axis of the instrument and the other at a nail or other fixed mark on the line.

It might be thought that this method must be much slower than ordinary ground taping. It undoubtedly is slower and less economical if the line lies in unbroken, flat, open country, or along a street; but it is usually more accurate than surface taping, because, for one thing, there is more control over errors arising from uncertainties about slope, alignment and temperature. Where the catenary scores heavily is in bush and forest country, because, not only is there a great saving in ground clearing, referred to in section 144, but, also, the surveyor or an assistant has not to run constantly up and down the line, straightening out the tape and freeing it from obstacles. Indeed, some surveyors who have become expert in catenary taping prefer to use this method in all circumstances, saying that they can make almost as good progress with it, even in flat open country, and that it gives them more confidence in the final accuracy of their work.

155. General Description of Catenary Method. When using the catenary method, it is best to set out the line beforehand, and use the results of the taping employed during setting out, as a check taping. However, the surveyor can, if he wishes, set out the line as he goes. The main thing is to have good stout pegs or posts driven on line at the points where each tape length will end. The posts

should stand about 30 inches above ground level, and the line should
be marked by nails projecting about an inch out of the top of each
post: the nails should be lined in by theodolite. Successive posts
must of course be within a tape length from each other. Instead of
the posts, it is possible to set up light tripods with reference marks
for the taping.

The general way of working will be understood from Fig. 82.

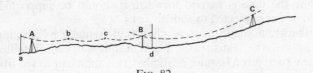

FIG. 82.

The theodolite is set up and centered at A, the beginning of the line,
and the first intermediate post or tripod is at B, a tape length away.
The tape is taken forward until a terminal mark is at B. The spring
balance is attached at the instrument end of the tape, and the handle
of the balance is fastened by a loop of string or leather thong to a
long pole *a*, which can be used as a lever to hold the pull steady.
The other end of the tape, near B, is similarly fastened to a pole, *d*,
which can be used to adjust the terminal mark of the tape to lie on
the nail at B.

If the tape is long, one or two intermediate supports may be
necessary, as indicated in the figure. A simple way to support a
tape without interfering with the natural catenary configuration, is
to hang it by about two feet of string from the end of a stick, rather
like a small fishing rod.

When everything is ready, the men at *a* and *d* prepare to take the
pull while those at *b* and *c* get the tape approximately into position.
The surveyor sights the instrument on to the nail at B, tightens
clamps, and signals his assistants to get the tape into position. The
man at *d* sees that the mark at his end is against the nail, while the
man at *a* applies the correct pull and adjusts the tape to lie lightly
against the end of the transit axis of the theodolite. Meanwhile,
the surveyor lines in the tape at *b* and *c*, by suitable signals to the
men there. As soon as everything is steady, the man at B signals
that the mark is against the nail, the man at *a* holds the correct
tension, and the surveyor reads the tape at the transit axis: if the
tape is not finely divided, the surveyor must use a short divided
scale to measure the small distance from the transit axis to a
tape division mark.

The surveyor must measure the slope from theodolite to nail, and record the temperature: the thermometer can be carried hanging freely in the air, by one of the assistants who hold the tape at intermediate points.

If a post has not been lined up at C, this should now be done from the theodolite, and the instrument can now be moved forward to C, and set up ready for the measurements of lengths BC and CD. It is best to arrange to apply the pull always at the theodolite end of the tape.

While the tape is carried forward, it should be supported and should not be subjected to sudden jerks.

In theory, a small error of alignment is made by holding the tape against the transit axis, but a simple calculation will show that any such error is quite negligible. It is sufficient to see that the theodolite is directed to the nail at the other end of the tape, and clamped.

156. Measurement of Odd-Length Bays. At the end of each line there will generally be a section less than a whole tape long. For measuring such a section, the pull can be applied to the tape by use of a clip as described above, otherwise the procedure is the same. The sag correction for each odd length can be computed, or taken from a specially prepared table or diagram.

157. Booking the Observations. Results can be booked in a manner similar to that used for ordinary surface taping. The only additional entries will be the short portions measured by the surveyor at the theodolite, but it is essential to make it quite clear whether these are to be added to or subtracted from the full tape length. Care must also be taken to get the odd-length sections recorded correctly. Each book should contain a very clear statement that the measurements were done in catenary, giving the pull applied and the weight of the tape. Full information about the method of supporting the tape at intermediate points should also be noted.

158. Fully Graduated Tapes. If a tape fully graduated to hundredths is used, the problem of setting a particular tape mark against the nail is avoided. The tape can be placed at random, and readings taken at each end, and booked in two columns, for 'back reading' and 'forward reading', with another column for the difference if desired. It is also easy to obtain a check by slightly moving the tape and taking another pair of readings. For this method, the distances between successive posts must be less than a tape length. A considerable amount of time is saved by not having to make precise adjustments of the position of the tape.

CORRECTIONS TO MEASURED LENGTHS

159. Corrections to tape measurements fall into two categories (1) corrections to convert tape readings to true lengths, and (2) corrections, perhaps better called modifications, to convert the true lengths into the distances required in computing results from the survey work.

160. Standardisation. A tape has a certain nominal length under given conditions of temperature and tension. However, a tape which has had a lot of use will probably be stretched. For instance, a so-called '200 ft.' tape may be found, when tested against a reliable standard under controlled conditions, to have a length of 200·007 ft. In this case, all measurements with the tape will be too small, and should be increased by 0·007 ft. for each tape length: on the assumption that the stretching is uniform, lengths less than 200 ft. should be corrected proportionally, that is by 7 parts in 200,000.

Another way to deal with a small standardisation error is to calculate the temperature at which the tape would be correct, and regard this as the standard temperature to be used in temperature corrections. Thus, the tape described above would be correct at a temperature about 5°F lower than the temperature at which it was originally correct.

Any large surveying organisation should have a permanent reference standard, or at least should keep at headquarters a standardised tape which is used only for comparison with working tapes.

161. Temperature. Increase of temperature causes a tape to increase in length and to give measurements that are too small. The change Δl is given by:

$$\Delta l = l \cdot c \cdot t,$$

where l is the length of the tape, t is the difference of temperature from the standard value, and c is the coefficient of thermal expansion. If t is in °F, the coefficient for steel is about 0·0000063 per degree, or in fractional notation about 1/160,000. In Centigrade degrees, the fraction is about 1/90,000.

Thus, suppose a line is measured with a steel 300 ft. tape which is of correct length at 96·4°F, and the measurement is 4987·24 ft., at an average temperature of 76·8°F. Then $t = 19·6$ and the correction Δl works out at 0·61 ft. which is to be subtracted because the working temperature was lower than the standard temperature.

The Table on the next page gives corrections for lengths from 100 to 1000 ft., and values of t from 10°F to 50°F.

TABLE OF TEMPERATURE CORRECTIONS
(Coefficient of Expansion = 0·000 0062 per 1° Fahr.)

$t°$ Fahr.	Lengths									
	100	200	300	400	500	600	700	800	900	1000
10	0·006	0·012	0·019	0·025	0·031	0·037	0·043	0·049	0·056	0·062
20	0·012	0·024	0·037	0·050	0·062	0·074	0·087	0·099	0·112	0·124
30	0·019	0·037	0·056	0·074	0·093	0·112	0·130	0·149	0·167	0·186
40	0·025	0·050	0·074	0·099	0·124	0·149	0·174	0·198	0·223	0·248
50	0·031	0·062	0·093	0·124	0·155	0·186	0·217	0·248	0·279	0·310

162. Sag. If a tape, standardised on the flat, is used suspended in catenary, the straight distance between the end points will be reduced by an amount called the 'sag correction'. In surveying practice, this correction is always small, and then it can be calculated, with sufficient accuracy, from the formula:

$$C = \frac{w^2 l^3}{24F^2}$$

where w = weight of tape in lbs. per foot run,

l = length of tape in feet, between marks,

F = pull applied at ends of tape, in lbs.,

C = sag correction, in feet.

The correction is to be subtracted, and the above formula applies when the ends of the tape are at the same level.

If the line is at slope θ, the tension cannot be the same at both ends, and the precise sag correction depends on whether the standard pull is applied at the lower or at the upper end. However, for slight and moderate slopes, the sag correction may be taken as $C . \cos^2\theta$. Precise formulas for sag corrections at all slopes are to be found in Vol. II of this work.

Some sag corrections for a tape weighing 10 oz. per 100 ft., at a tension of 10 lbs., are given in the table below:

TABLE OF SAG CORRECTIONS
(For tape weighing 10 oz. per 100 ft. under a pull of 10 lb.)

Span in feet	Sag correction. Feet	Span in feet	Sag correction. Feet	Span in feet	Sag correction. Feet
30	0·000	80	0·008	130	0·036
40	0·001	90	0·012	140	0·045
50	0·002	100	0·016	150	0·055
60	0·003	110	0·022	160	0·067
70	0·006	120	0·028	170	0·080

Since the sag correction depends on $(w/F)^2$, the sag for tapes of other weights and at other pulls can be obtained from this table by multiplying by the appropriate factor: thus, for a tape of 13·2 oz. per 100 ft. at tension 15 lbs., the sag for 150 ft. will be:

$$0·055 \times \left(\frac{13·2}{15}\right)^2 = 0·043 \text{ ft.}$$

In practical surveying, it will generally be inconvenient for the tape to hang more than 3 or 4 feet below the straight line between its ends: the drop is given by the formula $wl^2/8F$. If a tape has to be supported at intermediate points, with sag proportional to l^3, and evenly spaced supports dividing the tape into n sections, the sag for each section will be C/n^3, so the total sag for n sections will be C/n^2. Any intermediate supports must of course be free to move longitudinally so that the tension can remain uniform along the tape. See Section 155.

163. Tension. If the pull applied to the tape is not the same as that used during standardisation, a correction for the difference in the elastic stretching becomes necessary. The formula for the stretch of a tape depends on the cross-sectional area of the metal and on its Young's Modulus: neither of these quantities is likely to be known with any accuracy in practice, and it is better to determine the stretching experimentally. It is better still to avoid a correction, by using the proper tension.

The elastic stretch is easily found, at a suitable opportunity during a survey, while the tape is laid out, by applying several different tensions, say 5, 10 and 15 lbs., and noting the differences in the recorded lengths. A 300 ft. tape of cross-section $\frac{1}{8}$ in. by $\frac{1}{40}$ in. will stretch about half an inch under 10 lb. pull.

The above corrections will give the true length between the actual marks against which the readings of the tape were taken. As a rule, this length is not what is required for the purpose of the survey, and further changes have to be made.

164. Slope. It is usually required to find the projected horizontal length of a sloping line. The correction is given by $l(1 - \cos \theta)$, or l versine θ, where θ is the angle of slope and l is the length of the line. If special slope-correction tables are not available, the corrections are easily worked out with the aid of a table of versines such as is given in 'Chamber's Seven-figure Mathematical Tables'. Thus, the natural versine for an angle of $3°$ $20'$ is 0·00169, so the correction for a length of 300 ft. at this slope is $0·00169 \times 300 = 0·51$ ft. The correction is always subtractive. The formula can also be written as $2l \sin^2 \frac{1}{2}\theta$.

Sometimes, slopes are dealt with by finding the difference of height h between the end points, using level and staff. In this case, the slope correction is best worked out from the expression

$$\frac{h^2}{2l} + \frac{h^4}{8l^3}$$

which consists of the first two terms of a series. The second of these terms is inappreciable when h is less than about 14 ft. in 100 ft., that is a slope of about $8°$: it can therefore be neglected in most cases in practice.

165. Height above Sea-level. Owing to the curvature of the Earth, a line measured at a height above sea-level will be longer than the distance between the points that are at sea-level vertically below the ends of the line. The difference is only about 1/20,000 for each 1,000 ft. of height, and is usually ignored in ordinary survey work. It needs to be applied to precise geodetic bases and to extensive surveys that are controlled by a geodetic framework. For a full explanation of this correction, see Vol. II, Chapter IV.

STANDARDISATION OF TAPES, SPRING BALANCES AND THERMOMETERS

166. If the weight of a tape has to be determined, one method is to use the formula given above for the drop $-wl^2/8F$; this can also be written as $Wl/8F$, where W is the total weight of length l. The tape can be hung in horizontal catenary with one end against the transit axis of a theodolite, the telescope of which is directed to the other end: a levelling staff is placed at the centre of the tape, and the reading there, subtracted from the reading seen through the telescope, gives the drop. If h is the drop, it is easy to show from the formulas that $w = 8(Fh/l^2)$, and the sag C is $8h^2/3l$.

Really accurate taping work can only be achieved if tapes are regularly and carefully standardised, because any error in the length of a tape is cumulative in the final results. For this reason, it is desirable to keep at least one special tape for field standardisation purposes and to use this tape for nothing else.

A tape can be standardised at the National Physical Laboratory, which will issue a certificate of length guaranteed to 1/100,000, or if desired, to 1/1,000,000. When sending a tape to the N.P.L. for standardisation, it is desirable to send the spring balance with it, and to specify clearly the conditions under which the tape will be used—whether on the flat or in catenary, what tension will be used, and if possible an idea of the average temperature at which measurement is likely to be done.

The best way to compare standard and field tape is to measure between a pair of marks with both tapes.

A spring balance should be tested in the horizontal position, using two or three different weights, say 5, 10 and 15 lbs., hung by string over a pulley.

Field thermometers can be tested by placing them in warm water along with a thermometer of known accuracy, and comparing readings at intervals, as the water cools down. The range of temperatures likely to be encountered in the field should be covered in the tests.

ERRORS IN MEASUREMENTS WITH
STEEL TAPES

167. The principal errors to be guarded against in work with steel tapes are, as in the case of every other type of measurement, those of systematic or cumulative kinds. It is useful to classify errors into three groups:

(*i*) systematic errors that are proportional to the length measured,

(*ii*) errors that are always of the same sign but otherwise irregular,

(*iii*) accidental or random errors that tend to cancel out.

Mistakes, also, may be committed, but a proper procedure of measurement, such as double taping in different units, ought to shows up errors of this kind.

168. Systematic Errors. The effect of any error in standardisation, whether of the tape, the spring balance, or the thermometers, will be truly systematic. Thus, to introduce a systematic error of 1/20,000 a 200 ft. tape would have to be wrong by 1/100 ft., about $\frac{1}{8}$ inch, or the thermometer would have to have a constant error of 8°F, or the pull would have to be wrong by about 4 lb. if the tape had the cross-section mentioned above (Section 163).

If a tape is used in catenary, an erroneous tension will also cause a wrong value of sag correction to be applied. It is easy to see from the formulas that if C is the sag correction and dF is the error in the tension, the error in the sag correction will be $-2(dF/F) . C$: this may be larger than the effect due to stretching. If we suppose that the spring balance is giving a stronger pull than the reading on its scale, the tape will be excessively stretched and will give short measure: also, the calculated sag correction will be too large, since the pull used in the formula will be less than the actual pull, and the correction is subtracted, so the measurement will be short from this cause too.

Errors as large as those mentioned above in the thermometer or spring balance are avoidable and should not occur in practice. It is much more likely that temperature errors will arise from differences between the recorded temperatures and the actual temperatures of the tape.

169. Cumulative Errors. There are many possibilities here, but most of them are easily avoided. If the middle point of a 200 ft. tape is 1 ft. off line and the two 100 ft. sections are straight, an error of 1/20,000 will be generated. Obviously, a lack of straightness in horizontal or vertical direction will cause measurements to be too great. A similar effect will arise if the tape is lying on bumpy ground and the sag is ignored: the sag of a 100 ft. tape weighing 1 lb., under 10 lb. tension, is 0·04 ft. and if the tape is supported in only three equal sections the total sag will be 1/9 of this, which amounts to less than a fractional error of 1/20,000. Sag effects on a tape lying on rough ground are less than one might guess before working out some typical cases.

Another positive cumulative error will arise from ignoring slopes of lines that are nearly horizontal, Geometrically, the effect is exactly the same as that of non-alignment: a slope of 1/100 radian, about $\frac{1}{2}°$, will have a slope correction of 1/20,000 of the length. Thus, to maintain this accuracy, all slopes over $\frac{1}{2}°$ should be measured.

Obviously, the steeper the slope the more carefully it should be determined. The slope correction is $l(1-\cos\theta)$ and the differential of this is $l\sin\theta \,.\, d\theta$, or $h \,.\, d\theta$, where h is the difference of height between the ends. Suppose h is 20 ft., equivalent to a slope of about 6° in a length of 200 ft. Then an error of 1/100 ft. in 200 ft. will arise if $20 \,.\, d\theta = 1/100$, or $d\theta = 1/2000$ radian, which is less than 2 minutes of angle. Clearly, accurate taping on steep slopes demands careful determination of slope corrections, whether by vertical angles or by levelling.

An error that is very likely to be cumulative can arise from fluctuations in the temperature of the tape. In practice, a tape lying in sunshine is almost certainly hotter than the surrounding air, and an error of 8°F can easily occur. Suggestions have been made to determine the temperature of a tape from its electrical resistance, but the necessary apparatus would be cumbersome, and it would be simpler to use invar tapes or do high precision measurements only during overcast weather.

170. Compensating or Random Errors. These are the effects of fluctuations and uncertainties in the readings taken during a measurement, personal variations in the positioning of the tape, and suchlike. By definition, these errors tend to accumulate proportionally to the square root of the number of operations involved. It is in fact rather difficult to think of any kind of errors which, occurring randomly, could have, on average, a serious effect on any careful tape measurement work.

A graduated tape can be read to 0·001 ft. and a vertical angle to 1 minute or better, and random errors in these readings will presumably be generally much smaller than these amounts. Any estimate of the size of random errors is necessarily a kind of average—an average numerical value without regard to sign. Suppose a 200 ft. tape is laid down 25 times, thus measuring a distance of 5000 ft., nearly a mile. 1/20,000 of this is $\frac{1}{4}$ ft. or 3 inches. Because the probable effects of random errors are dependent on the square root of the number of operations involved, the estimate of random error of laying down the tape in the above case is one fifth of the amount last mentioned, that is 0·05 ft. – over half an inch. Since this is an estimate of random error, one must contemplate that on a few occasions an error of two or three times this amount will actually occur, in order to keep up the average, so to speak. In carefully conducted taping, errors of this order of size will not continually occur: an error of an inch or more would only occur as an occasional and isolated mistake, not belonging to the class of errors now in consideration.

PROPAGATION OF ERRORS IN TAPE MEASUREMENTS

171. It seems, from what has been written above, that the errors most likely to have dominant effects on linear measurements with steel tapes are those which tend to be cumulative, and most of them will make the final results too long. Errors must therefore be expected to be more or less proportional to the distances measured. This conclusion has bearing on the problem of adjustments involving measured lengths, particularly the closure of traverses.

Experience shows that an accuracy of 1/20,000 in steel tape measuring is easily achieved; three or four times this accuracy is not difficult.

EXAMPLES

(Take the coefficient of thermal expansion of steel to be 0·0000063 *per* °F, *or* 0·0000113 *per* °C.)

(1) A 200 ft. steel tape was used to measure a base line, and the tape readings totalled 1885·117 ft. The tape was tested and found to be actually 200·007 ft. long at the temperature at which the base was measured. What was the correct length of the line?

(2) A steel tape has true length of 199·994 ft. at 68°F. It is to be used at a

temperature of 83°F to set out a reference base exactly 200 ft. long. What tape reading should be set?

(3) A steel tape having correct length at 23°C was used to measure a base line and the recorded readings gave the total of 765·808 metres. The average temperature during the measurement was 15°C: what was the true length of the line?

(4) A traverse line was measured in three sections: (i) 296·22 ft. at slope 3° 44′, (ii) 156·34 ft. at slope 2° 09′ and (iii) 237·56 ft. at slope 4° 32′. What was the horizontal length of the line?

(5) A base was measured by a steel tape of correct length at 64°F. The recorded slope lengths, temperatures, and differences of height of the successive sections were:

Tape Reading	Temperature	Diff. of Height
186·217 ft.	48°F	− 8·27 ft.
285·554 ft.	50°F	− 3·55 ft.
266·409 ft.	50°F	− 2·96 ft.
147·338 ft.	51°F	+ 1·07 ft.
292·995 ft.	52°F	+ 4·25 ft.
201·444 ft.	54°F	+10·94 ft.

Calculate the correct horizontal length of the line.

(6) What is the sag correction for a 300 ft. tape weighing 3 lb. 4 oz. when used in catenary at tension 15 lb., with supports at 100 ft. and 200 ft.?

(7) A steel tape has a length 300·006 ft. at 68°F. At what temperature will the length be exactly 300 ft.?

(8) A steel tape weighing 1 lb. 9 oz. was standardised on the flat and found to have length 99·996 ft. at 68°F, tension 12 lb. It was used in catenary at the same tension to measure a horizontal base, at average temperature 77°F, and the readings on the successive sections were 99·105, 99·373, 98·976, 99·817 and 64·553. What was the correct length of the line?

(9) A 200 ft. steel tape is hung in catenary at a tension of 12 lb., and the centre point of the tape is found to hang 3·78 ft. below the straight chord. Estimate the weight of the tape, and the sag correction for a span of 100 ft. at the same tension.

(10) A tape is known to have a sag correction of 0·283 ft. at tension 15 lb. What will be the sag at tension 12 lb.?

(11) A 50 metre tape has weight 1·17 kg. What will be the sag correction if it is hung under tension 5 kg. with a support at the middle point?

(12) The tape mentioned in question 11 had correct length on the flat at 18°C. It was used to measure a base line in catenary supported as stated, at average temperature 11°C. There were three sections recorded as 49·907, 49·638 and 49·854, plus an odd length recorded as 21·505 metres in single catenary. What was the true length of the line?

(13) The slope length of a line is 407·22 ft. and the slope is 4° 53′. What is the horizontal length, and what error will be made if the slope is taken as 5° exactly?

(14) A section of a traverse line has slope length 95·517 m. and the difference of height between the ends is 17·24 m. What is the slope correction, and what error will be made if the height difference is taken as 17·2 m.?

REFERENCES

DYSON A. 'Catenary Measuring as applied to Mine Surveying'. *Chartered Surveyor*, May and June, 1957.

FROOME K. D. AND BRADSELL R. H. 'The N.P.L. Mekometer III'. *Conference of Commonwealth Survey Officers*, 1967.

JOHNSON, H. A. 'Surface taping with 300-ft. steel tapes'. *Australian Surveyor*, Vol. 15, No. 1, March 1954.

KELSEY J. 'The Tellurometer'. *Chartered Surveyor*, Apr. 1959.

MCVILLY R. B. 'Use of the Geodimeter on routine civil engineering projects'. *Survey Review*, Vol. XVIII, No. 138, Oct. 1965.

SAASTAMOINEN J. J. (Ed.) 'Electromagnetic Distance Measurement'. *University of Toronto Press*, 1967.

WRIGHT J. W. 'Electronic Distance Measurement in Land Surveying'. *Survey Review*, Vol. XVIII, No. 140, Apr. 1966.

CHAPTER III

CHAIN SURVEYING

---

172. Chain or linear surveying is that method of surveying in which only linear measurements are made in the field. It is a method well adapted to the survey of small areas with simple details, and its use can be extended to larger areas of open country if control points are previously established by triangulation, traverse or other instrumental surveys. Chain survey requires only very simple equipment. To the beginner, it forms a fitting introduction to the study of other methods, on account of its simplicity and the general applicability of many of its operations.

CHAIN SURVEYING–PRINCIPLES

173. Surveys with Straight Boundaries. The simplest possible survey is that of a triangular plot with straight boundaries. If the horizontal lengths and the relative positions of the three sides are recorded, the plan of the area can be drawn by the method of Fig. 1 (a). If, however, the area has more than three straight boundaries, it is no longer sufficient to measure the lengths of the sides only, as an infinite number of figures could be drawn satisfying the data. The field measurements must be so arranged that the plan can be drawn by construction of triangles: thus, if either diagonal of the four-sided field ABCD (Fig. 83) is chained, as well as the sides, the plotting can be performed without ambiguity, preferably by drawing the diagonal first and then constructing the two triangles.

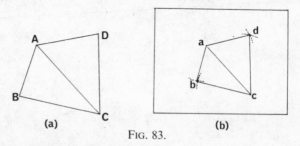

FIG. 83.

174. Check Lines. Consideration of Fig. 83 shows that a mistake made in measuring or plotting any of the five lines will not, as a rule, render the construction impossible, and may pass unnoticed. Such

130

mistakes will be brought to light, however, by measurement of additional lines, called check, proof or tie lines, which may not be needed for the surveying of detail, but which show the correctness of the work by fitting into their places in the drawing. Fig. 84 shows various ways in which the survey of Fig. 83 could be checked.

It is a good general rule that when crossing lines, like the diagonals of a quadrilateral, are surveyed, the point of intersection should be marked and the lengths of the two parts of each line recorded. When surveying is done far away from the drawing office, it is wise to make a rapid plot of chain lines before leaving the ground.

FIG. 84.

175. Offsets. As a rule, there will be irregular features and boundaries to be surveyed, so the work cannot be done solely by measurement of chain lines. The most rapid and commonly used system for locating such features is that of Fig. 1 (b), perpendicular offsets. Points may also be connected to the chain lines by tie-lines, as shown in Fig. 1 (a).

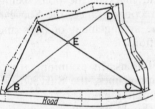

FIG. 85.

Fig. 85 represents a field with irregular sides. ABCD is a framework of lines lying alongside the features to be surveyed. From these lines, as many points can be fixed as are necessary to define the irregularities of the boundaries. In plotting, the triangles are first constructed, and on their sides short perpendiculars are erected in their proper positions and of correct lengths to scale. The boundaries, etc., are then drawn through the points so obtained.

176. Points to which Offsets are Taken. To survey a straight line from an adjacent chain line, it is sufficient to determine correctly the positions of both ends by offsets: these points being plotted, the straight line joining them represents the feature surveyed, and the measurement and plotting of additional offsets provides no further information (but it will provide a check).

In surveying an irregular feature, a sufficient number of offsets must be taken to enable the drawing to be accurately done: offsets should be taken to any point where there is a marked change of direction, otherwise the number of offsets must be determined by the experience and judgement of the surveyor, having regard to the scale at which the work is to be drawn.

Fewer offsets are required for surveying indefinite lines such as margins of woods, or features subject to variation such as shore-lines. In the survey of lines of regular curvature, like railway tracks, it is sufficient to take offsets at regular intervals.

CHAIN SURVEYING–FIELD WORK

177. Reconnaissance. On arriving at the field, the Surveyor should first of all walk over and thoroughly examine the ground, with the view to determining how he may best arrange the work. The importance of this step is sometimes overlooked by the beginner, but the utility of a thorough reconnaissance cannot be over-emphasised, the time spent being amply repaid in the greater ease with which the survey can be executed. The positions of stations can be selected and marked, the poles being used to test inter-visibility. During reconnaissance, the Surveyor should prepare a sketch showing the arrangement of lines and the numbering or lettering of the stations.

178. Selection of Stations. In examining the ground for a good arrangement of survey lines, the Surveyor should endeavour to meet the following requirements:

(1) Survey lines should be as few as practicable, and such as to form a geometrically sound framework.

(2) Triangles should be well-conditioned, and angles less than 30° should be avoided if possible.

(3) There should be an adequate number of check lines, and if these can be used to pick up some details, so much the better.

(4) Offsets should be kept short especially those to important features, as the survey of a lot of long tie-lines is wasteful of time.

(5) Lines should lie over the more level ground.

(6) Lines along roads with much traffic should be kept to one side: and if a road is to be crossed, the line should cross it at a good angle so that only a short measurement across the road will be necessary.

179. Marking Stations. Stations should be marked so that they can be found again at any time during the survey. In soft ground, wooden pegs about 18 ins. by $1\frac{1}{2}$ ins. square may be used: on roads or pavements, nails may be driven in flush and made conspicuous by dabs of paint or chalk. If it is likely to be necessary to re-locate a station again after a long interval, reference measurements from 3 or 4 surrounding points of detail should be recorded in the field book.

On the other hand, if a survey is small enough to be completed in a few hours, stations may be marked by cutting out small sods, pinning bits of paper down with twigs, or any such temporary devices.

180. Running Survey Lines. The routine of chain-surveying comprises the chaining of the lines and the location of details by offsets, ties, etc. Chain survey can be carried out by two operators, surveyor and leader, but it is probably more efficient to have a third, the follower, so that the surveyor can direct the work and write up the field book without interruption. See section 134.

The leader drags out the chain and places the forward arrow after being aligned by the follower. When offsets are to be taken, the chain is left lying on the ground and the leader holds the ring of the linen tape while the follower holds the tape-box and judges the chainage at which the offset is perpendicular, or proceeds to points suitable for measurement of tie lines to control long offsets.

The surveyor must be on guard against mistakes, and should generally be in a position to check the offset and chainage readings called out to him: he should see that all necessary offsets are taken.

Chainages at which the lines cross drains, fences, road-edges, etc., should be recorded.

The degree of accuracy to be observed in chain-survey depends on the scale at which the work is to be drawn. Assuming that 0·01 inch is the smallest distance that can be distinguished in plotting on paper, it is seen that measurement to the nearest foot is sufficient if the scale is 100 ft. or more to one inch, but at larger scales the measurements should be recorded to $\frac{1}{2}$ ft. or where necessary to 1/10 ft. In recording tenths in the field book, it is best to use the fractional notation, e.g. $8\frac{6}{10}$, rather than decimal points, which are easily erased or overlooked.

The maximum allowable length of simple offset will depend on the scale, and on the nature of the feature surveyed. It is found by experiment that the position at which a tape is perpendicular to the chain can be judged with an accuracy of about 5 % of the length of the offset. Thus, in surveying to $\frac{1}{2}$ ft. accuracy, offsets over 10 feet long should be checked by tie-lines, or positioned by using the tape to locate an isosceles triangle, as indicated in Fig. 86, and recording the middle point of the base. There is no difficulty about reading the length of the offset itself, and its accuracy depends on correct alignment of the chain. If many long offsets are needed, it is better to use an optical square to find the chainage points, rather than measure a lot of tie lines which take time and make the field book look very untidy.

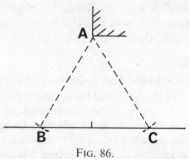

FIG. 86.

Offsets to buildings should be taken with special care. If corners of buildings are rectangular a few salient corners can be offseted and the other dimensions of the building recorded as direct measurements, sometimes called 'plus measurements', along the building itself. The surveyor must bear in mind that the corners of any building, however well constructed, may not be right-angles. It is useful to note the chainages at which the directions of walls, if produced, would cut the chain lines.

181. Note Keeping. The Field Book, about 9 in. by 5 in. and opening along the shorter edges of the pages, may be of strong plain paper, but more commonly each page has a pair of red lines ruled down the centre. The advantage of plain paper is that the surveyor may draw the pair of lines near one side if all the details to be surveyed are on one side of the line. The space between the lines is reserved for chainage distances. The survey of each line should be started on a fresh page, and booking proceeds from the bottom upwards. All offsets should be measurements from the chain line, and should be written close to the sketched detail, on the side towards the chain line.

Cross-referencing of lines to page numbers of the field book will greatly facilitate the draughtsman's work.

It is a useful check if each full chain length is written in the book, even if no offsets are actually taken at these points.

The field book should also contain (*i*) a sketch of the chain lines with letters or numbers, (*ii*) nature and location of the survey, (*iii*) numbers of the pages on which the work is recorded, (*iv*) dates on which the work was done, (*v*) names of members of the party, (*vi*) references to drawings on which the work is plotted, as soon as this is known.

A good field book is one which could be understood and drawn up by a draughtsman who has not seen the area covered by the survey.

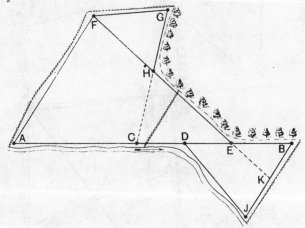

FIG. 87.—SMALL CHAIN SURVEY.

182. Examples of Chain Surveys. Fig. 87 shows a small survey of an irregularly shaped parcel of land. The noticeable features are (*a*) the long line AB as a backbone or base, (*b*) lines CH and EK as check lines, (*c*) the straight piece of hedge is fixed by surveying its ends from the lines AB and EF.

A farm survey is illustrated in Fig. 88. The central base-line AB is more suitably placed than any diagonal across the area would be.

The structure of survey lines is largely determined by the positions of the boundaries, but some choice for the positions of check lines is possible. The triangle PQN with proof line OQ is introduced to avoid long offsets to the stream.

A specimen field book for one of the lines of this survey is shown in Fig. 89.

CHAIN SURVEY PROBLEMS

183. To set off a Right-angle at a Given Point. In Figs. 90 and 91, the given point is A and several methods are possible. If A is at the end of a line, as in Fig. 90, the simplest method is to set up a right-

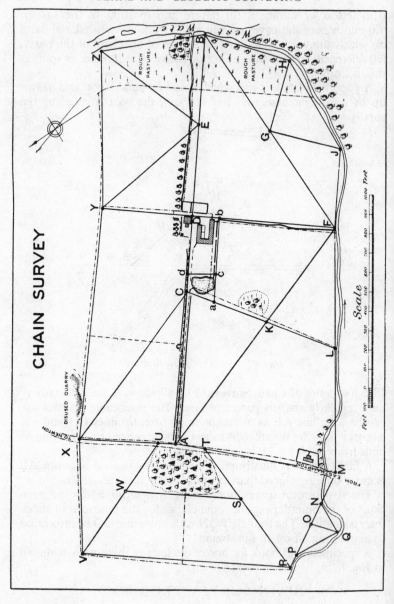

FIG. 88.

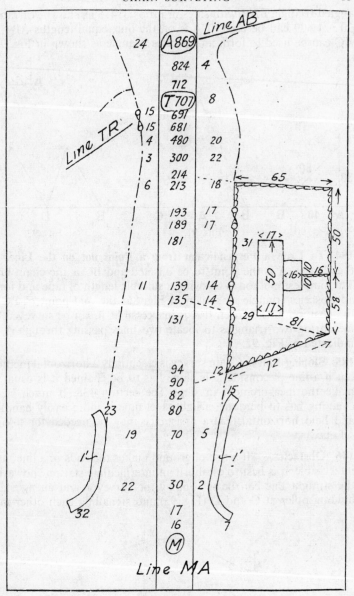

Fig. 89.—Double Line Booking.

angled triangle FAB with sides in ratios 3:4:5 as shown; or ratios 20 : 21 : 29 can be used. If A is on the line, equal lengths AB and AC can be used to form an isosceles triangle as shown in Fig. 91.

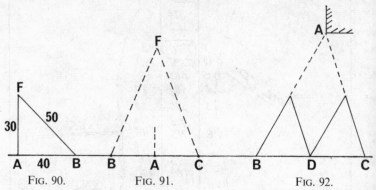

FIG. 90.   FIG. 91.   FIG. 92.

184. To Drop a Perpendicular from a Point not on the Line. In these cases, A is the point to be offseted and BC is the chain line. The simplest method is to swing a suitable length of tape and form an isosceles triangle as shown in Fig. 86; the mid-point of BC is the position of the offset. If A is inaccessible, it can be surveyed by measuring two triangles to locate two lines passing through A as indicated in Fig. 92.

185. Sloping Lines. Chain survey is essentially a horizontal method, but if a line of considerable slope has to be chained it is usual to make the measurement in steps. See section 138. If much steep chaining has to be done, a light steel tape is more easily handled and held horizontal. Extra assistance may be needed for taking offsets.

186. Obstacles.   If a rise of ground makes the ends of a line non-intervisible, it is best to establish an intermediate station: however, the straight line can be set out by the surveyor and an assistant holding poles at C and D (Fig. 93) and signalling each other into

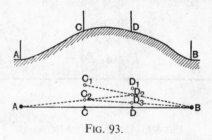

FIG. 93.

line with the visible terminal, B or A: the limiting position of straightness ABCD is rapidly reached.

An irregular area óf pond or woodland which cannot be chained through can be surveyed by a kind of traverse as indicated in Fig. 94. Measurements of all the sides of the small triangles like DBF enable the lines to be drawn up correctly; these small lengths should be measured very carefully.

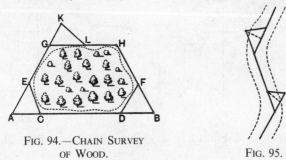

FIG. 94.—CHAIN SURVEY OF WOOD.

FIG. 95.

A long narrow area such as a highway through forest can be surveyed by a similar method, as illustrated in Fig. 95.

If a chain line crosses an obstacle that does not obstruct the sight, the distances across the obstacle can be found by various methods: see Figs 96 to 99, in which A and B are points marked on the chain line.

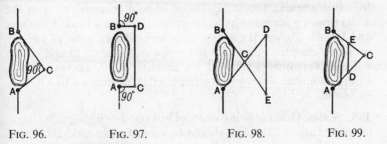

FIG. 96.   FIG. 97.   FIG. 98.   FIG. 99.

In Fig. 96 the sides of the right-angled triangle are measured and the length of AB can be computed. In Fig. 97 a rectangle is set out and the length of AB is equal to that of CD.

In Fig. 98, AC and BC are extended to double their lengths, then AB is equal and parallel to DE. In Fig. 99, D and E are the middle points of AC and BC, then AB is twice DE.

A line may, if necessary, be continued over an obstacle that obstructs sight by setting out a parallel line as indicated in

Fig. 100, but this sort of thing should be avoided in chain survey if possible.

If the obstacle is a wide river or anything. which prevents any measurement across it, there are several ways of finding the required distance. In Fig. 101, AC is set out perpendicular to the line,

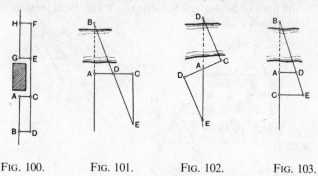

FIG. 100.　　　FIG. 101.　　　FIG. 102.　　　FIG. 103.

and D is its mid-point. CE is set out perpendicular to AC, and point E is located so that it is in line with B and D: then AB = EC. A rather similar method is indicated in Fig. 102. Fig. 103 shows a method with two perpendiculars whose ends are lined up with B: then AB = (AC . AD)/(CE − AD).

PLOTTING A CHAIN SURVEY

187. Instruments. The draughtsman should be provided with the usual drawing instruments which should include a Beam Compass and French Curves. If railways have to be drawn, a set of railway curves is essential: these are flat strips of wood or plastic cut to different curvatures: a useful set consists of 100 curves of radii $1\frac{1}{2}$ to 240 inches. A long steel straight-edge, of length 4 to 5 ft. may be useful.

188. Scales. Good drawing scales of boxwood or high-grade plastic, of lengths about 12 inches, should be used. A scale divided suitably for each map scale that is likely to be used should be available. Offset scales about 2 inches long, used as shown in Fig. 104 will also be required. A useful type of scale is one divided to chains and links on one edge, and feet on the other.

When Ordnance Survey Plans are employed as bases for plotting extensive amendments and new work, scales divided appropriately for the ratios 1/1250, 1/2500, and 1/10,560 will be needed. Furthermore, the decision to change over progressively from 1/10,560 to 1/10,000 has now been adopted by the Ordnance Survey.

189. Paper and Pencils. Good quality mounted drawing paper should be used for all important plans. Paper should be laid out flat in the drawing office for some days before being used, so that it can become stable in the prevailing atmosphere. Fairly hard pencils should be used, as they make fine sharp lines, but they should not be so hard as to cut the paper and interfere with subsequent inking-up. Softer pencils may be used for lettering and symbols that need to be conspicuous.

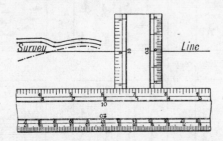

FIG. 104.—USE OF OFFSET SCALE.

190. Plotting. After determining the scale at which a survey is to be plotted, consideration must be given to the positioning of the work on the paper, and the plotting will then, as a rule, begin with the drawing of the longest chain line. Construction of triangles then proceeds, and the larger ones may have to be fixed by drawing arcs with the beam-compass. All chain lines should be drawn up and checked by the tie lines, before any detail is plotted. Stations may be marked by pricking, or with a hard pencil, as they will not usually be inked up: they may be shown as very small circles.

A chain survey may be controlled by points fixed in a triangulation, or other instrumental survey, and having rectangular co-ordinates. In this case, the drawing paper may be marked up with a grid of squares of suitable size having regard to the scale, and then the control points are first plotted from their co-ordinate values. Machines for plotting points from given co-ordinates are available in some offices.

Offsets may be plotted by marking points on the chain lines and drawing perpendiculars of correct lengths, but it is much quicker to use the offset scale as previously described (Fig. 104).

The draughtsman should be on guard against making simple mistakes like (*a*) plotting offsets at the wrong point or on the wrong side of the line, (*b*) omitting offsets, or even (*c*) plotting from the wrong end of a line.

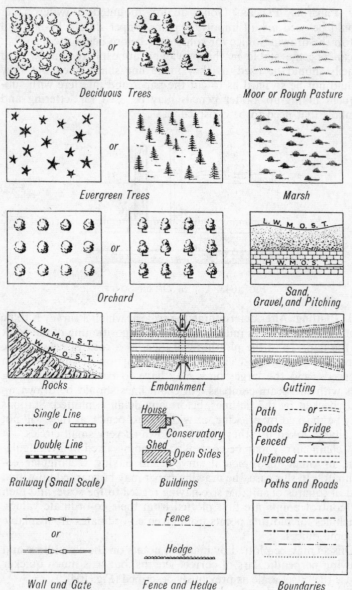

FIG. 105.—CONVENTIONAL SYMBOLS.

191. Inking, and Conventional Symbols. Black Indian Ink, obtainable in bottles, is used for most or all of the detail features, and it should be applied with a drawing pen, as an ordinary pen cannot give a line of uniform thickness. Lines of different thicknesses may be used for different kinds of features. Broken or dotted lines may also be used, and various types of divided line can indicate administrative boundaries.

To indicate special features, such as new buildings, red ink may be used.

Chain lines may be left in pencil, or inked in a fine pale blue or pink, so that they cannot be confused with surveyed features. Pale blue should be used if the lines are not wanted to appear in a printed copy.

Some conventionally adopted symbols for various features are shown in Fig. 105, and are self explanatory.

192. Colouring and Lettering. The less colouring, the better. If a large area is covered with colour-wash, the paper will be distorted. Narrow features like roads and streams may be coloured, brown or blue respectively, and areas of water may have a blue edging. Certain boundaries may have to be made prominent by means of coloured edging.

Neat hand-printing of letters and numbers is preferable to stencilling. Nowadays it is possible to get printed symbols on transparent films that can be stuck down on the plans. Title, scale and a North Point should appear on each plan.

If a plan is to be reduced in size when reproduced, the thicknesses of lines and sizes of symbols should be determined accordingly, and the scale should be indicated graphically (as in Fig. 88) and not by stating a numerical ratio.

193. Ordnance Survey Method. The Ordnance Survey has adopted enamel-coated metal plates, in place of drawing paper, for the initial plotting of large-scale plans, to avoid all distortion. Details of this method will be found described in 'Empire Survey Review', Vol. IX, No. 68 and Vol. X, No. 75.

CHAPTER IV

TRAVERSING

194. A traverse survey is one in which the framework consists of a series of connected lines, the lengths and directions of which are found from measurements. The system of fix involved is therefore that of Fig. 1 (c). When the lines form a circuit which ends at the starting point, the survey is termed a closed, or loop traverse.

A traverse that starts from a point already fixed in some survey system, and ends on another such point, is a controlled traverse, because it must be computed in such a way as to fit into the existing system.

A flying traverse is one not controlled as described above; this method is likely to be used in surveying an area in the form of a long strip, such as a road location survey or a pipeline. Special procedure should be adopted to ensure that a flying traverse is free from mistakes. See Section 243.

195. Scope. There is a considerable range in the character of traverse surveys according to the instruments used and the degree of accuracy necessary. In countries unsuited for triangulation, extensive theodolite traverses are required for the establishment of control points from which subsidiary surveys for the mapping of detail may proceed. In such primary traverse work great refinement is called for in both angular and linear measurements, and in certain countries the standard of accuracy now expected, and regularly attained, is of the order of about 1 in 70,000 to 1 in 100,000, a standard comparable with that of modern primary triangulation. The adoption of traversing as the primary framework for surveys of extensive areas has been greatly facilitated by the introduction of electromagnetic distance measuring instruments (Tellurometer, Geodimeter etc.) by which the traverse lines of length up to 20 or 30 miles can be accurately measured in a very short time. These instruments are described in Vol. II. In small surveys, particularly if they form closed figures, or if they are run between points whose relative positions have been otherwise determined, the precision aimed at may be considerably less, and at the bottom of the scale we may have, as in certain exploratory surveys, distances estimated from the rate of march, and directions taken by compass towards sound signals instead of to visible points.

In the present chapter the methods ordinarily used in general cadastral and large scale engineering surveys are considered, the applications of traversing in small scale mapping being dealt with in Vol. II, Chapter VII.

196. Comparison with Chain Surveying. Chain surveying requires only very simple and inexpensive equipment, but the use of angular as well as linear measurement makes traversing a more flexible method. If the traverse is the framework controlling detail surveys, it can be located along a suitable route where the necessary measurements can be carried out with maximum accuracy and speed; on the other hand, if the traverse is for detail survey with offsets, the lines can be located near the detail and the measurement of extra lines without offsets, merely to provide geometrical structure, is avoided.

Traversing can be done in certain types of country where chain surveying would be quite impossible.

BEARINGS

197. Direction. The directions of survey lines may be defined (*a*) relatively to each other, (*b*) relatively to some reference direction, or meridian. In the first case they are expressed in terms of the angles between consecutive lines, and in the second by bearings.

The use of a meridian of reference from which the directions of survey lines may be reckoned has so many advantages, particularly regarding the facilities afforded for checking and plotting, that this system is adopted in preference to the other. Only in small closed surveys involving few instrument stations should it be regarded as permissible to dispense with the establishment of a meridian.

The reference direction employed may be one of the following:

(*a*) True meridian.
(*b*) Magnetic meridian.
(*c*) Grid meridian.
(*d*) Any arbitrary direction.

198. True Meridian. The true or geographical meridian passing through a point is the line in which the Earth's surface is intersected by a plane through the north and south astronomical poles and the given point. The determination of its direction through a station involves astronomical observations, and is described in Vol. II, Chapter II.

The meridians converge from the equator to the poles, and consequently the true meridians through the various stations of a survey are not parallel to each other. All the survey lines, however, are to be referred to one meridian, *viz.* that through the initial station, or station at which the meridian has been established. The bearings of lines situated east or west of the initial station therefore differ from their azimuths, or directions from their respective local meridians. In consequence, a line common to two adjoining surveys is usually designated by different bearings in the two surveys. For ordinary small surveys the discrepancy is slight, and, when necessary, the correction for convergence (Vol. II, Chapter V) can be applied.

The direction of true meridian at a station is invariable, and a record of true bearing therefore assumes a permanence not otherwise possible. This is a matter of considerable importance for large surveys in unmapped or imperfectly mapped country. In engineering location surveys the adoption of true meridian may save much time in retracement of the lines during final location and construction, more particularly if the ground is rough or densely wooded. For small surveys, on the other hand, true bearings need be used only if they can be measured from a meridian already established.

199. Magnetic Meridian. The magnetic meridian of a place is the direction indicated there by a freely floating and properly balanced magnetic needle, uninfluenced by local disturbing forces. Magnetic meridian does not coincide with true meridian except in certain localities, and the horizontal angle between the two directions is termed the *Magnetic Declination*, or *Variation of the Compass*. (This term should not be confused with the changes of direction described in the next section.) The amount and direction of the declination is different at different parts of the Earth's surface: in some places the needle points west, and in others, east of true north. Its value at any place may be determined by making observation for true meridian, or may interpolated approximately by reference to published isogonic charts.

Isogonic lines, or isogons, are imaginary lines passing through points at which the magnetic declinations are equal at a given time. Those through places at which the declination is zero are termed agonic lines. Across Great Britain and Ireland the value of the magnetic declination in 1967 ranges from about 7° W. in E. Anglia to about 12° W. in the west of Scotland and the isogons have a bearing of roughly 15° to the east of true meridian.

200. Changes of Declination. The declination at any place is not constant, but is subject to fluctuations, which may be divided into: (1) regular or periodic variations; (2) irregular variations.

(1) This class of variation may itself be analysed into several components of different periods and amplitudes, but only two of

them, secular and diurnal variation, are sufficiently pronounced to merit attention by the surveyor.

(*a*) *Secular Variation* is a slow continuous swing having a period of several centuries. Thus, at London, previous to about the year 1657, the declination was easterly and decreased annually. In 1657 the needle pointed towards true north. Thereafter the declination gradually increased westwards until 1819, when a maximum westerly declination of about $24\frac{1}{2}°$ was attained. Since then the declination west has been decreasing until at present, 1968, its value at Greenwich is about $8\frac{1}{2}°$ W., with an annual decrease of about 5′.

While in this country the magnetic meridian has, within the recorded cycle, moved from one side of true meridian to the other, similar records in other places show that a complete swing may be performed on one side of the meridian and that the range and period of the oscillation vary in different localities.

(*b*) *Diurnal Variation* is an oscillation of the needle from its mean position during the day. The amount of this variation ranges from a fraction of a minute to over 12 min. at different places, being greater in high latitudes than near the equator, and more in summer than in winter at the same place. In the northern hemisphere, the needle is east of its mean position during the night, and attains its maximum easterly position at about 8 a.m. It then moves westwards, and is farthest west from its average position at about 1 p.m. The mean position of the magnetic meridian occurs in this country at about 10 a.m., and again between 6 and 7 p.m. The direction of the daily swing is reversed in the southern hemisphere.

(2) *Irregular Variations* are caused by magnetic storms. A variation within a quarter of an hour of as much as 5° has been recorded, but this is very exceptional, and variations exceeding 1° are rare.

On account of the secular variation it is always well, when magnetic bearings are used in a survey, to note on the plan the date of the survey, and, if possible, the magnetic declination and its annual variation on that date. Thus: 'Magnetic Declination (February 1958) 11° 35′ W. Decreasing $7\frac{1}{2}′$ per annum.'

201. Grid Line Meridian. In some countries the government survey and official maps and plans are based on one or more geographical meridians, each of which serves as an axis of a coordinate system covering a north-south strip of country restricted in width so that the errors arising from representing the round Earth on a flat map are kept within acceptable limits (see section 298).

The older maps of Great Britain were based on many different meridians, practically one for each county, but now the mapping of this country has one system based on the meridian 2° W. of Greenwich.

In a coordinate system defined in this way, a line on the map parallel to the central meridian defines the direction *grid north*, and a bearing reckoned from this direction is a *grid bearing*. The difference between true and grid bearing is the *convergence*; its precise value at a point depends on the mathematical structure of the coordinate system, but in most cases it is found quite accurately enough by multiplying the *sine* of the latitude of the point by the difference of longitude between the point and the central meridian.

202. Arbitrary Meridian. For small surveys, especially in unmapped country, any convenient direction may be assumed as a meridian. This artificial meridian is usually the direction from a survey station either to some well-defined and permanent point or to an adjoining station. It is desirable that its magnetic bearing should be known. An arbitrary meridian has the merit of being invariable, and its direction can be recovered when required if the station or stations defining it are permanently marked or fixed by ties from permanent objects. If it is subsequently found necessary, the bearings can be converted to true bearings by the establishment of a true meridian.

203. Designation of Bearings. Bearings are specified on either of two systems of notation: (*a*) the whole-circle system; (*b*) the quadrantal system. In the whole circle or azimuth method, bearings originate from north, which is marked 0° or 360°, and are measured clockwise from the meridian, through E., or 90°, S., 180°, and W., 270°.

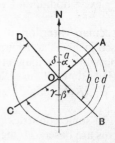

FIG. 106.

In the quadrantal system, they are numbered in four quadrants, increasing from 0° to 90° from N. to E., S. to E., S. to W., and N. to W. Thus, if O (Fig. 106) is a survey station, and ON the meridian through it, the bearings of the lines from 0 are:

Line	Whole Circle bearing	Quadrantal bearing
OA	a	N. α E.
OB	b	S. β E.
OC	c	S. γ W.
OD	d	N. δ W.

Quadrantal bearings are never reckoned from the E. and W. line, so that the letter which precedes the figure must be either N. or S.

204. Comparative Merits of Whole Circle and Quadrantal Reckoning. In the whole-circle method a bearing is completely specified by an angle, and the convention of reckoning clockwise from N. is so simple that the noting of the cardinal points as required in the quadrantal method must strike one as unnecessary trouble. The former system lends itself to the measurement of bearings on a continuously graduated circle as fitted to the theodolite. The fact that quadrantal bearings never exceed 90° is an advantage in extracting the values of their trigonometrical functions from ordinary tables, but the alternate clockwise and anticlockwise direction of increase of angle in the different quadrants is sometimes inconvenient and may very easily lead to mistakes being made. Thus, if the true bearings of Fig. 106 have to be converted to magnetic bearings, given that the declination of the needle is 16° W., each of the whole circle reckonings has to be increased by 16° (subtracting 360° if the result exceeds 360°), but in the quadrantal method the correction is positive in the 1st and 3rd, and negative in the 2nd and 4th quadrants. Care must be exercised that the appropriate cardinal points are applied to the resulting figures.

205. Back Bearings. The bearing of a line designated by the stations between which it lies is to be taken as that from the first station mentioned. Thus, in referring to the bearing of OA (Fig. 106), the sign of the direction is from O towards A. The bearing from A to O is termed the back, or reverse, bearing of OA, which latter may be distinguished as the forward bearing. By referring to a parallel meridian through A, it will be evident that the back and the forward bearings of the line differ by 180°.

In whole-circle reckoning, the back bearing of a line is obtained from the forward bearing by applying 180°. To apply 180°, add when the given bearing is less than 180°, and subtract when the given bearing exceeds 180°. In the quadrantal system, it is only necessary to change the cardinal points by substituting N. for S. and E. for W., and *vice versa*. Thus, if the bearing of a line AB is observed as 300°, or N. 60° W., the back bearing of AB, or the forward bearing of BA, is 120°, or S. 60° E.

206. Reduced Bearings. In finding the values of the trigonometrical functions of a whole circle bearing exceeding 90°, one must refer in the tables to the corresponding angle, less than 90°, which possesses the same numerical values of the functions. This angle is called the reduced bearing, and is that between the line and the part of the meridian, whether the N. or S. end, lying adjacent to it. It is therefore the angle used in the quadrantal reckoning of bearings.

To Reduce Whole Circle Bearings. The following rule is applied:
If the bearing lies between 0° and 90°, no reduction is required.

       „   „   90° and 180°, subtract it from 180°.

       „   „   180° and 270°, subtract 180° from it.

       „   „   270° and 360°, subtract it from 360°.

It follows that the reduced bearing of a line lying due N. or S. is 0°, while that of a line due E. or W. is 90°. Bearings need not be reduced in this way if tables giving trigonometrical functions for all directions up to 360° are available.

FIELD WORK OF THEODOLITE TRAVERSING

207. Field Party. The minimum party consists of the surveyor and two chainmen, and this is sufficient for small surveys, if no clearing is necessary. In large theodolite surveys, particularly where rapid progress is important, the party may consist of a chief, an assistant surveyor or instrument man, a note-keeper, two or more chainmen and a number of labourers, the number depending upon the object of the survey, the nature of the ground, and, in forest or bush country, the amount of clearing required.

With a large party, the chief of party directs the survey and, in particular, reconnoitres the forward ground, fixes the position of stations and of permanent marks, and sees to their being properly pegged and flagged. The assistant surveyor or instrument man, assisted by the note-keeper, is responsible for the angular observations and notes, and may also control the main taping. The labourers clear the lines of all bush and obstacles, fetch and carry poles, drive and make pegs, erect signals and permanent marks, support and carry the tapes, if long ones are used, and do any other unskilled or semi-skilled work required. In forest country, an ordinary traverse party may easily number between twenty and thirty men, the whole party being split up into a number of small gangs or sub-parties, each with its own particular work to do.

208. Equipment. For the linear measurements, the equipment is much the same as for chain surveying (section 135). In theodolite traversing, the steel tape is to be preferred to the wire chain because the errors arising from the use of the latter are much larger than those arising from angular measurements made with even the smallest theodolite. When the steel tape is used, the equipment may also include spring balance, tape clips, and thermometer.

As mentioned above, electromagnetic distance measuring instruments may be used in traversing with long lines.

The theodolite should be a transit, with either vernier or micrometer reading. An instrument reading to 20″ is sufficient for most purposes, but a more precise instrument may be needed for surveying long traverses, setting out tunnels, etc. It is sometimes useful to carry a prismatic compass for use on minor subsidiary traverses, where the same degree of accuracy as that needed on the main work is not necessary.

By including traverse or trigonometrical tables in the miscellaneous equipment, the accuracy of closed circuits, when these only consists of a very few sides, can be tested before leaving the field. As a general rule, however, computations will be done on the return of the party from the field.

209. Balancing the Accuracy of the Linear and Angular Measurements. In deciding on the equipment necessary, and the methods to be used for both linear and angular measurements, it is essential to bear in mind the standard of accuracy desired, and this is controlled by the object for which the survey is needed. When this point has been settled, the equipment and methods chosen should be such as to make the relative accuracy of the linear and angular measures about the same, and, in doing this, it is useful to remember the ratio of the linear displacement at the end of a line, subtended by a second, a minute, and a degree of arc, to the length of that line.

Thus approximately:

1 Second of arc corresponds to a displacement ratio of 1 : 206,300.

1 Minute of arc corresponds to a displacement ratio of 1 : 3,440.

1 Degree of arc corresponds to a displacement ratio of 1 : 57.

While remembering these figures, it must not be forgotten that angular errors tend to propagate themselves along a traverse, not directly as the number of stations, but as the square root of the number of stations, a point that will be referred to later, while, as we have already seen, errors arising from the linear measures tend to be roughly proportional to the length of the line.

210. Selection of Stations. If the traverse is being surveyed with the primary object of picking up detail, considerations of easy chaining and short offsets should, as in linear surveying, be given due weight. If, on the other hand, it is needed as a means of establishing control points, or for other work in which a relatively high degree of accuracy is required, the primary consideration should be to avoid short legs and to obtain as long sights as possible. In addition, it is well to endeavour to keep the lines of sight as high above ground level as possible and to avoid 'grazing rays', or rays which come very close to the surface of the ground, as, otherwise, shimmer or horizontal refraction, or both, will cause minor inaccuracies in observing the angles. The ground at stations should be firm to afford an unyielding support for the instrument, and should be moderately level rather than on a steep slope. Unless in exceptional circumstances, stations should not be established in situations, such as on roads, streets, or railways, where the observations will be delayed, and the instrument possibly endangered, by passing traffic.

It sometimes happens that difficulties on the ground prevent long legs being chosen for the linear measurements, but it is possible to sight a forward station some distance ahead on the line that the traverse has to follow. Thus, in Fig. 107, it is impossible, or at best

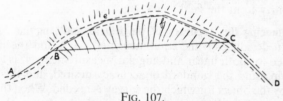

Fig. 107.

difficult or inadvisable, to carry the taping direct along the line BC, but B and C are intervisible. In this case, the taping line is made to follow the line B*ef*C. The angles at *e* and *f* are measured, as well as those at B to both *e* and C, and at C to both B and *f*. The portion B*ef*C is then treated as a subsidiary traverse to determine the length, and the length only, of the line BC, so that main bearings are brought forward along ABCD, which is now the main traverse, with a computed length for the line BC. In this way, errors in bringing forward bearings through the short legs B*e*, *ef*, and *f*C are avoided. Such a lay-out of a traverse line, where the taping line departs or deviates from the main angular line, may be called a 'deviation'.

In other cases, schemes of combined traverse and triangulation may sometimes be used to cover the ground quickly; if necessary, the bases for the triangles being short traverses. A 'traverse base' is one in which the distance between the terminal points is computed from a traverse run between them and it is sometimes used when the nature of the ground makes it impossible or inconvenient to measure a straight base. Much ingenuity may often be exercised by the surveyor in avoiding short traverse legs or in working round, or over, obstacles. Some of the methods available are dealt with in detail in the miscellaneous problems at the end of this chapter.

211. Marking of Stations. Stations may be marked by pegs as described in section 179, but more permanent marks are often required for some or all of the stations of a traverse. These may take the form of a bolt or spike set in a block of concrete or a large stone. Reference marks to aid in the recovery of the station point should be established in its vicinity, their distance and bearing from the station being measured. The notes should include a detailed description of the site of each station with particulars of the reference marks.

In some countries, the types of station pegs to be used in surveys for official purposes are specified by regulations.

212. Signals at Stations during Observing. When observing angles with the theodolite, the signals to which observations are taken can be of various kinds according as to whether legs are long or short. If legs are long, the signal may consist of a ranging pole, with a red and white flag at the top, carefully plumbed over the station mark and held in position by a labourer or supported by light wire stays or wooden struts. In the case of short legs, it may be possible to see, from the theodolite, the tack or mark in the centre of the peg or pillar, and, if this is so, the sight can be taken direct to the mark. If this is not possible an arrow, held with its point on the mark, may be visible. For legs that are not too long, a very good mark is a plumb bob with a piece of white paper threaded on the string, the sights being taken to the string and the paper serving merely to enable the position of the string to be found easily in the field of the telescope. The plumb bob can be kept fairly steady if it is suspended from a long pole, fixed firmly at a slight inclination to the vertical, and with its lower end on the ground. Otherwise, it may be hung from a 'bush tripod', made by lashing three poles together and fixing them so that they stand over the mark in the form of a tripod.

In all cases in traverse work, particular care should be taken to see that the signal is properly centered and plumbed over the station mark, and it should be sighted as low down as possible. Lack of proper care in plumbing signals is a frequent cause of minor error.

213. Special Equipment for Observing Short Legs. Very short traverse legs are often unavoidable in surveys in mines and tunnels, and in such work it is essential to avoid large centering errors.

Some makers therefore provide a special equipment for this kind of work which consists of a theodolite, three or more tripods, and at least two special targets which fit on the tripods and are interchangeable on them with the theodolite. When observations at a station are complete, the theodolite is taken off its tripod, without disturbing the latter, and then moved and placed on the tripod marking the forward station, from which the target has been removed. Meantime, the target at the rear station is taken off its tripod and placed on the original theodolite tripod, the forward target being moved on and placed on a tripod at the next forward station. In this way, centering errors are reduced to a minimum since the vertical axes of the theodolite and targets always occupy the same positions on the tripods. The targets are all provided with some sort of artificial illumination because in mines and tunnels work has to be done in darkness or in very subdued light.

214. Order of Field Work. If the party is small it may be preferable in minor surveys to complete the observation of the angles or bearings before taping is begun or *vice versa*. A full party on long traverses, however, is divided into two main groups; the first selects stations, clears the lines, erects permanent marks, and does the check taping, while the other follows and observes angles and does the main taping.

ANGULAR MEASUREMENTS

215. The accuracy of angular measurements in traversing depends on:

(*i*) the type of instrument used,

(*ii*) the procedure adopted in the use of it,

(*iii*) the care taken in centering the instrument and the observed signals.

If a small vernier theodolite is used and the bearings are obtained from one setting at each station, the accuracy may be no better than one minute; but several measurements of each angle with a micrometer theodolite on each face with change of zero, should give an accuracy better than 5 seconds. The methods used for linear measurements should have correspondingly similar accuracy to that of the angular measurement.

The methods of measuring the angles and bearings of a traverse may be divided into two classes:

(*a*) Those in which the angles at the different stations are measured directly, and the bearings subsequently calculated from the measured angles and the given bearing of an initial line.

(*b*) Those in which the setting of the theodolite is so arranged as to give direct readings of bearings.

As a general rule, the first method is to be preferred and is the one most generally used for long traverses, or where precision is required, while the second may be used for short traverses where great precision is not necessary and the traverse is either a closed one or ends on a line of known bearing. In engineering and cadastral surveying it is generally advantageous to arrange the angle measurement method so that at least one of the readings of the theodolite is approximately the true orientation of a line.

THEODOLITE OBSERVATION OF ANGLES

216. The horizontal angles measured at the several stations may be either (*a*) included angles, (*b*) deflection angles. An included angle is either of the two angles formed at a station by two survey lines meeting there. A deflection angle is that which a survey line makes with the preceding line produced beyond the station occupied, and its magnitude is the difference between the included angle and 180°.

The minimum routine to be adopted for the observation of included angles is given in section 85, and it is often sufficient in small surveys when the linear measurements are made by chain. When, however, it is decided that considerations of accuracy demand the use of the steel tape, each angle should be read 'face right' and 'face left' and, if a vernier instrument is used, both verniers should be read, not only to minimise the effects of instrumental error but also to provide a check against mistakes in reading. In the better classes of work, it is usual not to endeavour to set the instrument to read 0° or 360° exactly when the instrument is sighted on the back station, but to set it somewhere near that value and then, after the mark has been intersected with the cross-line, to read both verniers. The mark at the forward station is then sighted and intersected and both verniers again read. This gives one value for the angle. For the next observation, on the other face, it is as well to 'change zero', before the back station is resighted, by setting the verniers to read somewhere near 90° and 270°. Face having meantime been changed, the instrument is directed to the back station and the angle measured again. Changing zero in this way not only tends to avoid errors in measurement, since the two values of the angle are derived from two entirely different sets of figures, but it also tends to eliminate errors due to small periodic errors of graduation in the horizontal circle.

Included angles can be measured clockwise or counter-clockwise from either the back or the forward station, but it is well to adhere to the regular routine of measuring from the previously occupied station and in a clockwise direction, since the graduations of the theodolite circle increase in this direction. This does not necessitate that the telescope should always be turned clockwise, although it is better to wheel or swing it in a constant direction for observations on one face and in the reverse direction for observations on the other. Figs. 108 and 109 show that, in a closed polygon, angles

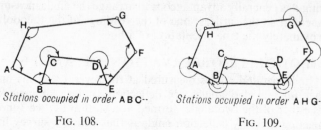

Stations occupied in order **A B C**--   *Stations occupied in order* **A H G**-

Fig. 108.   Fig. 109.

measured clockwise from the back station are either interior or exterior according to the direction of progress round the survey. Interior angles are obtained by proceeding counter-clockwise round the figure, but these will be exterior to subsidiary circuits as at C and D.

In measuring deflection angles, having bisected the mark at the back station by using the lower clamp and tangent screw and read one or both verniers, the theodolite is transited and is then pointed to the forward station, the upper clamp and tangent screw being used for this and to intersect the station mark. The verniers are again read, and the difference between the first set of readings and the second gives the angle of deflection. The measurement is either right- or left-handed from the production of the back line, and this direction must be most carefully noted in the field book. It is usual, when deflection angles are being measured, to set the horizontal circle to read zero when the back station is sighted, so that the reading when the forward station is sighted gives the angle of deflection directly.

Included angles are to be preferred to deflection angles. The latter are often used in surveys for railways, roads, pipe lines, etc., in which a series of traverse lines may make small deflection angles with each other, but they are open to the objections that right- and left-handed angles may be confused in booking or plotting or in working out coordinates and that the transiting of the telescope introduces possible errors of non-adjustment if observations are not made on both faces. Moreover, whatever other advantages the method of deflection angles may have, these largely disappear if

zero is changed between face right and face left readings. Consequently, for all but entirely specialised work in which deflection
angles are usually employed, it is preferable in every way to read
and book the angles of a traverse as the included angle, read clockwise from the back station.

217. Booking the Angles. The angles can be booked in an
ordinary field book which may also include the linear measurements. Fig. 81 will explain the method in the case of observations
by included angles, the angular observations being written at the
right of the page. With instrument set at face right over station 64
the reading on the circle when the telescope was sighted on station
63, the back station, was $0°\ 15'\ 20''$ on 'A' vernier and $180°\ 15'\ 40''$
on 'B'. These readings are set out as shown, minutes and seconds
only being recorded in the case of the 'B' vernier readings. The
mean is $0°\ 15'\ 30''$ and this is written out in full. When the instrument, still face right, was sighted at station 65, the forward station,
the reading on 'A' vernier was $196°\ 32'\ 40''$, and on 'B' the minutes
and seconds read $33'\ 00''$. The mean was therefore $196°\ 32'\ 50''$,
and all three sets of figures are written down as shown above
the corresponding readings to station 63. Hence, subtracting
the figures in the second column, the included angle at station 64,
as read with the instrument set face right, was $196°\ 17'\ 20''$. Face
was now changed and the horizontal circle set to a new zero near
$270°$ on 'A' vernier. When the back station was sighted, the readings
were $272°\ 12'\ 20''$ on 'A' vernier and (again omitting the degrees)
$12'\ 20''$ on 'B', the mean being $272°\ 12'\ 20''$. These are written down
as shown. Similarly, when the forward station was sighted, the
readings were $108°\ 29'\ 40''$ and $30'\ 00''$, the mean being $108°\ 29'\ 50''$.
Consequently, the second value for the included angle was $108'\ 29'$
$50''$ minus $272°\ 12'\ 20''$, which, after adding $360°$ to the first mean,
since it is numerically less than the second, gives $196°\ 17'\ 30''$. The
mean value of the included angle was therefore $196°\ 17'\ 25''$, and
this is shown underlined below the readings for the first observation.

In all cases, when observing included angles with the theodolite,
all observations should be worked out before a station is vacated,
because, if a mistake is made and noticed in time, it can be corrected
at once and there is no necessity to visit the station a second time
to take new readings.

*When using the method of included angles it is most important to
avoid readings taken to the forward station being booked in the place
usually reserved for booking readings taken to the back station*, as
otherwise the interior angle will be obtained instead of the exterior
one, or *vice versa*. In the example, it will be noticed that, as it is

usual to work up a page of a field book, the first observation taken—that to the back station—is booked at the bottom of the page, with the next observation—that to the forward station—booked immediately above.

Deflection angles can be booked in a somewhat similar manner, but, if the instrument is set to zero every time the back station is sighted and only one face used and one vernier read, the only entry will be the observed deflection angle. The figures *must* be followed in every such case by the letter 'R' or 'L', signifying a right-hand or left-hand deflection respectively, or else a small sketch made in the book to show clearly which direction the deflection takes.

218. Obtaining the Initial Bearing. In order to calculate the bearings of the different lines the bearing of one line at the initial station must be known or assumed. Thus, in Fig. 110, AB is the first line

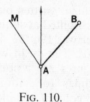

Fig. 110.

of the traverse and AM a line which is already marked out in some way on the ground and whose bearing is known. If bearings are to be referred to a true meridian, that of AM may be obtained either direct from astronomical observations, or, if the points A and M have been established by the government survey department, on application to the head or local representative of that department. The bearing of AM being known, the angle MAB is measured and from this the bearing of the first line, AB, of the traverse can be calculated.

219. Calculation of Bearings from Angles. Having obtained an initial bearing from which to start, the calculation of the other bearings of a traverse is very simple, and can be formulated as follows:

In Fig. 111 let B be the station at which the angle is measured and let the bearing of the line AB or BA be known. The point B may either be the first station or it may be any intermediate station on the traverse. If it is the first station, the point A will correspond to the initial reference mark M in Fig. 110. If B is an intermediate point in the traverse, BA will be a traverse leg, the bearing of which has already been computed.

Let c = the measured angle ABC *reckoned clockwise from* A.

d = the deflection angle.

(1) Bearing BC = Bearing BA + c,

(subtracting 360° if the result exceeds 360°).

(2) Bearing BC = Bearing AB ± d,

(using the + sign if d is clockwise from AB produced, and the − sign if d is counter-clockwise, as in Fig. 111; and adding 360° if the result is negative, and subtracting it if the result exceeds 360°).

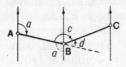

FIG. 111.

Quadrantal bearings should be converted to whole-circle reckoning before applying the formulae.

In the reverse process the formulae become:

(3) c = Bearing BC − Bearing BA,

(adding 360° if the result is negative).

(4) d = Bearing BC − Bearing AB,

(if d is positive, it is clockwise from AB produced).

When the bearings of a traverse have been calculated, and the bearing of each line obtained, the work should be checked by adding together the initial bearing and all the included angles or deflection angles. For included angles, the result should equal the bearing of the last line plus some multiple of 180°.

As an example, take the following:

Point	Observed included angle			Calculated bearing			Line
	°	′	″	°	′	″	
M		—		37	14	10	MA
A	224	15	25	81	29	35	AB
B	210	36	40	112	06	15	BC
C	135	14	10	67	20	25	CD
D	120	08	30	7	28	55	DE
E	167	42	35	355	11	30	EF
F							
	857	57	20				
	37	14	10				
	895	11	30				
3 × 180	= 540						
	355	11	30				

Here, the bearing of the line MA is 37° 14′ 10″ so that the bearing of the line AM is 217° 14′ 10″. Adding the included angle at A, the sum is 441° 29′ 35″, and, subtracting 360°, the bearing of the line AB is 81° 29′ 35″. Also, the sum of the included angles is 857° 57′ 20″, and this, plus 37° 14′ 10″, the bearing of MA, and less $3 \times 180° = 540°$, is equal to the bearing of the line EF.

The number by which 180° has to be multiplied in order to obtain the final bearing from the sum of the included angles and the initial bearing can be obtained by noting, for each station, the number of additions or subtractions of 180° or 360°. Thus, at A, 180° is added once, and 360° subtracted once, so that the net result is a subtraction of 180°. There are also similar subtractions of 180° at B, C, and D, and an addition of 180° at E. Hence, in all, 180° has been subtracted 3 times, so that 3 is the multiplier required. This, however, is not a very satisfactory check as far as an error of 180° is concerned, and, as the sum, provided too many angles are not included in it, can be relied on to check the minutes and seconds and the tens and units in the degrees, it is far better to make a rapid examination of the degrees column only. This should show at once whether or not a mistake of 180° has been made.

If deflection angles are used, it is best to tabulate right-hand deflections as positive and left-hand ones as negative. Thus, with the same angles as before but transformed into deflection angles:

Point	Observed deflection angle						Bearing			Line
	+			−						
	°	′	″	°	′	″	°	′	″	
M		—			—		37	14	10	MA
A	44	15	25		—		81	29	35	AB
B	30	36	40		—		112	06	15	BC
C		—		44	45	50	67	20	25	CD
D		—		59	51	30	7	28	55	DE
E		—		12	17	25	355	11	30	EF
F										
				116	54	45				
	74	52	05							
	37	14	10							
	112	06	15							
	116	54	45							
	355	11	30							

Here the plus and minus columns are added, and the sum of the plus deflection angles added to the initial forward bearing from

M to A and the sum of the negative angles subtracted from the result. This gives the bearing of the last line, and this acts as a check.

This method of computing can also be used as a very good check on bearings computed direct from the included angles by remembering that, if the included angle is greater than 180°, the deflection angle is right-hand or positive and equal to the included angle minus 180°, while, if the included angle is less than 180°, the deflection angle is left-hand or minus and equal to 180° minus the included angle. These are the rules for included angles measured clockwise from the back station, but equally simple ones can be devised for angles measured clockwise from the forward station.

THEODOLITE OBSERVATION OF BEARINGS

220. In this method the theodolite is oriented so that, when the telescope is sighted along an initial line, the reading corresponds with the bearing of the line. Hence, if the lower circle is kept fixed during observations, and the telescope is sighted along some other direction, the reading then gives the bearing of that direction.

A theodolite in which the upper and lower plates can be clamped together and given small relative movement with a slow-motion screw, must be used for the direct bearing method.

Three distinct systems of procedure are available:

(*a*) Direct method involving transiting of the telescope.

(*b*) Direct method without transiting.

(*c*) Back bearing method.

In every case, the routine at the initial station is the same and involves setting the instrument to the correct orientation. To avoid repetition in the following descriptions of the observations, it is supposed that a single observation only of each bearing is made, but the observations may be duplicated by being taken on both faces of the instrument.

221. Orienting on the True Meridian or on a Grid Line. Here, as in the case of finding bearings when using the method of observed angles, it is necessary that the first station occupied should be one end of a line whose bearing is already known, either from the results of astronomical observations, from the results of a previous survey, or from the data provided by the government survey department.

Referring to Fig. 110, let A be the first station of the traverse, AM the line of known bearing and AB the first leg of the traverse. To orient the theodolite by use of the bearing AM, the micrometer or one of the verniers is set to that bearing, and the upper plate is clamped. With the lower clamp slack, the telescope is then turned towards M. When M appears near the intersection of the hairs, the lower clamp is tightened, and the line of collimation is brought exactly on the signal by the lower tangent screw. This completes the orientation, and the reading is the bearing of the line along which the line of collimation is directed. Under these conditions, if the upper clamp is released, and the circle is kept fixed, the telescope will point towards north when the reading is set to zero or 360°. To observe the bearing of the survey line AB, the telescope is directed to B by the upper clamp and tangent screw. The bearing is then read on the vernier previously used, or on the micrometer.

222. Orienting on Magnetic Meridian. In orienting from magnetic meridian, reference is made to the compass on the theodolite. The compass box, if of circular form, is mounted on the upper plate (see section 72) while the trough pattern is either connected to the plate by being screwed to a standard or is attached to the lower or graduated plate. In the case where the compass is connected to the upper plate, the line joining the N. and S. graduations in the circular form, or that joining the zeros of the scales in the trough form, bears a fixed relationship to the line of collimation of the telescope, and is intended to be parallel to it. When the compass is attached to the lower plate, the line of zeros is parallel to the line of collimation only when the horizontal reading is 0° and 180°.

To orient, with either arrangement, the reading is first set to zero, the lower clamp being slack. The needle is then lowered upon its pivot, and the instrument is turned about the outer axis until the N. and S. graduations are brought opposite the ends of the floating needle. The lower clamp is then tightened, and the lower tangent screw is used to bring the zero graduation to exact coincidence with the point of the needle. The instrument is now oriented, since the line of collimation is directed towards magnetic north while the reading is zero. To observe the bearing AB, it is only necessary to set the line of collimation on B by means of the upper clamp and tangent screw, and note the reading. The result of the observation may be checked by a glance at the reading of the north end of the needle in the circular box form.

It is to be noticed that when the needle rests on its pivot, the instrument can be rotated about it without disturbing its direction.

The needle cannot be set properly if it is looked at from one side. The eye should be in the vertical plane of the needle as nearly as possible.

When the plate has been oriented, the telescope may be pointing south, in which case it must be transited before observing bearings.

While the needle of a trough compass is more sensitive than the shorter needle in a circular box, neither is capable of defining the meridian with the refinement with which angular measurements can be made by the theodolite. This is not always important, since in theodolite surveys, as distinct from compass surveys, the bearings of all lines after the first are measured from the bearing of the first line, and the needle is consulted at subsequent stations only as a check. Inaccurate orientation at the first station therefore merely turns the whole survey through a small angle, but does not distort it.

If the value of the magnetic declination is known, the needle may be used for approximately orienting to true meridian by setting off the declination. It is, however, better, as regards facility for applying the compass check, to adhere to magnetic bearings in the field, subsequently converting them to true bearings if required.

It must not be forgotten that, in practice, nearly every compass, whether of the trough or circular variety, has its own individual error, so that the magnetic axis does not coincide with the geometrical axis of the needle, and, in consequence, the magnetic meridian indicated by the instrument is not the true magnetic meridian. Individual compass errors of anything from half a degree to two or three degrees are quite common, and, in any particular instrument, the error may easily undergo a sudden change due to some outside influence upsetting the magnetisation of the needle and so altering the position of the magnetic relative to the geometrical axis. Hence, if absolutely correct orientation is desired, it is necessary to have the compass compared from time to time with a standard instrument or to test it on a line whose true magnetic bearing is known.

223. Orienting on Arbitrary Meridian. Having set the reading to zero· it is only necessary to sight the object or station defining the direction from A of the adopted meridian. When the signal is bisected by the use of the lower clamp and tanget screw, the circle is oriented. The upper clamp is then released, and bearing AB is observed as before.

224. Carrying Forward the Bearing. While the measurements in these methods are virtually those of the angles between the lines meeting at the various stations, the instrument is so manipulated that the required whole circle bearings are read directly. In the descriptions given in the next three sections, the use of a vernier theodolite is assumed, and it is necessary to distinguish between the two horizontal verniers. They are designated here as 1 and 2, the former being supposed used at the initial station. If they are not given distinctive marks on the instrument, they can easily be distinguished by their positions with respect to the vertical circle or a plate level.

In each of the three methods, after observing the bearing of AB from A, the lower clamp is released, and the instrument is carried to B with vernier 1 kept clamped at the reading obtained. A signal is left at A.

225. Direct Method with Transiting. (*a*) Set up and level the instrument at B. See that vernier 1 still records the bearing AB. If the plate has slipped during transfer of the instrument, correct the reading by the upper tangent screw.

(*b*) By using the lower clamp and tangent screw, sight back on A.

(*c*) Transit the telescope. The line of sight has now the same direction as it had at A, and vernier 1 still records the bearing AB. The instrument is therefore oriented.

(*d*) Bearing BC can now be observed by releasing the upper clamp and sighting C, the upper tangent screw being used in bisecting C. Vernier 1 now records the bearing BC.

(*e*) This reading being maintained on the vernier, the instrument is transferred to C, and the routine is repeated. On transiting at C, the instrument is returned to the same face as was used at A.

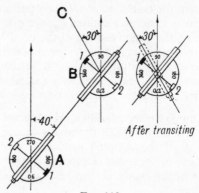

FIG. 112.

Example. Fig. 112 illustrates the case where the bearings of AB and BC are respectively 40° and 330°. It will be seen that, on taking the backsight BA, the verniers occupy the same positions relative to the telescope as at A, but, since the telescope is at 180° to its previous direction, the orientation of the circle is 180° different from what it was at A. This is neutralised on transiting the telescope. To bring the telescope into the position shown dotted, in order to sight C, it must be turned counter-clockwise through 70°, or clock-wise through 290°, either movement resulting in vernier 1 (shown in black) being brought opposite the reading 330° as required.

226. Direct Method without Transiting. The manipulation of the instrument at B is similar to that in the previous method, except that the telescope, instead of being transited after the backsight is taken, is turned directly on to C. The difference of 180° in the orientation of the circle at B from its orientation at A therefore remains uncompensated, and the reading of bearing BC on vernier 1 is 180° out. A correction of 180° has therefore to be applied to the reading or readings taken at B, adding the correction if the observed value is less than 180°, and subtracting if the reading exceeds 180°. At C the orientation, being 360° out, is correct, and the results need no adjustment. The application of 180° is therefore necessary only at the 2nd, 4th, 6th, etc., stations occupied.

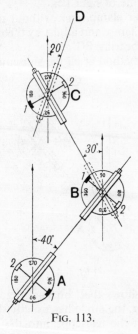

FIG. 113.

Example. Fig. 113 shows this system applied to the previous case. To sight C from B, the telescope must now be turned through 110° clockwise, or 250° counter-clockwise, and vernier 1 then reads 150°, which falls to be increased by 180°. Following the process to station C, let the bearing CD be 20°. After backsighting on B, the rotation necessary to bring the telescope into the position shown dotted, in order to sight D, is 130° counter-clockwise, or 230° clockwise, and the vernier then reads 20°.

Some surveyors adopt the routine of reading opposite verniers alternately to eliminate the 180° correction, but it is simpler to read one vernier throughout and note the correct values.

Notwithstanding that a number of successive instrument stations may lie in the same straight line, the correction must be applied at alternate stations.

227. Back Bearing Method. (*a*) Set up and level the instrument at B as before.

(*b*) *Before* sighting back on A, set vernier 1 to read the back bearing of AB, and fix the upper clamp.

(*c*) By using the lower clamp and tangent screw, sight back on A. The instrument is now oriented, since vernier 1 records the bearing BA, along which the line of sight lies.

(*d*) Release the upper clamp, and, without transiting, direct the telescope towards C. Clamp, and adjust by the upper tangent screw. Vernier 1 records the bearing BC.

(*e*) Apply the same method at all the subsequent stations.

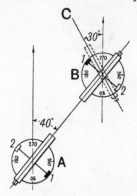

FIG. 114.

Example. Fig. 114 shows that, on back-sighting with vernier 1 set to the back bearing, the circle has exactly the same orientation at B as at A. To turn the telescope into the dotted position for sighting C, the same rotation is required as in the last case, but vernier 1 gives the required bearing directly.

228. Precautions in Carrying Bearings Forward. (1) It is necessary to guard against using the wrong clamp and tangent screw. The routine at all stations consists of the two steps; (*a*) orientation by backsighting; (b) measurement of a bearing or bearings. In back-sighting, only the lower clamp and tangent screw must be used to bring the intersection of the lines on to the signal, and these must not be touched again until the observations at the station are completed.

(2) If the telescope is set upright in carrying the instrument, it should be restored to its previous position before backsighting. Confusion may arise by inadvertently transiting.

(3) When several bearings are measured from one station, the round of bearings should be completed by observing the backsight a second time in order to detect possible movement of the circle. The observation to the station to be next occupied may then be repeated, so that the vernier may be left at the required reading.

229. Relative Merits of Methods of Carrying Bearings Forward. In point of speed there is little difference between the three methods, as the time occupied in the first two in checking the vernier reading before backsighting is not much less than that required for setting the back bearing. The first method is probably the most mechanical, but, if only single observations are made, the transiting of the tele-scope introduces possible errors of non-adjustment, and in respect of accuracy the others are preferable. On the whole, the second method is the most satisfactory. The necessity for applying 180° at every second station is not likely to lead to error, as an omission to apply the correction is easily traced.

Any method requiring re-setting of the verniers to specific read-ings at each station will be of somewhat inferior accuracy, since an error less than the least count of the instrument may be introduced at each re-setting. The back-bearing method is therefore the least satisfactory of the three.

If a micrometer reading instrument is used, there is no question of distinguishing between two verniers, and the procedures, with such an instrument, are much the same. Again, the second method is probably the most satisfactory.

230. Booking the Bearings. Bearings may be booked on the right-hand side of the field book in a manner somewhat similar to that already described for booking deflection angles. If only one vernier is read, there will only be a single entry—the observed bearing of the forward line. If both verniers are read, the readings to the back station should be booked as well as the means of these readings. On top of these, the readings to the forward station and their mean should be entered, and, from these results, a correction can be worked out to give the corrected forward bearing.

As the method of direct observation of bearings is usually used

168   PLANE AND GEODETIC SURVEYING

only for short unimportant traverses, mainly run for the survey of detail, it generally happens that the party is a small one and that detail has to be surveyed at the same time as the linear and angular measurements are made. In that case, the entries in the field book relating to the angular measures should be kept as low down on the page as possible, so as to leave plenty of room for entries relating to the survey of detail. Alternatively, a special book can be kept for the angular measurements.

The field notes should always include a sketch of the framework of survey lines, roughly to scale, so that the relative directions of the lines may be shown approximately correctly. The stations should be lettered and numbered on the sketch, and references given to the pages of the field books in which the measurements for each line or station are to be found.

LINEAR MEASUREMENTS

231. Various methods of measuring lengths have been described in Chapter II, and the particular one to be adopted should be chosen to suit the degree of accuracy required and the type of instrument available for the angular work. In general, however, a steel tape, and not a wire chain, is used in theodolite work, except, possibly, in the case where bearings are measured on one face only of a small vernier theodolite.

Distances may be measured by tachymetric methods, as described in Chapter XI, provided that the lines are short enough.

The horizontal lengths of lines are of course required for computing traverses. Methods of measuring slope are described in Chapter II. If the taping is done along the ground, the simplest method for measuring the slope will usually be to observe the vertical angle to a signal held at the other end of the tape at the same height above ground as the transit axis of the theodolite. If the tape is held against the transit axis, the vertical angle is recorded after sighting on the other end of the tape.

Where a line is longer than two tape lengths, the theodolite may have to be set up at intermediate points for measuring slopes.

With a self-reducing tachymeter, the horizontal distances are obtained directly from instrument readings.

232. Survey of Detail. Owing to the ease with which mistakes are made in traversing and the resulting need for concentration on the essential observations, it is advisable, on long or important traverses, not to throw any more subsidiary work on the main observing party than is absolutely necessary. Consequently, the survey of detail, the greater part of which can be done by offsets, should generally be done by the party which does the setting out and check taping. On the other hand, when the traverse is only a short minor one and the party is a small one, the detail survey can be done by it at the same time as the angles or bearings and distances are measured. For this, the radiation, or angle and distance, method is often useful for the survey of detail near instrument stations. In this, bearings to the various points are observed, and the lengths of the radial lines from the instrument are measured, possibly by use of the stadia lines in the telescope. Unimportant detail and distant and inaccessible points may be fixed by the intersection of bearings from two instrument stations.

In city surveying, where there is a large amount of detail, the best procedure is to survey only the frontages of the buildings during the running of the traverse lines. The miscellaneous detail can be subsequently located from the buildings and from subsidiary traverse lines projected where possible towards the back of the main buildings.

Tachymetric surveying is particularly advantageous in town surveys because the measurement of distances is not seriously impeded by heavy traffic.

SOURCES OF ERROR IN THEODOLITE TRAVERSING

233. Errors of linear measurement are dealt with in sections 167–171. Those to which the angular observations are liable may be treated as:

(1) Instrumental errors;

(2) Errors and mistakes in setting up and manipulating the instrument;

(3) Observational errors;

(4) Errors due to natural causes.

234. Instrumental Errors. The effects of residual errors of adjustment, as well as of non-adjustable errors, can be satisfactorily reduced only by the adoption of a system of multiple observations, either on the repetition or direction principle (Vol. II, Chapter 3), on both faces of the instrument, and on different zero settings of the graduated circle.

In employing any of the methods which have been described above for carrying forward the bearing, it is advisable to measure each bearing twice, face right and face left, and to read the opposite verniers at each observation. It is unusual to take more than one face right and one face left observation in ordinary traversing; but if the work is required to be of a precise character, and suitable precautions are taken in the linear measurements, a greater number of angular observations are taken for averaging. In such a case, the included angles between the lines should be measured, the bearings being deduced from the average values. In thus endeavouring to obtain results of superior accuracy with a small theodolite, the necessity for rigidity and stability in the instrument and tripod must be recognised.

The effects of certain instrumental errors on a single measurement of an angle are investigated in Chapter I (see sections 100 and 101).

235. Errors of Manipulation. (*a*) *Defective Centering*. If the centre of the instrument is not vertically over the station point, the angle or bearing is not measured from the point to which it is presumed the foresight has previously been taken. The angular error produced depends upon the error of centering, the lengths of sights, the magnitude of the angle being measured, and the position of the instrument relatively to the station and the points sighted, no error being introduced if the four points are concyclic. Under the worst circumstances, an error of centering of about $\frac{1}{6}$ in. may produce an error of 1 min. in the angle measured if the sights are only 100 ft. long, but centering would have to be at least 9 in. out to produce a similar error with sights a mile long. Considering the ease with which centering may be performed, the error is not likely to be appreciable in ordinary work.

The use of a special traversing outfit with interchangeable theodolite and signals on three tripods eliminates centering errors (see section 213).

(*b*) *Defective Levelling*. This means that the axis of rotation of the theodolite is not truly vertical, hence the angles actually measured are those projected on a plane slightly inclined to the horizontal plane through the instrument.

In Fig. 115, O is the centre of the instrument and OV the true vertical; OV' is the actual axis of rotation. The lines OTXYSO represent the horizontal plane, perpendicular to OV, and OTX'Y'SO represent the plane through O perpendicular to OV'. The line TOS common to the two planes is the axis of tilt and the direction of tilt is perpendicular to this line: let the tilt VOV' be θ, a small angle, we suppose.

An observed point P is at height H above the level of O : PM is the perpendicular from P to the horizontal plane and PM' is the perpendicular to the tilted plane: thus OM' represents the 'horizontal' setting of the theodolite when it is set on point P.

The horizontal displacement MM' is obviously perpendicular to

the tilt axis TS, and amounts to Hθ if θ is reckoned in radian measure. Thus, referring to Fig. 115 (b), the readings on the instrument
will be altered by angles like MOM′. If A is the direction of point
P referred to OT as zero, the component of the displacement MM′
perpendicular to the direction OM is MN = MM′ cos A =
H . θ . cos A. Therefore, the effect on the instrument reading is
(H . θ . cos A)/L, where L is the horizontal distance of P from O.
But H/L is tan E, where E is the elevation, or vertical angle, of P
as seen from O.

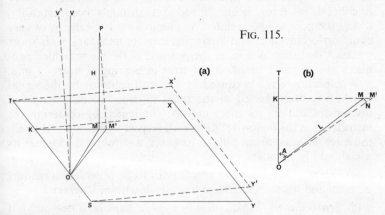

FIG. 115.

Thus, the effect of the dislevelment θ on the instrument setting
is θ . tan E . cos A, and the effect on a measured angle will be the
difference of the two effects at the settings between which the angle
is measured. In calculating the effects, vertical angles of depression
must be counted as negative, and it can be seen that the effects on
horizontal angles, though generally small, could be comparable with
θ itself if elevations and depressions of 20° or 30° are being observed.

(c) *Slip.* Results will be unreliable if the instrument is not securely
fixed to the tripod head, or if a centering head is not firmly clamped,
or if the tripod leg hinges are slack.

(d) *Using Wrong Tangent Screw.* If the upper tangent screw has
been used in backsighting, the mistake will be discovered by a
glance at the vernier before sighting forward, but the mistake of
turning the lower one in bisecting the forward signal is not evi-
denced until a check sight reveals a discrepancy.

A similar error will occur if the independent lower plate of the
theodolite is inadvertently moved during observations, through
accidental touching of its adjusting screw when it is not properly
protected.

236. Observational Errors. These consist of errors and mistakes in sighting and reading.

(*a*) *Inaccurate Bisection of Signal.* This may occur through defective vision or carelessness, particularly as regards the elimination of parallax. The bisection of a pole should always be made at the point of intersection of the lines, as the vertical line may not be truly vertical.

(*b*) *Non-verticality of Signal.* Poles should always be sighted as far down as possible. When the foot cannot be seen, there is every likelihood of error being introduced through non-verticality, unless the pole has been tested by plumb line. The error is of very common occurrence, and particular care is necessary with short sights. A deviation of 1 in. in the position of the point sighted produces at a distance of 100 ft. an error of bearing of nearly 3 min.

Whether the error is caused by non-verticality or by the signal not having been erected at the point intended to be used as an instrument station, its effect may be eliminated by centering the instrument at the forward station with respect to the point to which the foresight was taken. This is, however, a slipshod device and its use should be avoided.

(*c*) *Errors of Reading.* The precautions to be observed in reading verniers and micrometers have been dealt with in Chapter I.

(*d*) *Errors due to Displacement of Pegs or Signals.* If pegs are not very firmly driven, one may get displaced while work is in progress. For instance, after observations at a station have been completed, the peg marking it may be moved inadvertently, without the movement being noticed, before the observations to that peg from the next station are completed. The only way to guard against errors arising from this cause is to choose instrument stations on firm ground, drive pegs firmly and warn everyone concerned to use the greatest care when walking near, or moving round, a peg.

Similar care must be taken to see that no movement of signals takes place while observations are in progress.

When work ends for the day, the pegs required to commence from for the next day's work should either be carefully referenced or guard posts put around them to prevent people or animals walking into them.

(*e*) *Errors due to Wrong Booking.* A common source of error in traversing when included angles are observed is to book readings taken to a forward station as being taken to the back station or *vice versa*. This mistake, already referred to in section 217, is best avoided by adopting and keeping throughout to a standard system of booking.

237. Errors due to Natural Causes. (*a*) *Wind.* It is impossible to perform accurate work in a high wind because of the vibration of the instrument. If careful centering is required, the plumb bob and line must be sheltered. This is when the optical centering arrangement, or the rigid centering-levelling leg as fitted to some tripods (see section 42) can be very useful.

(*b*) *High Temperature.* On hot sunny days the heating of the ground during the midday hours causes warm air currents to ascend, and the irregular refraction produced gives rise to an apparent trembling of the signal near the ground. The effect is avoided if the line of sight is at all points above the disturbance, which may be taken as appreciable only within a height of 3 ft. from the ground.

The effect of the sun shining on one side of the instrument is to throw it out of exact adjustment, but in ordinary traversing uncompensated errors from this source are negligible.

(*c*) *Haze.* A hazy atmosphere increases the probability of inaccurate bisection of the signal, and may necessitate a suspension of the work.

FIELD CHECKS IN TRAVERSING

238. Various checks on a traverse are sometimes available in the field, and, when the opportunity offers itself, these should almost invariably be taken advantage of. They may be divided into two kinds:

(*a*) Those available when the traverse is a closed circuit.

(*b*) Those available when the traverse is unclosed or does not form a closed circuit.

In some cases, these checks are not only useful merely as checks, but the extra observations involved can also often be used in the office as additional data for adjustment purposes.

239. Field Checks in Closed Loop Traverse. Let AB . . . KA (Fig. 116) be a closed circuit. If there is no intervisibility except between adjacent stations, then, on proceeding from the initial station A and carrying the bearing forward to B, C, etc., all the bearings will have been determined when the observations at K are completed. The instrument should, however, be again set up at A, and the bearing AB read from a backsight on K. Any difference between 'he new observed value of bearing AB and that first determined represents the angular error accumulated in the circuit. More usually the bearings of both AK and AB will be determined at the first occupation of A, and the error is discovered at K by comparison of the value of bearing KA with that previously obtained for AK.

The magnitude of the angular closing error will reveal whether any gross mistake has occurred. If the error is appreciably greater than that to be expected from accidental errors of angle measurement, the observations must be repeated. It is therefore desirable to be able to detect a mistake as soon as committed, or at least to localise it, so that it may be discovered and eliminated without undue additional labour. The following methods of checking are directed to this end.

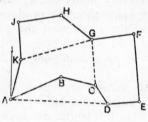

FIG. 116.

Double Observations. The method of obtaining each bearing from observations on both faces of the instrument is valuable because it reduces the effects not only of instrumental error but also of accidental errors of observation and it should almost invariably be used on extended traverses. At the same time, it must be recognised that, although it reduces the possibility of gross error very greatly, especially if care is taken to see that the degrees and minutes of the zero used for observations with the second face are different from those used in the case of the first face, it is not an absolute preventative of error. For this reason, other checks on the bearings should be used whenever they are available, not only to detect or prevent the possibility of gross errors, but also to provide a control on the growth of the purely unavoidable and accidental errors of observation.

Check taping helps to prevent the occurrence of gross error in the linear work, but, again, other checks should be taken advantage of when they can be obtained conveniently.

240 Cross Bearings. Possible mistakes may sometimes be localised if the survey is divided up by observing cross bearings between such non-adjacent stations as are intervisible. Thus, in Fig. 116, the signal at D being visible from A, a measurement of bearing AD is included in the observations at A. On reaching D, after having occupied stations B and C, a sight is taken on A, and, if the bearings DA and AD differ by 180°, the circuit observations up to D are checked. Cross bearings CG and GK similarly serve to verify the angular work from C to G and from G to K respectively.

Check bearings are also useful for the location of a mistake in the linear measurement. When the magnitude of the closing error shows that a mistake in chaining has occured, each of the compartments into which the survey is divided by the cross bearings can be tested for closure either by means of co-ordinates or by plotting. For example, in Fig. 116, the bearing of a line such as CG can be computed from the co-ordinates, or obtained from plotting, and the computed or plotted bearing compared with the observed bearing. It will then be seen which portion of the circuit is affected by the mistake.

Compass Check. If the theodolite is fitted with a circular compass box, a reading of the compass affords a rough but useful check on each observation. If the bearings are being referred to magnetic meridian, the compass readings should be the same as the observed bearings; otherwise the difference should be the constant angle between the magnetic and reference meridians. The compass is insufficient to check the bearings nearer than to about half a degree, but it serves to show up serious mistakes, and the check should not be neglected if opportunities for taking cross bearings are few. The trough form of compass is less direct in this respect, but serves to verify the orientation of the instrument in terms of magnetic meridian, except at those stations where the orientation of the circle is 180° from that at the initial station. The modern tendency, however, especially in the better class of instrument, is not to have a compass attached to the theodolite.

241. Summation Test for Angles. Adding together all the angles of a closed figure gives a useful check on the angular measurements. Assuming that included angles have been measured and that all are measured clockwise from the back station, the following two rules apply:

(*a*) If the closed figure always lies to the right of the observer as he walks around it in the direction followed by the survey, the measured angles are all the exterior angles of the figure, and their sum should be equal to $(2n+4) \times 90°$, where n is the number of stations or corners in the figure.

(*b*) If the closed figure lies always to the left of the observer as he walks around it in the direction followed by the survey, the measured angles are all interior angles, and their sum should equal $(2n-4) \times 90°$.

If deflection angles are measured, the difference between the sum of the right-hand and that of the left-hand angles should be equal to 360°

It will happen very seldom that the observed angles add up

exactly to their theoretical sum, and there will usually be a small excess or deficiency, which will be measured in minutes or seconds of arc. This represents the 'closing error' due to the unavoidable and accidental errors of measurement, and this closing error should be distributed or adjusted among the various stations, by methods to be described in the next chapter, before coordinates are computed.

242. Linear Measurement. The only reliable safeguard against error in taping or chaining is to have two independent measures of each line. As a counsel of perfection, the separate measurements should be of similar precision and done in opposite directions, on different dates, by totally different parties. However, in actual practice, it is usual to measure the check taping by less accurate methods than the main taping, the object being to prevent the occurrence of gross errors, such as dropped tape lengths, or of errors of the order of a foot or over, rather than the detection of minor errors of a fraction of a foot. If time does not permit of a regular check taping or chainage, check measurements, to detect the grosser type of error only, may be made by pacing or pedometer, or, if legs are short, by tachymeter. If legs are long, tachymetrical methods are no faster than ordinary check taping.

243. Field Checks in Unclosed Traverse. In the case of an unclosed traverse, opportunities may present themselves of observing check bearings as in a closed traverse. Thus in Fig. 117, illustrating part of an unclosed traverse, if E is visible from A, bearing AE may be observed from A, and, on reaching E, the orientation of the instrument and the angular work between A and E will be checked by a backsight on A. The forward and back bearings between E and H serve to check the carrying forward of the bearing from E to H. Mistakes in the linear work may also be discovered by means of these check bearings on computing the coordinates or on plotting.

In the case illustrated, the portions ABCDE and EFGH would best be treated as deviations (see section 210), the distances AE and EH computed, and the main bearings and coordinates brought forward along the lines AE and EH.

An alternative method consists in observing at intervals the bearing to a prominent object to one side of the traverse. Thus, in Fig. 117, the bearing of a point P is observed from stations such as A, E, and H. The co-ordinates of P are obtainable from the observed parts of figure ABCDEP. A computed value of bearing HP is derived from the coordinates of H and P, and agreement of this with the observed value serves to test the traverse from A to H. If the coordinates are not computed, the check may be applied on the drawing by trying if the rays from A, E, and H intersect in one point.

Practically, it will rarely be possible to maintain either of the above systems of checking throughout the complete survey.

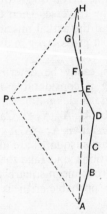

Fig. 117.

Double observation of each bearing and distance is therefore specially desirable in an unclosed traverse to avoid the carrying forward of a gross mistake. The traverse could be converted, in effect, into a closed traverse, by returning to the starting point along a slightly different route.

A thorough check is afforded in the case where the traverse is run between the stations of a reliable triangulation. Constant errors as well as inconsistencies in the linear measurements will contribute towards the closing error, which may be regarded as a real index of the quality of the work.

244. Astronomical Checks on Traversing. In a long traverse, astronomical observations of bearings can be used as checks. A bearing correct to one minute can be obtained from a simple observation of the Sun, provided the observer knows his Latitude and has the necessary information about the Sun's celestial position. The bearing so obtained is of course a true bearing, and proper allowance must be made if the survey bearing to be checked is referred to some other meridian system.

Methods for observing astronomical azimuth are described in Vol. II.

Observations of latitude and longitude are useless for checking ordinary traversing; they are employed only in connection with large geodetic survey operations, and then not as checks.

245. Closing Error of Bearings. A simple check on angles is obtained if a traverse is closed on itself, or if it starts and ends at points where known bearings are available.

Let r be the probable error of an observed angle and n the number of angles in the traverse. Then the 'probable misclosure' of bearing is $r\sqrt{n}$, and if the actual misclosure is more than about five times this amount, a gross error in angle measurement can be suspected.

The value of r can easily be determined for any particular instrument by making a considerable number–20 or more–of observations of the same angle on different zeros of the lower plate. Let the various values be $A_1, A_2, A_3 \ldots$ Calculate the mean value A, and the differences or residuals $A_1 - A, A_2 - A, \ldots$ Add together the squares of all the residuals (squares are always positive), divide by the number of observations, and take the square root of the quotient. The result is the 'root-mean-square' or 'standard deviation' of an observed angle; the 'probable error' is practically $\frac{2}{3}$ of the standard deviation.

The probable error should turn out to be of such size that about half the residuals are *numerically* smaller than it and the rest *numerically* larger.

The above operation determines the probable error of an angle observed once; if the angle is measured k times and the mean is calculated, the probable error of the mean is equal to the probable error of a single angle divided by the square root of k. It must be remembered, however, that an angle observed only once is affected by instrumental maladjustment errors which can only be cancelled out by the procedure of changing face.

OFFICE TESTS FOR LOCATING GROSS ERRORS IN TRAVERSES

246. When both ends of a traverse are known fixed points, and it is known that a gross error in bearing exists, the station at which it occurred can often be located by plotting or computing the traverse from each end, when the station having the same co-ordinates by each route will be the one where the error lies.

Another method, suggested by a writer in the *Empire Survey Review*, obviates the labour of plotting backwards, and is as follows. In Fig. 118, the correct line of the traverse is *Oabcdefgh*, but a gross error has been made in the measurement of the angle at *e*, with the result that, from *e* on, the traverse plots as *ef'g'h'*, so that the terminal point comes to *h'* instead of *h*. As the point *h* is a fixed one, its position is known and can be plotted. Join *hh'* and bisect it at k. At k erect a perpendicular, ke, to the line hh'. This perpendicular will then pass through e, the point at which the error was made. This follows from the fact that the triangle, ehh' is isosceles.

FIG. 118.

If the misclosure of a traverse is due to the existence of a gross error in taping, the exact leg in which the error occurred can sometimes be found by examining the bearing of the closing error. Provided that there is only one gross error and that it is much larger in magnitude than the normal accidental errors, the bearing of the closing error will be approximately the same as that of the leg in which the gross error exists.

PROPAGATION OF ERROR IN TRAVERSING

247. The propagation of error in linear measurement has been considered in Chapter II.

The accuracy of traversing is also affected by errors in observed angles. General formulae obtained by applying the theory of errors to traverses are very cumbersome, and it is only by making some simplifying assumptions that algebraic work on this problem can be reduced to reasonable proportions.

248. Propagation of Angular Error. If it is supposed that the random errors of angle measurement are independent of the lengths of the lines, or alternatively that all the lines are the same length, the probable misclosure $r \cdot \sqrt{n}$ mentioned above is the probable error of the bearing of the nth line as calculated directly from the initial bearing at the start of the traverse. In practice, as a rule, there will be a fixed bearing at the end, and the bearing misclosure will be distributed back through the traverse, and in this case it is naturally to be expected that the maximum residual error will occur at the middle of the traverse. In fact, it can be shown that the probable error of the adjusted bearing at the middle is $\frac{1}{2}r\sqrt{n}$. There will be a small additional contribution from the given initial and closing bearings which are held fixed but which cannot in reality be free from errors.

In a compass traverse in which the bearings of the lines are independently observed the errors do not depend on the number of lines in the traverse.

249. Propagation of Error in Position. In Fig. 119, OM is the true position of a line of length l, bearing A. A small alteration a in the bearing will shift the end of the line from M to N and the length of MN will be practically la if a is reckoned in radian measure. The bearing of MN will be $A+90°$. It is easily seen that the displacement MN amounts to $(-la . \sin A)$ in X coordinate and $(+la . \cos A)$ in Y coordinate.

Investigation of the effects of errors in a traverse with lines of differing lengths and bearings will obviously lead to very heavy algebra, but useful estimates can be obtained if the traverse is considered to consist of lines of equal lengths and constant bearing. In such a traverse the angle errors will result in displacements perpendicular to the direction of the traverse, and linear errors will give displacements along the same direction as the traverse.

As regards the linear errors, if we suppose that the probable error of measurement of a line of length l is proportional to l, it can be represented as kl where k is a small fraction. Then, if there are n lines in the traverse, the probable error of position at the end of the traverse will be $kl\sqrt{n}$.

If θ is the error of the first angle of the traverse, its effect will be that due to a deflection θ over the whole length of the traverse, that is $nl\theta$, θ being reckoned in radian. If the error θ is in the second angle the deflected part of the traverse has length $(n-1)l$ and the displacement will be $(n-1)l\theta$. And so on; finally, an error θ in the nth angle will deflect only the last line and the effect will be $l\theta$.

Now, if we take θ to be the probable error of a measured angle, the probable displacements due to random errors in all the angles must be combined by the usual statistical 'sum of squares' rule, and the result is that the probable displacement perpendicular to the direction of the traverse will be

$$[n^2l^2\theta^2+(n-1)^2l^2\theta^2+\ldots 4l^2\theta^2+l^2\theta^2]^{\frac{1}{2}}$$

or

$$l . \theta(1^2+2^2+3^2+\ldots+n^2)^{\frac{1}{2}} = l . \theta[\tfrac{1}{6}n(n+1)(2n+1)]^{\frac{1}{2}}.$$

The displacements due to angular and linear errors are in perpendicular directions and the total 'probable displacement' is

$$[k^2l^2n+\tfrac{1}{6}\theta^2l^2n(n+1)(2n+1)]^{\frac{1}{2}}.$$

For example, consider a traverse of 20 lines each 1000 ft. long. Suppose $k = 1/10,000$ and $\theta = 1/40,000$, practically 5 seconds. We have

$$
\begin{array}{ll}
k . l = 0{\cdot}1 & \theta . l = 0{\cdot}025 \\
k^2l^2 = 0{\cdot}01 & \theta^2 , l^2 = 0{\cdot}000625 \\
k^2l^2n = 0{\cdot}2 & \tfrac{1}{6}\theta^2l^2n(n+1)(2n+1) = 1{\cdot}79
\end{array}
$$

The total probable error is thus $\sqrt{1\cdot99} = 1\cdot41$ ft., or fractionally 1/14,200 of the total length.

In this case the contribution of the linear errors is comparatively insignificant and emphasises the necessity to pay particular attention to the angle observation in traversing. To make the effects of angle errors comparable with those of the linear errors, the probable error of a measured angle would have to be about 2 seconds.

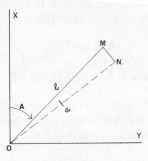

FIG. 119.

250. The foregoing investigation has given a formula for the probable error of position at the end of a 'flying' traverse not controlled by closure on fixed values at the end. In studying the errors of a controlled traverse, allowance must be made for the fact that the bearings are adjusted before positions are computed and then the positional misclosures are adjusted. As before, it is to be expected that the maximum probable error of position will occur at the middle of a controlled and adjusted traverse.

Provided that the methods of adjustment are clearly specified, it is possible to calculate the required probable errors. Some results for cases of traverses of simple geometrical shapes and legs of equal lengths have been worked out by Mr. F. Yates in *Records of the Gold Coast Survey Department Vol. III.*

MISCELLANEOUS PROBLEMS IN THEODOLITE SURVEYING

251. By means of the theodolite, obstacles to ranging or chaining may be more rapidly and accurately surmounted than by the use of chain surveying equipment only. The examples given below also include some problems frequently encountered in the course of setting out works. While all possible cases cannot be included, the solutions given will suggest the geometrical methods to be applied in particular circumstances.

252. Obstacles. *Obstacles which Obstruct Ranging.* (*a*) When an intermediate point can be selected from which the ends are visible, the line may first be ranged by eye (section 131). On setting the theodolite in this line, it may be adjusted to the exact line by trial and error, until on sighting one end station and transiting, the line of sight is found to cut the other station.

(*b*) From A (Fig. 120) project a random line AC'B' by the theodolite in the estimated direction towards B, leaving a point C' visible from A. Measure AB' and the offset B'B. Compute the angle

$$a = \tan^{-1}\frac{BB'}{AB'},$$

and, with the theodolite at A, set off *a* from AC', and establish points on AB.

(*c*) Select a point C (Fig. 121) from which A and B are visible. Measure AC, CB, and *c*. Solve for *a* and *b*, and set off one or both of these angles at A and B respectively.

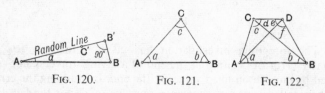

FIG. 120.   FIG. 121.   FIG. 122.

(*d*) If AB is very long, select two intervisible points C and D (Fig. 122) from both of which A and B are visible. Measure angles *c*, *d*, *e*, and *f*. Solve triangle ACD for AC and AD, taking CD as unity, and from triangle BCD obtain BC and BD. Now solve triangle ACB for *a* and *b*, and check by solving triangle ADB. By orienting on C, set off *a* and *b* from A and B respectively, and establish points on AB. This method will not be very satisfactory if CD is short, but will give a direction good enough for clearing a line from A to B.

(*e*) If intervisible points C and D cannot be obtained, run a traverse from A terminating on B, and compute the bearing of AB. Set off this bearing from A.

As a general rule, when any of the methods (*b*), (*c*), (*d*), or (*e*) are used to run a straight line between two points A and B which are not intervisible, and the line has to be set out on a computed bearing or angle, the line laid out will not strike the end point exactly. This is due to small errors in observation which cannot be avoided. Thus, in Fig. 120, the line actually laid out on the ground follows the direction AB' instead of the desired direction AB. With care, however, B' should be very close to B. The line AB' is chained and the distance BB′ measured, when the length of the offset at any intermediate C′ can be calculated from the formula

$$\text{Offset at C}' = \frac{\text{BB}' \times \text{AC}'}{\text{AB}'}$$

Consequently, permanent marks should not be put in on a line of this description until the distance BB′ has been obtained and any necessary offsets calculated.

Obstacles which Obstruct Chaining. (*a*) The methods of section 186, involving the use of right angles, are available, the theodolite being employed to erect perpendiculars.

(*b*) When both points, A and B (Fig. 123), are accessible, set out an equilateral triangle by laying off *a* and *b* each = 60°, obtaining the intersection C. Measure AC, and check by BC. Alternatively, set out any triangle ABC, and measure a sufficient number of the parts to enable the triangle to be solved for AB.

(*c*) When B is inaccessible, set out AC perpendicular to AB (Fig. 124). Measure AC and *c*, and solve for AB. Alternatively, measure out AC in any convenient direction, observe the angles at A and C, and solve for AB.

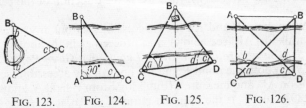

FIG. 123.   FIG. 124.   FIG. 125.   FIG. 126.

(*d*) When B is inaccessible and invisible from A, select two intervisible points C and D (Fig. 125) from which A and B are visible. Measure CD, AC, and AD and angles *a*, *b*, *c*, and *d*. Solve triangle BCD for BC and BD, then solve triangle ABC for AB, and check from triangle ABD. Time is saved in the measurements if C, A and D are collinear.

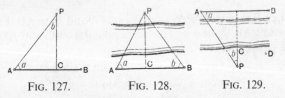

FIG. 127.   FIG. 128.   FIG. 129.

(*e*) When A and B are both inaccessible, select two points, C and D (Fig. 126) from which both A and B are visible. Measure CD and angles *a*, *b*, *c*, and *d*. Solve triangle ACD for AC and AD, and BCD for BC and BD. Obtain AB by solving triangle ABC, and check from triangle ABD.

253. Perpendiculars and Parallels. *To Set Out a Perpendicular to a Given Line from a Given Point.*

(*a*) When both the point P and the line AB are accessible, select a point A on AB, and measure a (Fig. 127). At P sight A, and set off $b = (90° - a)$; then PC is the perpendicular. Alternatively, measure a and AP, and compute AC.

(*b*) When P is inaccessible, select two points A and B on AB (Fig. 128). Measure AB, a, and b. Then

$$AC = \frac{AB \tan b}{\tan a + \tan b}$$

or, in a form suitable for logarithmic calculation,

$$AC = \frac{AB \cos a \cdot \sin b}{\sin (a+b)}$$

(*c*) When P is accessible, but AB is not, deduce a (Fig. 129) by observations from P and a point D, as in Fig. 126. From P sight A, and set off $b = (90° - a)$; then PC is the direction of the perpendicular.

(*d*) When both P and AB are inaccessible, set out any accessible line parallel to AB (see below), and proceed as in (*b*) above.

To Set Out a Parallel to a Given Line from a Given Point.

(*a*) When both the point P and the line AB are accessible, select a point A on AB, and measure a (Fig. 130). From P sight A, and set off $b = (180° - a)$. PC is the required parallel.

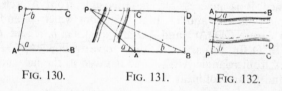

FIG. 130.   FIG. 131.   FIG. 132.

(*b*) When P is inaccessible, select points A and B on AB (Fig. 131), and measure AB, a, and b. The perpendiculars AC and BD to the required parallel are :—

$$\frac{AB}{\cot b - \cot a} = \frac{AB \sin a \sin b}{\sin (a-b)}.$$

(*c*) When P is accessible, but AB is not (Fig. 132), deduce a by observations from P and a point D, as in Fig. 126. From P sight A, and set off $b = (180° - a)$; then PC is the required parallel.

254. The Traverse Base and its Use. It is sometimes inconvenient or difficult to measure the direct distance between the two ends of a line which it is proposed to use as a base of a triangle for fixing a third point. In such a case a 'traverse base' may be used. This is a base the length of which is obtained by computation from a traverse run between the terminal points.

In Fig. 133, it is desired to connect the traverse *abcd* to the point C, the ground between *d* and C being unsuitable for linear measurements. C can be seen from the points A and B, but the ground between these points is not suitable for the direct measurement of the line AB. In this case, the subsidiary traverse A*gdef*B is observed, and the length and bearing of AB computed. If A and B are intervisible, the angles CAB and CBA and, if possible, ACB can be observed, and the triangle ABC computed. If A and B are not intervisible, the bearings of the lines AC and BC can be obtained from the traverse by measuring the angles *g*AC and *f*BC, and, these, together with the computed bearing and length of AB, give all the essential data for the solution of the triangle.

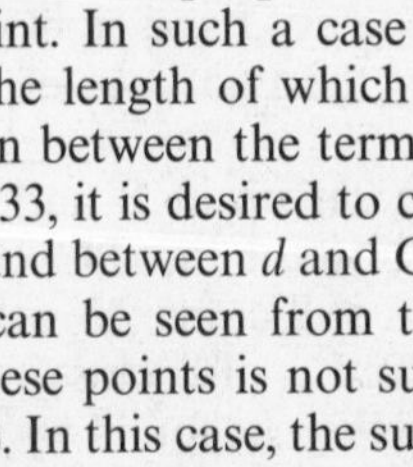

FIG. 133. 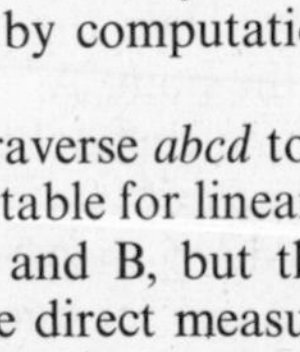   FIG. 134.

255. Referencing Marks and Beacons. When there is any possibility of temporary or permanent marks being moved and it is desired to be able, at any future time, to replace them in the exact positions in which they were originally, they should be carefully 'referenced'. There are several ways of doing this. In a town, several careful linear measurements to the corners of permanent buildings or to other clearly defined points of a permanent nature will serve to fix the position of the point. In other places the following is a suitable method.

Let A, Fig. 134, be the point which is required to reference. Set up the theodolite at A and choose a permanent feature B, or put in a mark in such a position that it is unlikely to be moved. Sight on B and then set out a mark at a point C, likewise in a position where it is not likely to be moved, and on the straight line BA or BA produced. Now put in two similar marks D and E on the straight line DAE and at a suitable angle with the line BAC.

Measure and record the distances AB, AC, AD, and AE. These measurements will be sufficient to enable the point A to be re-established, but it is better to do this by theodolite. If two theodolites are available, set one at B and one at D and sight on C and E respectively. Then the intersection of the two lines of collimation will be at the point A.

If only one theodolite is available, set up at B and sight on C and then line out two marks, *b* and *c*, on the line BC, close to, and on either side of, the position where A will come. Now set up at D, sight on E, and set out two pegs, *d* and *e*, also close to A and on either side of it. The position of A can then be found by stretching a string from *b* to *c* and one from *d* to *e*, when the intersection of the two strings will give the position required.

This method is very commonly used on construction work to enable points that are liable to be moved while construction is in progress to be recovered at any time.

For very important points, it is advisable to put in three or four marks, instead of two, on each line or else to set out additional lines. Failing this, linear measurements from two points will suffice to recover the position fairly accurately, though not so accurately as by the use of the theodolite in the manner described.

COMPASS TRAVERSING

256. Directions observed with a magnetic compass cannot be relied on closer than about 10 minutes, representing a fractional error of about 1/350: thus the precision of compass surveying is in no way comparable with that of theodolite work.

The advantages of compass surveying lie in:

(*i*) the instrument is simple, and is more portable and less expensive than a theodolite,

(*ii*) the work is much quicker because the elaborate procedure of setting up and levelling at each station is not required. The bearing of a line can be observed at any point on it. If high speed is required, the bearings of all the lines can be found by occupying only every second station.

257. Free Needle Surveying. A reading of a magnetic compass is a direction referred to the magnetic meridian through the compass station: if these directions are recorded and used as the bearings for computing the traverse, the method is called *free needle surveying*.

As mentioned in Chapter II, any such bearing may be affected by local attraction, that is, it may be referred to a direction which is out of sympathy with the general direction of magnetic meridians in the area covered by the survey: further, any magnetic reading may be influenced temporarily by diurnal variation or magnetic storms.

Local attraction will usually show up by making the forward and reverse bearings of a line differ by an angle other than 180°: allowance for these effects can be made before computing the traverse, see section 262.

258. Fixed Needle Surveying. A compass may be used as a theodolite. If, at a station, the bearings to the previous and next stations are measured, their difference is the included angle: this will be free from effects of local attraction and variations. Bearings of lines are obtained from the initial bearing by adding successive angles, and the method is in principle just the same as in theodolite traversing, but of course much less accurate. This method is called *fixed needle surveying*.

With an instrument of good quality, fixed needle surveying is more accurate than the other method, except perhaps where the traverse has a large number of very short lines, and then the accumulation of random errors in the included angles may cause the calculated bearings to differ markedly from the directly observed magnetic bearings. In this case, a bearing closure correction can be applied as for a controlled theodolite traverse.

The Wild TO Compass Theodolite is quite suitable for compass traversing: this instrument, referred to in section 72, has a graduated circle which can be placed in magnetic orientation by magnets attached to it.

259. Linear Measurement. In compass traversing, the linear measurement need not be done to high accuracy. Rapid measurement with a linen tape or wire-link chain is more than good enough, and little time need be spent in lining up. In reconnaissance surveying with a small prismatic compass, it will often be sufficient to measure distance by pacing.

If much traversing is to be done, with rather long legs, a rope, say 500 ft. long with distinctive knots or tags at every 10 ft. may help to speed up the work: this is dragged along by an assistant to each forward station and the surveyor then records the distances and takes the bearings.

Slopes of lines may be ignored unless very steep.

If offsets are taken, a much greater length may be allowed than in chain surveying, and short offsets can be estimated by eye. Detail near the traverse can be fixed by taking bearings from two traverse stations.

260. Application of Compass Surveying. The portability of the compass makes it specially suitable for reconnaissance and exploratory surveys, where rapidity of observation is important. Short compass traverses may, however, be introduced with advantage between stations of a theodolite survey for the rapid survey of detail which need not be recorded with a high degree of precision. The time saved by using the compass in place of the theodolite is greatest when survey lines have to be short on account of obstructions or irregularities of detail. The method is therefore suitable for tracing streams, particularly in wooded gorges, irregular shore lines, clearings in woods, etc.

On engineering sites nowadays, the use of the compass will often be prohibited by the presence of massive machines, pipes, girders, reinforcing bars, etc.

261. Recording Compass Traverse Surveys. If a detail survey is being done from a compass traverse, the offsets and distances can be recorded in much the same way as for a chain survey. In addition, there will be the bearings observed at each station, and these should be booked in a distinctive manner so as not to be confused with other numbers. Care must be taken that the direction to which a bearing refers is clearly evident, and particularly that there is no possible confusion between forward and reverse values.

On the other hand, if the traverse is simply a record of compass readings and distances, the observations can be written line by line in four columns giving the numbering of the line, the forward and back bearings, and the length.

262. Local Attraction. If, in a compass traverse, there are discrepancies between forward and reverse bearings of the same line, the observations must be examined to find the best way to eliminate the discrepancies without making large alterations.

As a preliminary, a line in which the two bearings differ by exactly 180° is selected, and it is assumed that there is no local attraction anywhere on this line: thus, bearings taken at either end of this line are accepted as recorded. A simple method of adjustment is illustrated in the following example:

The readings have been estimated to the nearest $\frac{1}{4}°$, and the bearing of each line has been observed twice, the first letter in the designation of the line showing the station of observation. On examining the values of the observed bearings, it will be seen that only in the case of the line between C and D are the forward and back bearings consistent. It may therefore be taken that stations C and D are *both* free from local attraction. Consequently all bearings observed at C and D are correct, and their values are transferred to the third column. Now, since the correct bearing of DE is 229°, that of ED must be 49°, but, as it was read as 48°, station E is influenced by local attraction, and a correction of $+1°$ must be applied to all readings taken at E. The corrected bearing of EA is therefore $136\frac{1}{2}°$, and consequently that of AE must be $316\frac{1}{2}°$, giving a correction for station A of $-2\frac{1}{2}°$. On applying this correction, and obtaining the adjusted value $70\frac{1}{4}°$ for bearing AB and $250\frac{1}{4}°$ for BA, the correction at station B is found to be $-1\frac{3}{4}°$, so that bearing BC should be $347\frac{1}{4}°$, which agrees with the reading CB taken at the unaffected station C.

Line	Observed Bearing	Correction	Corrected Bearing
	°	°	°
AE	319	$-2\frac{1}{2}$	$316\frac{1}{2}$
AB	$72\frac{3}{4}$	$-2\frac{1}{2}$	$70\frac{1}{4}$
BA	252	$-1\frac{3}{4}$	$250\frac{1}{4}$
BC	349	$-1\frac{3}{4}$	$347\frac{1}{4}$
CB	$167\frac{1}{4}$	0	$167\frac{1}{4}$
CD	$298\frac{1}{2}$	0	$298\frac{1}{2}$
DC	$118\frac{1}{2}$	0	$118\frac{1}{2}$
DE	229	0	229
ED	48	$+1$	49
EA	$135\frac{1}{2}$	$+1$	$136\frac{1}{2}$

It will be realised that this process is equivalent to referring the whole traverse to the magnetic meridian at the line first chosen as free from local attraction.

In a long traverse, the above method may lead to large corrections of a systematic nature, and it may be that the line chosen as reference was affected by local attraction at both ends; in such a case, a constant correction to all the calculated bearings can be applied so as to get as good as possible a general agreement throughout the traverse.

263. Sources of Error. Apart from local attraction effects, errors will arise from a sluggish needle due to wear of the pivot, and from parallax in the reading of the needle against the scale. The compass may have a constant error due to irregular magnetisation, but this will cause no trouble if the same instrument is used throughout the survey.

264. Estimate of Accuracy of Compass Traversing. In free-needle surveying, each observed bearing is independent of the others, and displacements due to random errors will be expected to have statistical effects similar to those of random errors of chaining. If θ is the probable error of an observed bearing of a line of length l, the probable displacement produced is θl, and over the whole of a straight traverse of n lines each of length l the probable displacement is $\theta l \sqrt{n}$ transversely to the general direction of the traverse.

Let $\theta = 1/400$ radian, about 9 minutes of angle, and suppose that the traverse has 25 legs each of length 400 ft. Then $\theta l = 1$ ft. and $\theta l \sqrt{n} = 5$ ft. This is comparable with the probable error of 10,000 ft. of linear measurement with a chain. In practice, systematic errors may be more serious.

REFERENCES

LAWES A. V. 'Cadastral Traverses in the Gold Coast'. *Empire Survey Review*, No. 58, Oct. 1945.

LEWIS E. M. 'Precise surveying at Llanwern'. *Chartered Surveyor*, Sept. 1962.

LYTHGOE R. 'Surveying problems encountered during the sinking of Parkside Colliery'. *Chartered Surveyor*, May 1963.

McCAW G. T. 'Routine cadastral traversing'. *Empire Survey Review*, No. 36, Apr. 1940.

REES E. S. 'The basic surveys and plans of a mining project'. *Chartered Surveyor*, March 1960.

RUSK A. J. 'A method of booking and reducing traverse angles'. *Empire Survey Review*, No. 104, Apr. 1957.

TURPIN R. D. 'A study of the use of the Wild Telemeter D.M.1 for a closed traverse'. *Surveying & Mapping*, Vol. XIV, No. 4, 1954.

CHAPTER V

OFFICE COMPUTATIONS

265. The office computations consist of the numerical calculations required to put the results of the field measurements into a form in which they can be used for whatever purpose the survey is required, and they generally form an intermediate stage between the field work and the plotting of the plans. In some classes of work, chain surveying for example, there is very little, if any, computing to be done; but in others, theodolite traversing for instance, the total time spent on the computations may be almost equal to that spent on the field work. In this chapter we will deal first of all with some fundamental points in computing and then more particularly with the computation of traverses. Other special computations will be dealt with in later chapters as the need arises.

In dealing with the field operations, much emphasis has been laid on the necessity for great care at every stage in order to prevent mistakes, and the same considerations apply, perhaps with even greater force, to computing. In a long traverse, the computational work is exceedingly heavy, and, for the most part, monotonous; and only those who have much experience of this class of work know how fatally easy it is to make a mistake, and how difficult it sometimes is to detect or locate it when it has been made. This fact must never be overlooked, and the computer must constantly be on guard to see that mistakes do not occur, or that, if they do, they are detected and corrected at once or before it is too late.

In some computations, automatic checks are available and, whenever this is so, they should always be used. We have had one simple example of this type of check in the case of the computation of bearings, where adding together all the included angles and the initial bearing checks the final bearing. In other computations, simple and effective checks of this kind are not always available, and, in that event, the only real check is at least one complete re-computation.

A re-computation to be effective must be as independent of the original one and as complete as it is possible to make it. It is not sufficient to go over the original work and to tick off figures as being correct. If possible, the check computation should be done by a different person from the one who did the original, and it should start at the very beginning–that is, with the measurements recorded in the field books. All reductions or corrections made from, or to, the figures recorded in the field books should be re-computed and made anew, and it is only after this has been done that the figures should be abstracted to the forms or sheets on which further calculations are to be made. Computations should be arranged so that the copying of figures from one place to another is reduced to the minimum, or if possible eliminated.

Sometimes alternative methods or formulae can be used, and, if so, full advantage should be taken of them. For instance, if the coordinates of two points are given, and the bearing and distance between them is required, the distance can be worked out either from the sine or the cosine of the bearing after that has been obtained. Both formulae should be used, and the results compared, although, as will be pointed out later, it may be better, when the two results show a *very small* difference, to accept those from one formula in preference to those from the other.

Neatness and uniformity in method are as essential in computing work, as they are in keeping the field books. The use of a special paper, divided by faint blue rulings into squares of one-quarter-inch side, so that each square holds two digits, will do much to assist in keeping work neat; and working as much as possible on standard forms, preferably printed, will help considerably to prevent mistakes being made. All figures should be legible and clearly set out, and, where alterations are necessary, incorrect figures should not be erased but should be crossed out, and the correct figures written above them. Also, it is well to do all important computing work in ink, and not to do it in pencil first, and then ink the figures in later. The different computation sheets should be numbered in order, and references made from one sheet to the other and to the field books, whenever this is necessary to enable the survey to be followed easily by an independent computer or draughtsman. Finally, when the work is complete and it is desired to preserve it, the different sheets would be put together in their proper order, indexed and filed.

266. Mathematical Tables and other Aids to Computing. The quantities entering into the computations of surveying work are, almost exclusively, distances and angles, and the angles usually enter the calculations through their sines, tangents, or other trigonometrical functions.

Computations may be done either in natural numbers or with the aid of logarithms. In any case, books of mathematical tables are required, and for computation in natural numbers a calculating machine of some sort is essential. Simple machines are now available at prices less than £100. Some kinds of survey computation are considerably speeded up by being done on a machine, so its use may amount to a genuine economy.

Simple computing machines are of two types:

(1) Machines which are primarily *adding* machines. A number tapped out on the keys is added into an accumulated total by pulling a lever (or pressing a button if the machine is electrically operated), and the numbers put in are typed on a roll of paper. At any stage, the total accumulated may be typed out by using the appropriate key, and the accumulated total cleared or left in, as desired. Subtraction is equally possible.

FIG. 135.—THE OLYMPIA ADDING MACHINE
(By courtesy of Olympia Business Machines Co. Ltd.)

(2) Machines which also operate by adding (or subtracting) numbers set by keys or levers, but which have provision for moving the accumulator section so that a number can be added (or subtracted) at any position within a certain range, usually some 7 to 10 places. This means that the machine is easily used for multiplication or division. The simpler types of *calculating* machine do not have printing out mechanisms, so it is not possible to check the numbers actually put into them, and this may be a disadvantage. Calculating machines with print-out are available but of course they are comparatively expensive. Three simple machines are illustrated in Figs. 135, 136, 137.

As for mathematical tables, there are many available. The well-known 'Chamber's Seven-Figure Mathematical Tables' is a very comprehensive collection: it gives, to seven decimal places, the logarithm of every number from 10,000 to 108,000, the natural sine, tangent, etc., and their logarithms of every angle at intervals of 1′ from 0° to 90°. This book also contains many other tables of use in mathematical and astronomical work, and in navigation.

For logarithmic trigonometrical functions only, the 'Shortrede' tables, published by C. E. Layton Ltd. give seven-figure logarithmic values at every second of arc. Peters tables give natural trigonometrical functions at intervals of 10″, to six figures.

FIG. 136.—THE CURTA CALCULATOR.
(By courtesy of Automatic Office Machines Ltd.)

Separate books of natural and logarithmic six-figure tables are also published by W. and R. Chambers Ltd.: these give the trigonometrical functions at 1′ intervals and many other numerical tables.

Five-figure tables giving trigonometrical functions at $\frac{1}{2}$′ intervals as well as logarithms of numbers, are available, and there are numerous publications of four-figure tables of logarithms, antilogarithms, natural and logarithmic trigonometrical functions, squares, reciprocals, etc.

For computation of low-order traverses with fairly short legs, special Traverse Tables are obtainable. These contain natural sines and cosines multiplied by the factors 1, 2, 3 . . . 9, usually at 1′ intervals of arc. See section 292.

267. Precision of Computation. It is clear that the precision of computation, that is the number of significant figures retained through the calculations, should be related to the actual accuracy of the field observations. There is no point in computing a compass traverse with seven-figure logarithms, and it would be ridiculous to use a slide-rule to compute a traverse surveyed with theodolite and steel tape.

FIG. 137.—THE PRECISA CALCULATOR.
(By courtesy of Muldivo Calculating Machine Co. Ltd.)

As a rough rule, work using angles observed to single second accuracy requires six-figure, perhaps seven-figure computation. Ordinary survey work with a theodolite reading to 10″ and lengths measured to 0·1 ft. can be calculated to five-figure accuracy.

Small corrections, such as slope corrections in traverses, can be calculated with four-figure tables or on a slide-rule: if a survey operation involves the calculation of numerous small quantities, it may be economical to prepare a nomogram or some other kind of diagram from which the required numbers may be taken by direct reading.

268. Rounding Off Figures. When rounding off figures it is well to keep to a definite system so that the laws of chance operate and not the personal caprice or unconscious bias of the computer. If the computation is to n significant figures, the nth figure should be kept as it is if the figures following it are less than 5 in the $(n+1)$th place. If the figures following the nth significant figure are greater than 5 in the $(n+1)$th place, the nth figure should be increased by one unit. Thus, in working to four decimal places, 16·453238 is rounded off at 16·4532, but 16·453261 becomes 16·4533. If, however, a figure like 16·45325 is obtained, the computer might call this either 16·4532 or 16·4533. In such a case, the following is a usual and convenient rule to adopt: when the nth significant figure is an even number followed by a 5, keep it as it is and discard the 5, but, if it is an odd number followed by a 5, add one unit to it and discard the 5. Thus, 16·45325 becomes 16·4532, but 16·45335 becomes 16·4534.

269. Average Error of Last Figure. When a number is rounded off, the maximum residual error is half, and the average is a quarter, of a unit in the last place retained. Thus, the average error of the entries in a four-figure table may be taken as 0·000025.

A similar rule holds with regard to reading a scale. For example, if a tape graduated to hundredths is being read to the nearest tenth, any reading between, say, 94·65 and 94·75 would be called 94·7. In that case, the probable error of any single reading is $\pm 0·025$. This means that the real reading is as likely to lie within the interval 94·675 − 94·725 as it is to lie outside it.

When a column of quantities, each with average error of say 0·000025 is added up, the rounding-off errors must be assumed to occur randomly, so the average error of the total can be estimated as

$$0·000025 \times \sqrt{\text{number of quantities making the total}}$$

In the case of logarithmic calculation, a given error in a logarithm represents a particular *fractional* error in the corresponding number. The differential relation is

$$\Delta(\log_{10}x) = \frac{\Delta x}{x} \times 0·4343, \quad \text{or} \quad \frac{\Delta x}{x} = 2·3026 \times \Delta(\log_{10}x):$$

thus, if $\Delta(\log_{10}x)$ is 0·000025, $\Delta x/x$ is 0·00005756 or $\frac{1}{17,400}$. This is the average fractional error in obtaining a quantity from its ordinary four-figure logarithm.

270. Choice of Formula for Computing. Many formulae can be put into several different forms, and, while one form may be suitable for one method of computation, it may not be suitable for some other method, although another form will. To take a simple case, if a and b are given and $\sqrt{(a^2-b^2)}$ is required, the expression $\sqrt{(a^2-b^2)}$ is not very suitable for logarithmic computation as the solution involves looking up three logarithms and three anti-logarithms. However, if we put it into the form $\sqrt{((a-b)(a+b))}$ it is only necessary to look up two logarithms and one anti-logarithm. On the other hand, if a table of squares which goes to sufficient places, and not a table of logarithms, is available, the original form might be the more convenient one to use.

Nearly all formulae used in survey work are most conveniently expressed when put into a form suitable for logarithmic computation, and, in this form, provided they do not contain fractional indices, they are usually suitable for computation by machine, though, when a machine is used, it is best to examine the formula and see if it is expressed in the form most suitable for machine computation.

Most of the formulae involving the trigonometrical functions, when derived from first principles, are obtained in a form unsuitable for logarithmic computation, and anybody who has studied trigonometry will be familiar with the transformations necessary to put, say, the formulae for the solution of triangles into forms that are suitable for computation by logarithms.

In many cases an unsuitable formula can be transformed into a suitable one by the choice of an auxiliary angle. Thus, if we want to compute $A \cdot \cos\theta + B \cdot \sin\theta$; where A, B, and θ are known or given, we can find an auxiliary angle ϕ from $\tan\phi = B/A$ and we then have:

$$A\cos\theta + B\sin\theta = A\left(\cos\theta + \frac{B}{A}\sin\theta\right) = A(\cos\theta + \tan\phi\sin\theta)$$

$$= \frac{A(\cos\phi\cos\theta + \sin\phi\sin\theta)}{\cos\phi} = \frac{A\cos(\phi-\theta)}{\cos\phi}$$

Here, the expression on the right is easier to handle by logarithms than is the expression on the left. Again,

$$\sqrt{A^2+B^2} = A\sqrt{1+\frac{B^2}{A^2}} = A\sqrt{1+\tan^2\phi} = A\sec\phi$$

where $\tan\phi = B/A$. The angle ϕ is found from $\log\tan\phi = \log B - \log A$ and then $\sqrt{(A^2+B^2)}$ from $\log\sqrt{(A^2+B^2)} = \log A - \log\cos\phi$. When A and B are large numbers, this is a much easier method than taking the square root of the sum of their squares by ordinary arithmetic.

In both these cases, ϕ is an auxiliary angle so chosen as to put the expression into a form suitable for easy computation by logarithms.

Another case where choice of formula is important is where angles are either small or else near 90° in value. The logarithm of the sine changes most rapidly and irregularly and that of the cosine most slowly and regularly when the angle is small, and the converse is true when the angle is almost a right angle. If an angle is to be found from a calculated value of its logarithmic sine or cosine, then the sine is preferable if the angle is less than 45°. Better still, use the logarithmic tangent, when there is no such restriction. Special procedure for interpolation must be observed, however, if the angle is very small: see Section 272. If the trigonometrical function is being used for calculating something else, the argument goes the other way: thus the length between two points A and B can be found from either of the formulae:

$$l = \Delta X . \sec \alpha ; \quad l = \Delta Y . \operatorname{cosec} \alpha$$

where α is the bearing of the line and ΔX and ΔY the coordinates of one end referred to the other end. If α is small, it is better to accept the value computed from the cosine in preference to that computed from the sine. On the other hand, if α is large and near 90° in value, it is better to accept the value computed from the sine. (For an example, see section 278).

271. Circular Measure and Small Angles. The unit for *circular measure* of an angle is a *radian*, which is the angle of a sector of a circle given by an arc of length equal to the radius: since the total circumference of a circle of radius R is $2\pi R$, there are 2π radians in 360°. From this relation it is easily found that a radian is equal to 206 264·8 ... seconds, or nearly 3437·5 minutes.

Thus, to convert any angle to radian or circular measure, it should be expressed in seconds or minutes and divided by the appropriate one of the above numbers.

Many formulae printed in works on surveying contain the term 'sin 1‴'; this is to be taken as standing for the fraction 1/206264·8 ... Strictly 'sin 1‴' is not equal to this fraction but it is quite near enough for all practical purposes.

The practical value of circular measure is that for small angles the *sine*, *tangent* and *circular measure* are about equal. For example, all three can be taken as equal, to 5 decimal places, for any angle up to about ·1°.

The angle 'subtended' by an object seen at a distance can readily be estimated: for instance, a 10 new penny coin (dia. 1·1 in.) at a distance of 100 yards subtends an angle of 1/3600 radian, which is very nearly one minute of angle.

272. Logarithmic Sines and Tangents of Small Angles. A glance at the logarithmic tables shows that when angles are small their log . sines and log . tangents change very rapidly, and simple interpolation will not give correct values. There are several ways of dealing with this problem:

(1) Interpolation by a formula using second-order and if necessary higher differences: this is a cumbersome method and should be avoided.

(2) Interpolating in natural value tables, where the entries change quite smoothly, and then converting to logarithm if required. The only possible disadvantage here is that the order of accuracy depends on the number of significant figures in the natural value and, for example, the sines and tangents in the seven-figure tables have only 5 significant figures for angles below about 35 minutes.

(3) Use of Maskelyne's Rules as explained in the Introduction to Chamber's Seven-figure Tables.

(4) Use of the 'S' and 'T' functions on the last page of Shortrede's Tables.

273. Solution of Plane Triangles by Sine Formulae. As an example of convenient arrangement of computation and of suitable independent checks, take the case of the solution of a triangle by the ordinary sine rule. Let the data be as follows:

Station	Observed Angles			Adjustment	Adjusted Angles		
	°	′	″	″	°	′	″
A	47	15	50	− 10	47	15	40
B	68	35	10	− 10	68	35	00
C	64	09	30	− 10	64	09	20
	180	00	30	− 30	180	00	00

Length AB = 4527·27 feet.

Here, the observed angles exceed the theoretical sum of 180° by 30″. Hence, in order to adjust the angles, 10″ is subtracted from each angle and the adjusted angles, which are the ones to be used in the computation, are set out in the last column. The logarithmic computation is now arranged as follows:

$$\begin{aligned}
\log AC &= \underline{3{\cdot}670\,529} \quad AC = 4683{\cdot}05 \\
\log \sin B &= 9{\cdot}968\,926 \\
\log \operatorname{cosec} C &= 0{\cdot}045\,767 \\
\log AB &= 3{\cdot}655\,836 \\
\log \sin A &= \underline{9{\cdot}865\,965} \\
\log BC &= 3{\cdot}567\,568 \quad BC = 3694{\cdot}61
\end{aligned}$$

The different logarithms of the known quantities are arranged in the order shown, when the sum of the first three gives log AC, and that of the last three gives log BC. This is better than the following, more common, arrangement:

$$\begin{aligned}
\log \sin B &= 9{\cdot}968\,926 & \log \sin A &= 9{\cdot}865\,965 \\
\log \operatorname{cosec} C &= 0{\cdot}045\,767 & \log \operatorname{cosec} C &= 0{\cdot}045\,767 \\
\log AB &= \underline{3{\cdot}655\,836} & \log AB &= \underline{3{\cdot}655\,836} \\
\log AC &= 3{\cdot}670\,529 & \log BC &= 3{\cdot}567\,568
\end{aligned}$$

as it not only saves extra writing but it avoids an extra possibility of copying error, due to having to write log AB and log cosec C down twice instead of once. A very similar arrangement will be found on Fig. 141 for working out the latitudes and departures of a traverse.

If the coordinates (section 274) of the points A and B are known, those of C will usually be required. In this case, the bearing of the line AB, if not already given, can be computed, and, from this and the given angles, the bearings of the lines AC and BC can be obtained. These, together with the values of log AC and log BC obtained from the solution of the triangle, give sufficient data to enable the coordinates of C to be computed from both A and B and the check therefore consists in seeing if the coordinates computed from each point are the same. If they are, the solution of the triangle is correct. This, therefore, not only gives an independent check on the solution of the triangle, but it also gives checked coordinate values for the point C. The check should be exact, because the angles of the triangle have been adjusted to 180°.

If the coordinates of A and B are not known, or if those of C are not required, the solution can be checked by the test:

$$AB = AC . \cos A + BC . \cos B,$$

and we also have:

$$AC . \sin A = BC . \sin B.$$

Thus, in our example:

$$\begin{aligned}
\log AC &= 3{\cdot}670\,529 & \log BC &= 3{\cdot}567\,568 \\
\log \cos A &= \underline{9{\cdot}831\,651} & \log \cos B &= \underline{9{\cdot}562\,468} \\
\log AC \cos A &= 3{\cdot}502\,180 & \log BC \cos B &= 3{\cdot}130\,036
\end{aligned}$$

$$\begin{aligned}
AC \cos A &= 3178{\cdot}19 \\
BC \cos B &= \underline{1349{\cdot}08} \\
AB &= 4527{\cdot}27 \text{ (Check.)}
\end{aligned}$$

Similarly, we also see that AC . sin A = BC . sin B is satisfied, but the first is the better and more complete test.

The art of computing by machine is in avoiding, as far as possible, copying numbers down from the machine to paper until the final required values are to be recorded.

Consider the triangle solution given above: the data for machine computation are:

$$AB = 4527 \cdot 27 \text{ ft.}$$
$$\text{cosec } C = 1 \cdot 111\ 135$$
$$\sin A = 0 \cdot 734\ 454$$
$$\sin B = 0 \cdot 930\ 950$$

On a simple calculating machine, set in the number 4527·27 and multiply by 1·111 135: the product to seven significant figures is 5030·408 on the product register. Set this number on the keys and subtract it from the product register as a check on the setting. Clear the other registers and multiply by 0·734 454 obtaining BC = 3694·60 on the product register. Now operate so as to change the counter register to 0·930 950 and get AC = 4683·06 on the product register. Note that only the two required quantities are written down after the data, and it is not actually necessary to write down the three trigonometrical values because each is used on the machine as soon as it is obtained from the tables.

An alternative start to the computation is to use sin C (0·899 981) and divide AB by it, but then there is no possibility of using the setting check mentioned above.

It is not even necessary to interpolate the exact values for cosec C, etc., before computation: after setting 4527·27, multiply by 1·111 187, which is cosec 64° 09′, clear the counter register, and subtract 4527·27 × 0·000 052 which is the effect of the interpolation for 20″. The other interpolations can be operated on the machine similarly.

RECTANGULAR COORDINATES

274.   In survey work it is usual to define the position of a point with reference to two lines, drawn through some convenient point, at right angles to each other. These reference lines are known as the 'axes of coordinates', and the point of their intersection is known as the 'origin of coordinates'.

In Fig. 138 let OX and OY be the two reference lines and O the origin. Let P be the point whose position is to be defined with reference to O. From P draw PM perpendicular to OX and PN perpendicular to OY. Then the distance OM = PN represents the 'X' coordinate' of P and ON = PM represents the 'Y coordinate'.

If these two distances are known, the position of P is completely defined with reference to the origin, because this position can be plotted by laying off OM = NP = X along OX and then drawing MP = ON = Y perpendicular to OM. Distances measured from O in the directions opposite to X and Y, that is along OX^1 and OY^1, are reckoned as negative distances and give negative coordinates. Thus, the point P$_1$ is defined by the distances OM^1 and ON, corresponding to a negative value of X and a positive value of Y.

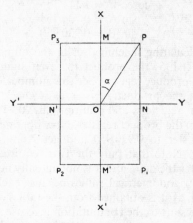

Fig. 138.

The position of P can also be defined by means of the length OP = L and the angle XOP = α, because P can be plotted by laying off the line OP at the angle α from OX and then cutting off the distance OP = L. If P lies in the first quadrant, as at P, the angle α lies between 0° and 90°, but if P is in the second quadrant, as at P$_1$, α lies between 90° and 180°. Following up this line of reasoning we can make the following table:

P lies in	Sign of Coordinates		Limits of α
	X	Y	
First Quadrant	+	+	0° to 90°
Second Quadrant	−	+	90° to 180°
Third Quadrant	−	−	180° to 270°
Fourth Quadrant	+	−	270° to 360°

From Fig. 138 it follows that:

$$X = L \cos \alpha; \quad Y = L \sin \alpha$$
$$L = X \sec \alpha; \quad L = Y \operatorname{cosec} \alpha.$$

These are the equations giving the relations between the co-ordinates X, Y, and the length L and the angle α.

We also have:

$$\tan \alpha = \frac{Y}{X}, \quad L^2 = X^2 + Y^2.$$

The X axis of coordinates is usually taken either as the meridian through the origin O or parallel to some standard meridian, and in that case the angle α becomes the bearing referred to the meridian of origin.

The term *grid bearing* may be used if it is necessary to emphasise the fact that the bearing is referred to an axis of a coordinate system, not necessarily to True North.

275. Latitudes and Departures. In Fig. 139 the points P, Q, R are plotted in the coordinate system with origin at O and axes OX,

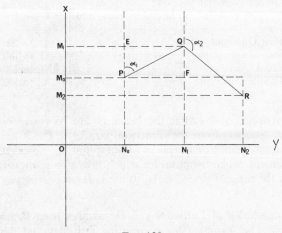

FIG. 139.

OY. Suppose that the coordinates of the three points are respectively $(X_0 Y_0)$, $(X_1 Y_1)$ and $(X_2 Y_2)$. Draw lines through the points and parallel to the axes as shown.

The position of Q in relation to P is expressed by the differences of coordinates $X_1 - X_0$ and $Y_1 - Y_0$, which may be written, for brevity, ΔX, and ΔY. These differences are of course independent of the position of the origin O provided the directions of the axes are not changed. The differences of coordinates between Q and R are likewise $X_2 - X_1$ and $Y_2 - Y_1$, and these may be written ΔX_2 and ΔY_2 respectively. Expressed numerically, the ΔX_2 in the Figure will be negative.

The series of lines may be continued beyond R.

Let the bearings of the successive lines be $\alpha_1, \alpha_2, \alpha_3 \ldots$ and their lengths $l_1, l_2, l_3 \ldots$: then the coordinates of the nth point will be:

$$X_n = X_o + l_1 \cos \alpha_1 + l_2 \cos \alpha_2 + \ldots l_n \cos \alpha_n,$$
$$Y_n = Y_o + l_1 \sin \alpha_1 + l_2 \sin \alpha_2 + \ldots l_n \sin \alpha_n.$$

If the points P, Q, R ... are the stations of a traverse, the quantities $\Delta X_1, \Delta X_2 \ldots$ are usually called *latitudes*, and the quantities $\Delta Y_1, \Delta Y_2 \ldots$ are called *departures*.

When expressed numerically, the latitudes and departures have signs depending on the quadrant in which the bearing lies, and in fact we can draw up a table very similar to that already given for defining the signs of the coordinates.

Position of line with respect to initial point of line	Sign of latitude	Sign of departure	Limits of α
First Quadrant	+	+	$0°$ to $90°$
Second Quadrant	−	+	$90°$ to $180°$
Third Quadrant	−	−	$180°$ to $270°$
Fourth Quadrant	+	−	$270°$ to $360°$

It is assumed above that the bearings are reckoned clockwise $0°$ to $360°$ starting from the direction parallel to the X axis of the coordinate system. For purposes of computation, however, it may be convenient, when using tables which give trigonometrical functions for the range $0°$ to $90°$ only, to use *reduced bearings* as defined in section 206.

276. Computation of Latitudes and Departures from Bearings and Distances. The formulae for computing latitudes and departures have already been derived above and are

$$\text{Latitude} = \Delta X = l \cdot \cos \alpha$$
$$\text{Departure} = \Delta Y = l \cdot \sin \alpha$$

In order to illustrate the most convenient method of computation we will take a numerical example in which

$$l = 1745 \cdot 28, \text{ and } \alpha = 164° \ 15' \ 20''.$$

Hence the end point of the line lies in the second quadrant so that the latitude is minus and the departure plus. Also, the reduced bearing is 15° 44′ 40″. For logarithmic calculation arrange the work as follows:

$$\begin{aligned}
\log \Delta X &= 3{\cdot}225\,2576 & \Delta X &= -1679{\cdot}80 \\
\log \cos \alpha &= 9{\cdot}983\,3924 \\
\log l &= 3{\cdot}241\,8652 \\
\log \sin \alpha &= 9{\cdot}433\,5252 \\
\log \Delta Y &= 2{\cdot}675\,3904 & \Delta Y &= +473{\cdot}58
\end{aligned}$$

Log l is written down in the position shown. Above it is written log cos α and below it log sin α. The addition of the first two logarithms gives log ΔX, and the addition of the last two gives log ΔY. Having found log ΔX and log ΔY, ΔX and ΔY are written down, with their proper signs, as shown.

This arrangement of the computation is much better than the usual one of writing down log l twice, with log cos α written under one set of figures and log sin α under the other. Having to write down log l twice means that the chances of the occurrence of copying errors are increased. Also, the arrangement shown saves one column in the latitude and departure sheets (Fig. 141) when, as is usual, the computation is done and shown on these. This may be an advantage when, for filing purposes, it is desired to restrict the size of the latitude and departure sheets so that they are no larger, or very little larger, than ordinary foolscap size.

It will be noticed that, following the usual practice in printed tables, the negative characteristics in the log sin and log cos are increased by 10, and an appropriate multiple of 10 is subtracted from sums of characteristics.

For machine computation, the l is set on the keys and is multiplied successively by cos α and sin α. Alternatively, if the latitudes and departures are computed separately, it is possible to set the coordinate of the initial point on the machine's product register, and 'turn in' successive values of $l . \cos \alpha$ (or $l . \sin \alpha$, if departures are being computed), adding or subtracting according to the quadrant of the bearing, and then the product register will show successive coordinates, which must be copied down if the machine has no print-out facility.

277. Check Computation of Latitudes and Departures. Traverse Tables may be used to check computations of compass traverses, or as a rough check on computation of theodolite traverses.

If a calculating machine is available, the best check of traverse computation is to make one computation by natural numbers and one by logarithms.

A very effective method of checking, little known or used in England, is carried out by dividing the length of each traverse leg by $\sqrt{2}$ and adding 45° to each bearing. Using these modified distances and bearings, the calculation of a line yields modified 'latitude' and 'departure', C and S, given by:

$$C = l.\cos(\alpha + 45°)/\sqrt{2}, \quad S = l.\sin(\alpha + 45°)/\sqrt{2}.$$

It is easily shown that:

$$S + C = l.\cos\alpha = \Delta X, \quad \text{and } S - C = l.\sin\alpha = \Delta Y,$$

and these provide simple checks on the original latitudes and departures. The calculation of the modified traverse can be done on the same type of prepared form as that used for the proper traverse, and the duplication of mistakes is avoided because the lengths and bearings used in the check computation are quite different from those used in the original computation.

Noting that $\Delta X + \Delta Y = 2S$ and $\Delta X - \Delta Y = 2C$, B. Goussinsky has suggested that either of the above formulae would be a sufficient check, and it would not then be necessary to compute *both* S and C. Another good check is obtained by verifying the relation:

$$\tan\tfrac{1}{2}\alpha = \frac{\Delta Y}{\Delta X + l}$$

278. Computation of Bearing and Distance from Coordinates. If we are given the coordinates of two points we can compute the bearing and distance between them, this being the reverse problem of the previous one. The formulae are:

$$\tan\cdot\alpha = \frac{\Delta Y}{\Delta X} \qquad l = \frac{\Delta X}{\cos\alpha} = \frac{\Delta Y}{\sin\alpha}$$

ΔX and ΔY are obtained as the differences of the X's and the Y's of the two points respectively. The quadrant in which the bearing lies depends on the signs of ΔX and ΔY.

$$
\begin{aligned}
X_B &= 4379\text{·}28 & Y_B &= 7849\text{·}36 \\
X_A &= 6186\text{·}32 & Y_A &= 7764\text{·}28 \\
\Delta X &= -1807\text{·}04 & \Delta Y &= +85\text{·}08
\end{aligned}
$$

$$
\begin{aligned}
\log \Delta Y &= 1\text{·}929\,8275 \\
\log \Delta X &= 3\text{·}256\,9678 \\
\log \tan \alpha &= 8\text{·}672\,8597
\end{aligned}
$$

$$180 - \alpha = 2° 41' 44''; \ \alpha = 177° 18' 16''$$

$\log \Delta X = 3\text{·}256\,9678$	$\log \Delta Y = 1\text{·}929\,8275$
$\log \cos \alpha = 9\text{·}999\,5192$	$\log \sin \alpha = 8\text{·}672\,3654$
$\log l = 3\text{·}257\,4486$	$\log l = 3\text{·}257\,4621$
$l = 1809\text{·}04$	$l = 1809\text{·}10$

Bearing AB = 177° 18' 16''; Bearing BA = 357° 18' 16''

The bearing from A to B must be in the second quadrant because ΔX is negative and ΔY positive in going from A to B.

This example has been chosen purposely to illustrate the point referred to in section 270 regarding the choice of formulae when angles are small, or are near 90° in value. Here, the reduced bearing is a very small angle so that its log tangent and log sine are changing very rapidly. In this case, the value to be taken for l is that computed from the cosine–1809·04. In order to obtain better agreement between the values computed from the sine and cosine it would have been necessary to work out the bearing to decimals of a second, instead of to single seconds, but, as the example is taken from an ordinary survey, and not from geodetic work, so that geodetic accuracy is not required, it is hardly necessary to go to such extremes.

The formula $l = \sqrt{(\Delta X)^2 + (\Delta Y)^2}$ is useless for logarithmic calculation but can be operated quite simply on a machine of adequate capacity if a table of squares is available. Consider the example: $\Delta X = -2215\cdot75$, $\Delta Y = +3049\cdot91$. The sum of the squares of these two numbers is obtained on a calculating machine by straightforward multiplication without clearing the product register after the first squaring: the result is 14211499·0706. Reference to a table of squares shows that the square root of this number is 3769 to four significant figures. The number on the product register is now divided by 3769 by the continued subtraction process; the quotient is 3770·628. The correct square root is $\frac{1}{2}(3769 + 3770\cdot628) = 3769\cdot81$ to two decimal places.

The bearing of the line is best obtained separately, by dividing the smaller by the greater of ΔX and ΔY: in the above case, the result is 0·726 497 which is the tangent of 35° 59′ 54″. Considering the signs and relative values of ΔX and ΔY it is clear that the bearing of the line is 125° 59′ 54″.

279. Check Computation of Bearing and Distance by Use of Auxiliary Bearings. Bearings and distances can also be computed from given coordinates by the use of auxiliary bearings. Being given the coordinates of the two points, we know ΔX and ΔY. Then:

$$S + C = \Delta X \qquad\qquad S = \tfrac{1}{2}(\Delta X + \Delta Y)$$
$$S - C = \Delta Y \qquad\qquad C = \tfrac{1}{2}(\Delta X + \Delta Y)$$

$$\tan(\alpha + 45°) = S/C$$

and

$$l = \frac{\sqrt{2} \cdot C}{\cos(\alpha + 45°)} = \frac{\sqrt{2} \cdot S}{\sin(\alpha + 45°)}.$$

THEODOLITE TRAVERSE COMPUTATIONS

280. The computations of a traverse involve the following operations:

(1) Application of all necessary corrections to the measured lengths.

(2) Reduction and meaning of measured angles in the field books.

(3) Abstracting measured lengths and angles on to the coordinate sheets.

(4) Computation of bearings.

(5) Adjustment of bearings if traverse is either a closed loop or ends on a line of fixed bearing.

(6) Computation of reduced bearings.

(7) Computation of latitudes and departures.

(8) Computation of coordinates.

(9) Check computation.

(10) Adjustment of coordinates if the traverse is either a closed one or ends on a point whose coordinates are fixed and known.

Traverse computation should be carried out on a prepared form, which can be designed to suit the methods of survey and computation in use: some ingenuity must be exercised in order to keep the width of the form within manageable limits. A check computation should always be done, preferably by another computer who should start at the beginning, that is from the original observations.

A design of form suitable for logarithmic calculation is illustrated in Fig. 141, and the computations entered on it relate to the first few lines of the traverse shown in Fig. 140. The traverse started at

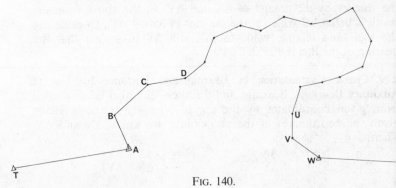

Fig. 140.

the point A which was already fixed into the coordinate system and had coordinates:

$$+218\,763 \cdot 55, \ +135\,227 \cdot 81.$$

The first angle measured was α_1, between the first line of the traverse and the fixed line from A to T: the given bearing of this line was 253° 16′ 30″. Successive observed included angles are the angles at the traverse stations on the left-hand side in relation to the order of computation.

This traverse had 20 stations including the terminal points, and therefore 19 lines. The last measured angle was α_{20} at the fixed point W the coordinates of which were given: this angle is included between the last line of the traverse and the fixed line WZ the bearing of which was given.

The traverse illustrated was one of fairly high order, in which the linear measurement was done with a steel tape and the angles were measured to 05″ with a small micrometer theodolite.

281. Abstracting Lengths and Angles. Calculation of mean values of observed angles, and computation of true horizontal lengths of lines are best done in the field books. The final values are then entered in the appropriate places on the traverse computation form, and the entries should be checked, preferably by another computer.

282. Computation and Adjustment of Bearings. Starting from the given bearing of AT, it is easy to calculate a bearing of each line of the traverse, and these are the bearings written in the Bearing column against the symbols of the traverse leg. The last calculated bearing is that of the line WZ; but the bearing of this is already known and the value calculated through the traverse will usually be different, since it includes all the errors in the observed angles.

In the traverse illustrated, the calculated bearing of WZ was 1′ 42″ less than the fixed value. This discrepancy is to be spread over the 20 observed angles, and the first adjustment operation is to increase the first calculated bearing by 5″, the second by 10″ and so on. These adjustments, and the corrected bearings, are shown in the Bearing column. The extra 2″ has been put into the last angle: normally, any reasonably uniform distribution of the angle misclosure will suffice.

The next column contains the lengths and the reduced bearings, written in for convenience of computation as explained in section 275; if Shortrede's Tables are to be used, reduced bearings are unnecessary, and this column could be dispensed with if the lengths are put in the bearing column.

283. Computation of Latitudes and Departures. The latitudes and departures are computed in the space provided in the next column of the form. The example shows a logarithmic computation but the space can also be used for computation by machine, the only entries

TRAVERSE COMPUTATION

From .... Field Books .... Computer ....

To ....   original/check

Point	Line	Bearing — Angle	Adj.	Length L — Reduced Bearing β	log ΔX log cos β log L log sin β log ΔY	X ΔX — Adj.	Y ΔY — Adj.	Point
	AT	253° 16′ 30″				218 763·55	135 227·81	A
A		78 58 45			3·271 100 9·946 959 3·324 141 9·667 946 2·992 087	+ 1 866·81	− 981·94	
	AB	332 15 15	+ 5″	2 109·31 — 27° 44′ 40″		− 0·05	+ 0·09	
B		251 17 50			3·128 731 9·860 172 3·268 559 9·838 244 3·106 803	220 630·36 + 1 345·03	134 245·87 + 1 278·80	B
	BC	43 33 05	+ 10″	1 855·92 — 43° 33′ 15″		− 0·10	+ 0·17	
C		219 56 20				221 975·39	135 524·67	C

(conclusion of traverse computation above)

U	203 10 10					U
			3·244 319	−0·78	+1·36	
			9·998 005	211 308·45	154 945·94	
UV	185 27 45	+90″	3·246 314	−1 755·17	−168·62	
			8·980 587			
		1 763·25	2·226 901			
		5° 29′ 15″				
V	129 56 40					V
			3·209 885	−0·82	+1·43	
			9·852 745	209 553·28	154 777·32	
VW	135 24 25	+95″	3·357 140	−1 621·38	+1 597·04	
			9·846 175			
		2 275·83	3·203 315			
		44° 34′ 00″				
W	147 33 50					W
				−0·88	+1·53	
				207 931·90	156 374·36	
WZ	102 58 15	+102″		(207 931·02)	(156 375·89)	
	(102 59 57)					
		Total length				
		35 646·13				

Fig. 141.

in this case being the natural cosine and sine of the angles, as the length of the line is already written in. It is quite possible, using a machine, to omit the computation column altogether.

As the values of the latitudes and departures are found, they are entered in the next columns, their signs, of course, being settled by the values of the bearings entered in column five. It would usually save time if the computer, before he works out latitudes and departures, examines the bearings and enters the correct signs in the latitude and departure columns. This will save him having to think about signs when looking out anti-logarithms or reading off multiplication results from the computing machine.

One very common type of error in traverse computations is to enter a latitude or departure with the wrong sign, or else to enter a latitude in a departure column and *vice versa*. Errors of this kind can easily be seen by inspection and, as each sheet is compiled, it is well to examine the latitudes and departures to see that all are entered correctly with the correct sign. If the reduced bearing is less than 45° the latitude will be numerically greater than the departure, but if the reduced bearing is greater than 45°, the latitude will be less than the departure. Hence, it is easy to see if a latitude and departure have been interchanged.

284. Computation of Coordinates. When the latitudes and departures have been calculated, the coordinates are easily obtained. If the initial point A has definite coordinates, the values of these should be entered in the top line in the coordinate columns on the form. If the initial point is not a fixed one, so that there are no definite values for it, zero coordinate values in X and Y could be assigned to it; but it is better to assign arbitrary positive values— each a multiple of 1,000—so that there will be no negative coordinates in the whole survey. This is equivalent to choosing an arbitrary 'false origin', in such a position that it lies to the south and west of any point included in the survey. The values assumed for the coordinate values of A then become the coordinates of this point referred to axes parallel to the true axes through A, but intersecting at the false origin.

Having written down the coordinate values of A in the appropriate columns, the latitude and departure to the point B are now added algebraically to these values and the results are the coordinates of B. These are written down and the latitude and departure from B to C added algebraically to them to give the coordinates of C. Proceeding from point to point in this way, we arrive at the coordinate values of the last point entered on the sheet.

Before going on to the next sheet the work already completed should be checked. To do this, add together the positive latitudes and the negative latitudes and find their algebraic sum. To the result add the X coordinate of the first point on the sheet, when the figure obtained should be exactly equal to the X coordinate of the last point on the sheet. Similarly, check the last Y coordinate from the algebraic sum of the sums of the positive and negative departures. This test is a very simple one, and it should be used for every sheet as soon as the other work on it is completed. It may help with the arithmetical checks to have separate columns for positive and negative latitudes, and for positive and negative departures; also separate columns for coordinates. This makes the computation form much wider and wastes a lot of space on the paper, however.

285. Check Computation. When a rough check against gross error only is needed, or it is known that a gross error exists and it is required to locate it, a check by traverse tables, or by some other approximate method will often suffice; but, if the traverse is an important one, a thorough and complete check, of a standard of accuracy not less than that of the field work, is essential. For example, if the traverse were measured in order to establish the alignment of a long tunnel, the ends of which were not intervisible, an error of only a very few feet might have serious consequences and cause considerable expense.

Assuming that a complete and thorough check is required, it can be done by the auxiliary angle method, or by use of a calculating machine if the original computation was done with logarithms. In either case, none of the data should be copied from the original computation form, which should be put out of sight, and all the distances and included angles should be taken again from the field books. If the auxiliary bearing method is to be used, the initial and final fixed bearings are increased by 45°. The check computation can be done on the same type of form as that used for the original computation, but it is important to write 'CHECK' boldly on each sheet of the check computation.

286. Adjustment of Traverses. If the traverse starts and ends on points whose coordinates are fixed and known it will be found that the coordinates obtained for the end point will differ slightly from the fixed values. The differences so found are caused by an accumulation of small errors and are unavoidable. It is, however, desirable to adjust the whole traverse so that the differences between the computed and the fixed coordinates of the end point are eliminated and the closing error is smoothly distributed throughout the various stations of the traverse.

A traverse may be adjusted by applying corrections to the latitudes and departures, or directly to the coordinates, or else by means of corrections applied to the individual lengths and bearings. Since it is usually the coordinates that are ultimately required, and not the distances and bearings from which they are derived, it is generally more convenient to use corrections that can be applied directly to the coordinates.

Of the methods available for the adjustment of traverses there is none that is really simple and at the same time theoretically sound. Probably the one most commonly used, especially for the adjustment of compass traverses, is that due to Bowditch. This method depends on the assumption that the probable errors in length and bearing produce equal displacements at the end of the leg and that the probable errors of the linear measurements are proportional to the square roots of the lengths of the legs. From these two assumptions it follows that the probable errors of the bearings must be proportional to the reciprocals of the square roots of the lengths of the legs and, as we have already seen, neither this, nor the assumption regarding the probable errors of the linear measurements, is a sound one.

From a purely theoretical point of view, the Bowditch rule is more suitable for the adjustment of compass traverses than it is for that of theodolite traverses, because it has been worked out for the case in which bearings are observed directly, as in a compass traverse, whereas, of course, in a theodolite traverse it is angles, and not bearings, that are observed. Moreover, the corrections obtained are corrections to the latitudes and departures and one disadvantage of the rule is that these corrections may cause excessive disturbance to the bearings, though, if coordinates alone are needed, this disadvantage does not make itself apparent, nor does it matter much. However, for ordinary work, as opposed to precise traversing for primary framework purposes, any elaborate method of adjustment, which involves a considerable amount of labour, is hardly justified, so that Bowditch's rule can quite safely be, and usually is, employed in nearly all normal cases of traverse adjustment.

The Bowditch rule is:

$$\text{Correction to } \frac{\text{latitude}}{\text{departure}} =$$

$$\text{Closing Error in } \frac{\text{latitude}}{\text{departure}} \times \left(\frac{\text{length of corresponding side}}{\text{total length of traverse}} \right)$$

the sign varying with the sign of the closing error at the end.

In the example given on the form, the total length of the traverse is 35,646 ft. and the misclosures are $+0.88$ and -1.53.

Then, according to Bowditch's rule, the correction to the X coordinates will be $-0.88/356.5 = +0.00247$ ft. per 100 ft., and to the Y coordinates $+1.53/356.5 = 0.00429$ ft. per 100 ft.

The first step is to add each length to the sum of the lengths before it, so as to get the total distance of each point from the initial point of the traverse, these total lengths being entered under the length of each leg. In the example, take the leg BC. The total distance of the point C from the beginning of the traverse is 3965 ft. Rounding off to the nearest 10 ft., this is 40 in hundreds of feet. Hence the corrections to the coordinates of C are $-40 \times 0.00247 = -0.10$ ft. in X and $+40 \times 0.00429 = +0.17$ ft. in Y. These corrections are entered above the coordinates of C. The corrected coordinates of each point may be written down in red to make them conspicuous.

It will be noticed that in this example we have applied the corrections to the coordinates directly, not to the individual latitudes and departures. This is by far the more convenient procedure, and it is obvious that it amounts to the same thing as correcting the latitudes and departures, as, in computing the corrections, we have used the total length from the beginning of the traverse up to the point being adjusted, instead of the length of the corresponding leg.

If the corrections are small, they may be derived very quickly either by slide rule or even graphically by the method outlined in section 293 in connection with the adjustment of compass traverses.

The Bowditch rule can easily be amended to fit the assumption that the probable errors of the linear measurements are proportional to the length of the leg and the probable errors of the angular measurements proportional to the reciprocals of the lengths of the legs. In this case, the factor which multiplies the closing error in latitude or departure is the square of the length of the side divided by the sum of the squares of the lengths of all the sides. This, of course, involves more arithmetical work than the ordinary rule, and, considering that the assumption made regarding the probable errors of the bearings is still unsound, it is doubtful if the extra labour is worth while.

Theodolite traverses may be adjusted by making the correction to a latitude (or departure) equal to the closing error in latitude (or departure) multiplied by the latitude (or departure) of the leg to be adjusted and divided by the arithmetical sum of all the latitudes (or departures). This rule is purely empirical, and there is no sound theoretical foundation for it.

It is also possible to devise methods of adjustment that do not alter the bearings first obtained from the bearing closure adjustment: that is, the corrections to the latitude and departure of each line are proportional to the latitude and departure.

In all probability, the simplest method so far devised of adjusting the bearings and lengths of a traverse to fulfil the conditions of closure in bearing and in position, and which has a reasonably sound theoretical justification behind it, is one described by Mr. H. L. P. Jolly in the *Empire Survey Review*, Vol. IV, No. 28, April 1938. This method involves corrections to each bearing and length so that corrections to the coordinates have to be worked out subsequently and are not obtained directly. The labour is therefore much greater than that required for the application of the Bowditch rule, so that, whatever defects this latter rule may have from a theoretical point of view, it probably remains the most practical one so far as the type of traverse considered in this volume is concerned.

287. Adjustment of a Network of Traverses. It often happens that the framework of a large survey consists of a network of traverses with junction points at which three or more traverse surveys meet at a common station. In such a case, it would be possible to make a simultaneous adjustment based on the principles of probability, but such a calculation is very cumbersome and not justified for ordinary survey work.

If the network is self-contained without any point having previously determined coordinates, the simplest procedure is generally to calculate and adjust a peripheral closed circuit, then adjust, one by one, a series of cross traverses, successively dividing up the area into smaller and smaller portions. Similarly, if the network is based on trigonometrical stations already fixed, the traverses connecting fixed points should be adjusted first, followed by traverses connecting already adjusted traverse stations.

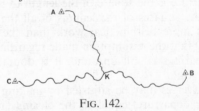

FIG. 142.

A slightly more elaborate procedure can be carried out, and will be described with reference to Fig. 142. A, B, C are three points of given coordinates, and the traverses AKB and CK have been surveyed. The procedure described above would require first the adjustment of AKB to fit the points A and B; this would give 'final' coordinates for K, and then the traverse CK would be adjusted to fit the coordinates of C and K.

The more elaborate method would be to calculate the co-ordinates of K starting from each of the points A, B and C: firstly, the bearings of the three lines meeting at K would be calculated through from the three starting points, these bearings would be adjusted so that the three differences between them agreed closely with the three angles observed at K, the bearing adjustments would be distributed back through the traverses, and the three versions of the coordinates of K calculated.

Suppose the three X coordinates so obtained are X_A, X_B and X_C; then the final value for the X coordinate of K could be a weighted mean of these three. The weights would be dependent on the relative lengths of the routes AK, BK, and CK. It would be reasonable, for instance, to take the weights as inversely proportional to the lengths of the routes, L_A, L_B and L_C say. Then the final X coordinate of K would be:

$$\left(\frac{X_A}{L_A}+\frac{X_B}{L_B}+\frac{X_C}{L_C}\right)\Bigg/\left(\frac{1}{L_A}+\frac{1}{L_B}+\frac{1}{L_C}\right).$$

Similarly for the Y coordinate.

The coordinates so obtained for K would be taken as fixed, and the three traverses adjusted to fit them.

288. Corrections of Latitudes and Departures for Small Corrections in Bearings and Distances. If, after the latitudes and departures have been computed, small corrections have to be applied to the bearings and distances in a traverse, and new corrected coordinates are required, it is not necessary to re-compute the latitudes and departures from the corrected bearings and distances. The small corrections δx and δy are:

$$\delta x = \delta l \,.\, \cos \alpha - l \,.\, \delta \alpha \,.\, \sin \alpha \,.$$
$$\delta y = \delta l \,.\, \sin \alpha + l \,.\, \delta \alpha \,.\, \cos \alpha \,.$$

or, if $\delta \alpha$ is expressed in seconds of arc:

$$\delta x = \delta l \,.\, \cos \alpha - l \,.\, \delta \alpha \,.\, \sin 1'' \,.\, \sin \alpha \,.$$
$$\delta y = \delta l \,.\, \sin \alpha + l \,.\, \delta \alpha \,.\, \sin 1'' \,.\, \cos \alpha \,.$$

Hence, if δl and $\delta \alpha$ are small corrections to the length and bearing of a line, δx and δy will be corrections to be applied to the latitude and departure of the line.

289. Corrections of Bearings and Distances for Small Corrections in Latitude and Departure. As we have already seen, Bowditch's rule involves corrections applied to the latitudes and departures or, what is more usual, directly to the coordinates, and these corrections mean disturbances to the lengths and bearings. In certain cases, notably in land surveying work, it is often necessary to insert or to tabulate the corrected bearings and distances on the plans. Here, again, it is not necessary to re-compute new bearings and distances from the new coordinates, as it is much simpler to work out small corrections to the original values.

By multiplying the first of the equations given in the last section by cos α and the second by sin α and adding we obtain

$$\delta l = \delta x \cos \alpha + \delta y \sin \alpha.$$

Similarly, by multiplying the first equation by sin α and the second by cos α and subtracting, we get

$$\delta \alpha = \frac{\delta y \cos \alpha - \delta x \sin \alpha.}{l \sin 1''}$$

Here, δx and δy are the known corrections to the latitude and departure of the leg, and δl and $\delta \alpha$ (in seconds) are the required corrections to length and bearing. If, as is usual, corrections for misclosure are applied to the calculated coordinates, δx and δy are to be taken as the differences of the coordinate corrections at the two ends of the line.

290. Traverse Closure by Triangle. If a traverse is closed on an inaccessible or non-occupied point, shown at S in Fig. 143, the measured length of QR and the angles α and β are used to calculate the length of QS or RS by sine formulae, and this line is then treated as the last line of the traverse. There can be no bearing closure in this case, but if point S can in fact be occupied a bearing

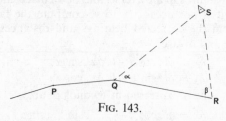

FIG. 143.

closure might be obtainable. This arrangement at the end of a traverse is convenient if the lines QS and RS would be difficult to measure. See section 254.

COMPUTATION OF COMPASS TRAVERSES

291. The computation of a compass traverse may be done on the same form as for a theodolite traverse, with very slight modification.

The 'Bearing' column may be amended to 'observed bearings' and used for the forward and reverse bearings of the lines: the next column can be used for any adjustment of bearings to take account of local attraction influences (see Section 262). It is suggested that the first computation of the traverse should be done with the aid of traverse tables and the arithmetic written in the computation column: the rest of the form is filled up and the adjustment carried out as for a theodolite traverse.

A check computation can be done by logarithms, calculating machine, or a spiral slide-rule, provided that the checking computer refers to the original bearings and lengths, not to those copied on to the first computation forms. If the traverse runs between points already fixed, a graphical check by drawing the traverse with protractor and ruler may be considered sufficient.

292. Traverse Tables. A traverse table simply gives the values of the natural cosines and sines of angles from 0° to 90°, each multiplied by 1, 2, 3, etc., up to 9 units. To find the latitude and departure of a given line, the tabular values for the reduced bearing are extracted for each digit of the given length, and, on properly placing the decimal points, the required coordinates are given by summing the parts.

Example. The following figures are extracted from a traverse table, for bearing 36° 41′.

1		2		3		4		5	
Lat.	Dep.	Lat.	Dep.	Lat.	Dep.	Lat.	Dep.	Lat.	Dep.
0·8019	0·5974	1·6039	1·1948	2·4058	1·7922	3·2078	2·3896	4·0097	2·9870

If it is required to determine to the first place of decimals the latitude and departure of a line having this reduced bearing and of length 534·1 ft., the additions are:

	Lat.	Dep.
500	400·97	298·70
30	24·06	17·92
4	3·21	2·39
0·1	0·08	0·06
534·1	428·3	319·1

After their numerical values have been determined, the latitudes and departures are given their proper signs and are entered in the appropriate columns, the remainder of the form being completed in exactly the same manner as that already described for the case of theodolite traverses.

293. Adjustment of Compass Traverses. After the coordinates have been obtained, a compass traverse can be adjusted by Bowditch's rule, which, from a theoretical point of view, is really more suitable for the adjustment of compass than for theodolite traverses. This adjustment can easily be done graphically as follows.

Set out a line A*a*, Fig. 144 (b), equal to the total length of the traverse and mark off the distances A*b*, *bc*, *cd*, *ef*, *fa* equal to the lengths of the different legs. At *a*, draw *a*A′ equal to the closing error in latitude. Join A A′, and at *b*, *c*, *d*, *e*, and *f* draw *b*B, *c*C, *d*D, *e*E, *f*F, parallel to *a*A′, and to meet the line A A′ in B, C, D, E, and F. Then the lengths of the different ordinates give the corrections in latitude to the coordinates of the points corresponding to *b*, *c*, *d*, *e*, and *f*. Similarly, repeat the process for the departures, using the closing error in departure instead of the closing error in latitude.

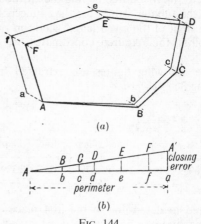

(a)

(b)

FIG. 144.

If the traverse has been directly plotted, not computed, the adjustment can likewise be performed graphically. In Fig. 144 (a) let A *b c d e f a* be a traverse which closes back on itself at A. When the traverse is plotted, the end point will not coincide exactly with A, but will be at the point *a*, and A*a* will represent the closing error. Set off, as before, the line A*a*, Fig. 144 (b), equal to the length of the traverse, and at *a* draw *a*A′ perpendicular to A*a*, but make *a*A′ equal to the total closing error A*a* in Fig. 144 (a). Then the magnitudes of the closing error at *b*, *c*, *d*, *e*, and *f* are equal to the ordinates *b*B, *c*C, *d*D, *e*E, *f*F. In Fig. 144(a) draw lines through *b*, *c*, *d*, *e*, and *f* parallel to A*a*, and along these set off distances *b*B, *c*C, *d*D, *e*E, *f*F, equal to the corresponding ordinates on the lower diagram. Then A B C D E F A represents the adjusted traverse.

In the case of lower order compass traverses–e.g. ones in which bearings are measured with a small hand compass instead of with a large one used on a stand–it will be quite sufficient to distribute the closing errors in latitude and departure evenly among the co-ordinates of the traverse, without reference to the lengths of individual legs. Thus, suppose that the closing error in latitude is 150 ft. and that there are 50 legs. Then, the adjustment to the X co-ordinate of the end point of the first leg would be 3 ft., that to the X coordinate of the end point of the second leg would be 6 ft., and so on, no notice at all being taken of the lengths of the legs.

MISCELLANEOUS PROBLEMS IN RECTANGULAR COORDINATES

294. Transformation of Coordinates. It occasionally happens that it becomes necessary to transform coordinates referred to one set of axes to values referred to another. This case sometimes occurs when a new survey is tied to an old one or it is desired to transform coordinates based originally on the magnetic meridian to others based on the true meridian as axis of X, or on to the grid meridian of a national mapping system.

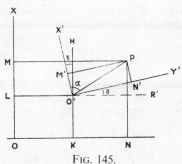

Fig. 145.

In Fig. 145 let O'X', O'Y' be the old axes of coordinates and OX, OY the new. Through O' draw KO'H and LO'R parallel to OX and OY respectively. Then angle HO'X' = γ = angle between old and new axes of X. Through P draw PM', PN' perpendicular to OX' and OY' and PM, PN perpendicular to OX and OY. Then the old coordinates of P are O'M' = x, and O'N' = y, and the new ones are OM = X, and ON = Y. Also, let co-ordinates of O' referred to OX and OY be OL = X_o and OK = Y_o. Let X'O'P = α

be the angle made by O'P with O'X'. Then:

$$X = X_0 + O'P \cos(\alpha - \gamma)$$
$$= X_0 + O'P \cos \alpha \cos \gamma + O'P \sin \alpha \sin \gamma$$

But O'P $\cos \alpha = x$,　O'P $\sin \alpha = y$,

$$\therefore X = X_0 + x \cos \gamma + y \sin \gamma \quad \dots\dots\dots\dots\dots\dots(A)$$

Similarly

$$Y = Y_0 + O'P \sin(\alpha - \gamma),$$
$$= Y_0 + y \cos \gamma - x \sin \gamma \quad \dots\dots\dots\dots\dots\dots(B)$$

and these are the formulae required.

For the reverse problem, multiply equation (A) by $\cos \gamma$ and (B) by $\sin \gamma$ and subtract (B) from (A). Then

$$x(\cos^2\gamma + \sin^2\gamma) = (X - X_0) \cos \gamma - (Y - Y_0) \sin \gamma$$
$$\therefore x = (X - X_0) \cos \gamma - (Y - Y_0) \sin \gamma \quad \dots\dots\dots\dots(C)$$

Similarly, by multiplying (A) by $\sin \gamma$ and (B) by $\cos \gamma$ and adding, we get

$$y = (X - X_0) \sin \gamma + (Y - Y_0) \cos \gamma \quad \dots\dots\dots\dots(D)$$

295. Computation of the Cut of a Fixed Straight Line on a Grid Line or Sheet Edge.　　When work is being plotted on a very large scale, it often happens that a straight line, whose position and bearing are known, or can be computed from the known co-ordinates of the two ends, cuts across several sheets, and it is required to plot the line on each sheet. The best way of doing so is to compute the positions of the points where the line, or the line produced, cuts the edges of each sheet.

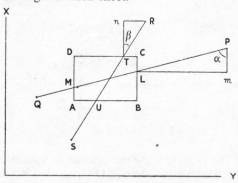

FIG. 146.

In Fig. 146 let ABCD be the sheet, the edges of which are parallel to the coordinate axes OX and OY, and let PQ be the fixed line joining the points P and Q. The coordinates of one point and the bearing of the line are known, or else the coordinates of both points are known, so that the bearing can be computed. Let the line cut the sheet edges BC and AD in L and M respectively. We require the distances BL and AM, or the coordinates of L and M.

Through L draw the line Lm parallel to OY, and through P draw Pm parallel to OX cutting Lm in m. Then angle LP$m = \alpha =$ the known bearing of PQ $- 180°$, and P$m = Lm \cot \alpha$. But L$m = Y_P - Y_B$, where Y_P is the Y coordinate of P and Y_B that of the sheet edge BC, which is known. Hence P$m = (Y_P - Y_B) \cot \alpha$. But P$m = X_P - X_L$, where X_L is the X coordinate of L, and therefore

$$(X_P - X_L) = (Y_P - Y_B) \cot \alpha.$$

But the X co-ordinate of B is known, because AB is a sheet edge. Hence BL is known. Similarly AM can be calculated, and therefore a straight line can be drawn through M and L, and this straight line will lie on the straight line PQ.

When the positions of L and M have been calculated by the formulae just given, the work can be checked by the relation:

$$X_L - X_M = AB \cot \alpha.$$

AB will usually be a round number of thousands, so that, when natural cotangents are used, the multiplication is a very simple and easy one.

Similarly, for the line RS, R$n = n$T $\tan \beta$,

$$\therefore (Y_R - Y_T) = (X_R - X_T) \tan \beta,$$

so that CT can be calculated. Also, another similar point U can be found on the sheet edge AB, so that a line joining T and U lies on the line PS. The check on this, of course, is: $(Y_T - Y_U) = CB \tan \beta$.

296. Graphical Intersection. The above method of computing the positions of the cuts of lines, of fixed bearing and position, on lines parallel to the coordinate axes can often be used in minor triangulation to determine, graphically, the most probable position of a point which is to be fixed by intersection from two or more points.

This case can best be illustrated by an example: A point P has been observed from four points A, B, C and D, and the coordinates of each point, and the observed bearings to P, are given in the following table:

Point	X	Y	Observed bearing	Approximate distance
A	37346·3	37469·3	212° 22′ 10″	16900
B	35180·8	26070·5	168 51 10	12300
C	27361·4	17761·7	111 39 00	11500
D	25676·4	36016·4	251 19 40	8000

The first thing to do is to obtain approximate coordinates for the values of the intersected point P, and a preliminary solution of the triangle BCP, obtained by drawing it on a large scale, showed that the coordinates of P were somewhere round about X = 23120, Y = 28450. These figures are only required to give an indication of the lines to be chosen as grid lines on which to compute the cuts. The grid lines chosen are:

$$X = 23110 \qquad X = 23130$$
$$Y = 28440 \qquad Y = 28460$$

Now compute the cut of the bearing $A - P$ to the lines $X_G = 23110 \cdot 0$ and $X_G = 23130 \cdot 0$.

$$
\begin{array}{rl}
X_A & = 37346 \cdot 3 \\
X_G & = 23110 \cdot 0 \\
\hline
& 14236 \cdot 3
\end{array}
\qquad\qquad
\begin{array}{r}
37346 \cdot 3 \\
23130 \cdot 0 \\
\hline
14216 \cdot 3
\end{array}
$$

$$
\begin{array}{ll}
\log 14236 \cdot 3 & = 4 \cdot 153\,397 \\
\log \tan 32\,22'\,10'' & = 9 \cdot 802\,001 \\
\hline
& 3 \cdot 955\,398
\end{array}
\qquad
\begin{array}{ll}
\log 14216 \cdot 3 & = 4 \cdot 152\,787 \\
\log \tan 32\,22'\,10'' & = 9 \cdot 802\,001 \\
\hline
& 3 \cdot 954\,788
\end{array}
$$

$$
\begin{array}{rll}
\therefore & Y_A - Y_G = & 9024 \cdot 0 \\
& Y_A = & 37469 \cdot 3 \\
\therefore & Y_G = & \overline{28445 \cdot 3}
\end{array}
\qquad
\begin{array}{rll}
\therefore & Y_A - Y_G = & 9011 \cdot 3 \\
& Y_A = & 37469 \cdot 3 \\
\therefore & Y_G = & \overline{28458 \cdot 0}
\end{array}
$$

(Check: $20 \times \tan (32° \ 22') = 20 \times 0 \cdot 6338 = 12 \cdot 7$: $9011 \cdot 3 + 12 \cdot 7 = 9024 \cdot 0$.)

Hence the line from A, on a bearing of 212° 22' 10", cuts the line X = 23110·0 where Y = 28445·3, and the line X = 23130·0 where Y = 28458·0. We therefore have sufficient information to plot the points where the line cuts the two grid lines that have been chosen so that they are very close to the final position of P.

Compute the cuts of the other bearings on suitable grid lines and form the following table:

Line	Grid line and position of cut	Grid line and position of cut
A − P	X = 23110·0 Y = 28445·3	X = 23130·0 Y = 28458·0
B − P	X = 23110·0 Y = 28449·0	X = 23130·0 Y = 28445·1
C − P	Y = 28440·0 X = 23122·8	Y = 28460·0 X = 23114·8
D − P	Y = 28440·0 X = 23116·0	Y = 28460·0 X = 23122·8

A piece of squared paper of suitable size is now taken, the chosen grid lines plotted on it and the lines corresponding to the lines of observed bearing plotted from the data given in above table. These lines intersect, two by two, in six points and each of these points gives one set of values for P. The position finally adopted for the point P should be as close as possible to all the lines, with due regard to the different distances from which the bearings were obtained; the final position should be chosen, where there is any conflict, to be nearer the shorter lines. This is because the linear displacement at the end of a given line, for a given error in bearing, is proportional to the length of the line, and it is assumed that the probable error of a bearing is the same for all lines.

From the diagram, the best values for the coordinates of P are taken as:

$$X = 23118 \cdot 2 \qquad Y = 28448 \cdot 8.$$

The approximate distances from the point P to the points A, B, C, and D, required for estimating the position of P in the graph, can be obtained quite accurately enough for this purpose by measuring on a fairly large-scale plot of the points.

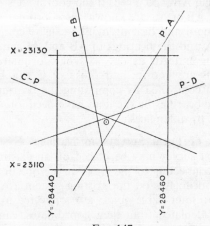

Fig. 147.

This method of graphical solution can also be adapted to the problem of resection—that is when a point is fixed by angles observed from it to three or more fixed points. Observations to three points, involving two angles, are sufficient for a 'fix', but, of course, a better result is obtained if more than three points are used. In that case, an ordinary computed solution is somewhat laborious and the graphical method is much quicker. This method is fully described in

Chapter 7 of the *Textbook of Topographical Surveying*, 4th Edn., (H.M. Stationery Office). See also sections 510–513.

297. Solution of Triangles by Coordinates. It sometimes happens that a triangle has to be solved when two sides and the included angle are given, and, in that case, a solution by coordinates is quicker than the usual method given in text books on trigonometry.

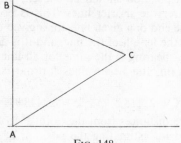

FIG. 148.

In Fig. 148 let the angle A and the lengths of the sides AB and AC in the triangle ABC be given. If the coordinates of A and the bearing of either AB or AC are known, the coordinates of B and C, and then the length and bearing of BC, can be computed. If the coordinates of A and the bearings of AB and AC are not known, take A as origin and AB as one axis of coordinates. Then the coordinates of B are known, and those of C can be computed and the solution completed by computing the bearing and length of BC. When the bearings by the lines have been found, the angles can be obtained by differences.

298. Connection of Local Surveys to National Systems of Co-ordinates. A local survey may be placed on an existing system of coordinates by connecting to the local survey one or more points for which the coordinates on the existing system are available or obtainable. If two intervisible points are connected in, the bearing between them, calculated from their coordinates, can be used as an initial bearing for the new survey; such a bearing is a 'grid bearing' as described in section 274.

A coordinate system covering a large area must be based on some 'map projection', which is simply a method of representing the curved surface of the Earth on a plane surface. It is inevitable in any map projection that some distortion of lengths and angles must occur, that is, the shape and size of any real geometrical figure on the Earth cannot be exactly reproduced to scale on a plane.

Many national mapping systems, including that of Great Britain, are based on a projection belonging to a class of projections characterised by the property that angles are correctly represented; such projections are called *orthomorphic* or *conformal*. The projection now used by the Ordnance Survey for Great Britain is the Transverse Mercator Projection. The meridian of 2° West is represented by a straight line which is the OX axis for coordinates. The mathematical origin of coordinates is the point on this meridian at latitude 49°N: after computation referred to this origin, the Y coordinates are increased by 400,000 metres so as to avoid having any negative coordinates over the area covered, and the X coordinates are decreased by 100,000 metres.

This projection is orthomorphic, but there is an inevitable increase of scale from the central meridian outwards each way. What this means, in practice, is that the length of a line calculated from the coordinates of its end-points will differ from the actual length on the ground by a small amount depending on the distance from the central meridian, that is on the Y coordinate. The various numerical constants used in the formulae for coordinate calculation have been chosen so that the scale is too small by 1/2500 on the central meridian, correct at a distance of about 180 km. (112 miles) east or west of this meridian, and too large further away, amounting to about +1/2500 at the extreme longitudinal limits of the country.

In connecting a local survey to the national system therefore, all measured lengths should be altered to 'projection' lengths by application of the appropriate scale correction formula. Strictly, small corrections to bearings are called for by the projection formulae, due to the fact that a line which is 'straight' on the ground is represented in the coordinate system by a line which is very slightly curved; the corrections for this effect are negligible for all ordinary purposes.

A surveyor wishing to connect his work to a national system should ask the government survey organisation for information and advice. As regards Great Britain, the Ordnance Survey will supply coordinates and descriptions of trigonometrical stations and other coordinated points in any specified area, on application and for a small fee. Information on the theoretical basis of the British coordinate system is to be found in the publication: *Constants, Formulae and Methods used in the Transverse Mercator Projection*, H.M. Stationery Office, 1950.

PLOTTING TRAVERSES

299. Methods of Plotting Survey Lines. One end of a survey line being located on the paper, the line can be plotted either by setting off its direction and length, or by fixing the position of its terminal point by coordinates. The first system, termed the angle and distance method, involves the laying down of an angle which, according as a reference meridian has or has not been adopted for the field observations, will be either (*a*) the bearing of the line, (*b*) the angle, included or deflection, from the preceding survey line. In the coordinate system, the direction and length of the line are used in the calculation of coordinates, and are not directly plotted.

300. Angle and Distance Methods. The more commonly used angle and distance methods of plotting a line are:

    (*a*) By protractor.
    (*b*) By the tangent of the angle.
    (*c*) By the chord of the angle.

By Protractor. The protractor furnishes the most rapid method of laying down angles, but, while the degree of accuracy afforded is sufficient for many purposes, the best results cannot be expected, particularly in plotting lines much longer than the radius of the protractor. With the ordinary simple pattern, of about 6 in. diameter, angles cannot be protracted nearer than to about 10 or 15 min., which accords with the accuracy of free needle readings, but not of theodolite work. Protractors of 9 to 12 in. diameter with vernier reading to 1 min. are more suitable for plotting theodolite surveys, but in using such protractors it is doubtful if full advantage can be taken of the vernier owing to the probability of error in centering. A very useful and inexpensive form of protractor has the graduations printed on cardboard, and may be obtained in diameters of from 12 to 18 inches. The cardboard inside the ring of graduations is cut out, and lines can be drawn at correct bearings by use of a parallel ruler.

By Tangents. In this method, the angle θ to be plotted is set out by geometrical construction with the aid of a table of natural tangents. From the station point a base length, scaling a round number of units, is set off along the given side of the angle, and from the end of the base a perpendicular is erected of length = base $\times \tan \theta$. The line joining the station to the point thus obtained includes θ with the given side. The length of base should be such that the hypoteneuse obtained exceeds the length of the survey line to be plotted along it. This method should not be used if the angle is much greater than 45°; if it is so, the complementary angle can be set out from a line drawn perpendicular to the initial line.

With care, the tangent method is capable of higher accuracy than direct protracting.

By Chords. In this geometrical method, an arc of any convenient radius r, preferably 10 or 100 units, is first swept out with the station point as centre. A second arc, centered at the point in which the first cuts the given side, is then described with radius equal to the chord subtended by the angle to be plotted. The line joining the station to the intersection of the arcs defines, with the given side, the required angle. The chord length $= 2r \times \sin \frac{1}{2}\theta$. Various mathematical tables give the lengths of chords corresponding to unit radius. If the angle exceeds a right angle, the construction should be applied only for the part less than 90°, as thereafter the intersections become unsatisfactory. The accuracy attainable is not materially different from that of the tangent method.

301. Plotting Angles and Bearings. The manner of applying the above methods to the plotting of a survey line from the preceding one will be evident. Results of angular measurements, noted in tabular form, should always be accompanied by a sketch, to which the draughtsman must make constant reference to avoid setting off angles in the wrong direction.

In plotting bearings, two methods may be distinguished.

(1) A meridian is drawn in a convenient central position on the sheet, and at an assumed point on it the bearings of all the lines to be plotted are set out from it. These directions are then transferred to their proper positions, as required (Fig. 149).

(2) Each bearing is plotted from a meridian through the origin of the line and parallel to the reference meridian (Fig. 150).

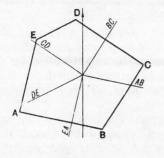

FIG. 149.

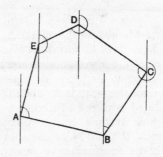

FIG. 150.

The first method is the more rapid. By using a circular protractor, all the bearings can be plotted from one setting of the protractor. In the tangent method the same bases can be used for various bearings, while in applying the chord method the several chord lengths can be marked on one circle. Great care is necessary in transferring the directions: a heavy parallel ruler gives the best results.

In the second method there is the same need for accuracy to secure parallelism between the successive meridians.

302. Plotting by Coordinates. The coordinate method is recognised as the most accurate and useful method of plotting traverse lines from bearings. Two systems are used.

(1) Each station is plotted from the preceding one by setting out the latitude and departure of the line between them in directions respectively parallel and perpendicular to the reference meridian.

(2) Each station is plotted independently by ascertaining its position relatively to two assumed coordinate axes respectively parallel and perpendicular to the reference meridian.

In plotting closed surveys by either method, the closing error, if appreciable, should first be eliminated suitably, and the adjusted values of the latitudes and departures used.

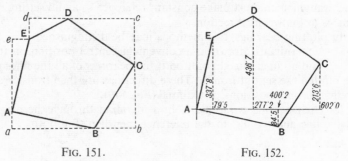

FIG. 151.   FIG. 152.

First Method. In this method, direct use is made of the latitudes and departures as shown in Fig. 151. Having assumed a point to represent the first station A and a line *ae* showing the direction of the reference meridian, B is plotted by setting off in the proper directions the latitude A*a* and departure *a*B of line AB. C is similarly located from B, and so on.

Second Method. By coordinate values. The origin may be the origin of the survey or the south-west corner of the sheet. If the latter, the coordinates of this corner can be subtracted from the coordinates of the different points falling on the sheet, so that the new coordinates of these points are referred directly to the sheet edges.

Fig. 152 shows a survey plotted with respect to axes through the most westerly traverse station. Here, there is a negative X coordinate for the point B. As a general rule, however, it is better to avoid negative coordinates, if necessary by choosing a 'false origin' to the south and west of any point likely to fall in the survey.

303. Precautions in Coordinate Plotting. To obtain good results, pencil lines should be fine, and coordinate lengths should be very carefully scaled. Inaccuracy in the directions of reference lines is the most likely source of error, particularly in the second method, which requires the plotting of longer lines than the first. Accuracy depends largely upon the care taken to secure perpendicularity between the reference axes. Long perpendiculars should be set out by the use of beam compasses rather than by set-square. When a survey covers a large area of paper, so that the station coordinates are long lines, it is better to construct a rectangle to enclose the survey. Before using this rectangle, its accuracy should be verified by checking the equality of its diagonals. Stations are then plotted from the nearest sides of the rectangle by subtracting, if necessary, the tabular coordinates from the lengths of the corresponding sides of the rectangle. Distances to be scaled can be still further reduced by subdividing the bounding rectangle into smaller rectangles, from the sides of which the plotting can be performed.

304. Relative Merits of the Two Coordinate Systems. The second method is much more commonly used than the first, but each has advantages. Errors of scaling are revealed in either system by measuring the length of each line as soon as its end points are plotted, but this test does not afford an absolute check on the bearings. In the first method, small errors of plotting are carried forward, and in a closed survey the closing error due to inexact plotting is discovered, since the coordinates used have been adjusted. In the most refined plotting this is an advantage, the cause of the discrepancy being searched for and eliminated. Since small errors are not accumulated in the second method, it is to be preferred for plotting unclosed traverses.

305. Advantages of Coordinate Plotting. (1) The table of coordinates exhibits the extent of the survey, and the position of the point representing the origin of coordinates may be selected so that the survey will fall centrally on the sheet.

(2) By calculation of the coordinates the closing error can be ascertained, and adjusted if necessary, before plotting commences. In angle and distance methods, plotting must be completed before the closing error is discovered, and then it is not known to what extent the error is due to inaccurate field work or plotting.

(3) The plotting of the stations of a large survey can proceed simultaneously on different sheets or on different parts of the same sheet.

306. Coordinate Plotting Devices. Precise plotting of rectangular coordinates can be mechanised by use of a specially fitted table. On one of the longer edges of the table is fitted an accurately straight guide, along which can be moved a slide which carries another straight guide perpendicular to the first. A carriage with a pencil moves along the second slide. Scales fitted to the two slides enable the positions of the moving slide and the pencil to be set precisely, with the aid of clamps, fine motion screws and verniers.

The plotting table is in fact a large precision-built version of the scale and offset arrangement illustrated in Fig. 104.

EXAMPLES

(*N.B. X co-ordinates are positive northwards, Y co-ordinates are positive eastwards.*)

(1) The length of a line AB is 751·1 ft. and its bearing from A to B is 303° 11′. Calculate the latitude and departure of the line.

(2) The coordinates of point A are $X_A = +841·5$, $Y_A = 139·2$ and the coordinates of point B are $X_B = +1890·7$, $Y_B = +488·4$. Calculate the length and bearing of the line AB.

(3) The bearing of a traverse line is 109° 33′ and its length is 716 links. By mistake, the computer uses 119° 33′ as the bearing. What errors in co-ordinates will result from this mistake?

(4) A traverse line AB has length 344·7 ft. and bearing 71° 09′. The co-ordinates of A are +3135·7 and +8709·2. Calculate the coordinates of B. At a point 241 ft. along the line, measuring from A, a right-angle offset of length 44·7 ft., on the left side of the traverse line, is measured to point C. Find the coordinates of C.

(5) In a triangle ABC the side AB has length 712·8 and the angles at A and B are 62° 15′ and 51° 42′ respectively. Solve the triangle and find the perpendicular distance of C from the line AB.

(6) The coordinates of points A, B and C are as follows:

	X (ft.)	Y (ft.)
A	7316	2709
B	7931	3898
C	6601	2945

Calculate the sides, angles and area of the triangle ABC.

(7) The coordinates of points A and B are:

	X	Y
A	4619	13815
B	6107	13076

Bearings from these points to another point Q are:

AQ	251° 06′
BQ	212° 51′

Find the coordinates of Q.

(8) The bearings and lengths of the lines of a closed compass traverse are given below:

AB	115 °	344 ft.
BC	95½	429
CD	345	202
DE	306	236
EF	319½	320
FG	245	439
GA	152½	244

The magnetic declination is 8½° West. Convert the bearings to true meridian and calculate the coordinates of the stations, taking the coordinates of point A to be +1000, +1000. Adjust the misclosure.

(9) A determination of the distance between two mutually visible points A and E is required, but cannot be made with sufficient accuracy by direct measurement. Obtain the distance from the following notes of a traverse run from A to E, using the line AE as local meridian:

Line	Bearing	Distance in feet
AE	360° 00′	
AB	346 18	2386·4
BC	73 57	583·2
CD	296 33	401·8
DE	18 21	1156·4
EA	180 00	

(10) A theodolite traverse ABCDEF starts from fixed point A at which the bearing to another fixed point M is known to be 199° 56′ 50″. The measured lengths, and the angles measured clockwise from the back station, are given below:

	Angle		Length	
A	225° 17′ 40″			
B	263 58 10	AB	719·3 ft.	
C	168 04 30	BC	1025·7	
D	231 09 00	CD	695·2	
E	98 28 20	DE	917·0	
F	112 33 30	EF	1247·5	

Point F is a fixed point and the last angle was measured to a bearing known to be 39° 28′ 40″. Calculate the bearings and adjust the angular misclosure. Calculate latitudes and departures and adjust the coordinate misclosure, using the following coordinates for the fixed points A and F:

	X	Y
A	16528	22411
F	14169	25119

234   PLANE AND GEODETIC SURVEYING

(11) An unclosed traverse ABCDE is surveyed along a shore-line for the purpose of setting out sounding lines. Taking the first line AB for local meridian, the bearings and lengths of the lines are found to be:

AB	00° 00′	426 ft.
BC	18 55	386
CD	69 10	511
DE	43 40	384

It is decided that the sounding lines shall be parallel lines on bearing 130° in the above system, and 300 ft. apart, the first line being at point A. Find the points of the survey lines where the sounding lines should be set out.

(12) From a common point A, traverses are conducted on either side of a harbour, as follows:

(1)			(2)		
Line	Length in feet	Bearing	Line	Length in feet	Bearing
AB	875	74°	AE	348	192°
BC	320	109	EF	436	160
CD	1064	82	FG	521	97
			GH	1683	89

Calculate the distance from H to a point K on GH due south of D, and the distance DK.

(13) It is required to ascertain the distance from A to an inaccessible point B invisible from A: a straight line CAD is run, with AC = 240 ft. and AD = 190 ft., and angles ACB and ADB are found to be 64° 10′ and 72° 40′ respectively. Calculate the distance AB.

(14) Given that the coordinates of two points A and B are:

	X	Y
A	37 842·10	46 616·19
B	32 384·74	46 593·72

find the bearing and distance B to A, and check by the method of auxiliary bearings.

(15) In the triangle ABC the coordinates of the points B and C are:

	X	Y
B	23 225·49	14 362·18
C	33 179·18	16 143·74

and the observed angles are:

at A	77° 18′ 20″
B	31 28 50
C	71 12 35

Find the coordinates of the point A, which is approximately west of point C.

(16) Observations are taken to two inaccessible points C and D from a measured base AB which is 1320 ft. long. The angles observed are:

CAD = 73° 57'   DBC = 86° 08'
DAB = 22 23   CBA = 32 22

Find the distance between C and D.

(17) Two stations A and B in a large survey are not intervisible. They have the coordinates (in metres):

$X_A = 12412$ $Y_A = 18253$
$X_B = 13559$ $Y_B = 18891$

In a compass traverse run from A to B the observed bearings and lengths were:

AP	104°	417 m.
PQ	87	602
QR	30	520
RS	350	438
SB	303	591

Compute the compass traverse and adjust it to the fixed values of A and B, and estimate the amount of magnetic declination with respect to the co-ordinate system.

(18) The coordinates of three consecutive stations of a theodolite traverse are:

X	Y
5000·0	5000·0
5532·1	5940·8
6897·5	6349·4

From these points, the bearings to an intersected point are, respectively, 350° 23', 297° 22' and 243° 10'.
Find the most likely position for the intersected point.

(19) The following traverse is run from A to E, between which points there occur certain obstructions:

Line	Distance in feet	Bearing
AB	426·7	38° 20'
BC	518·1	347 55
CD	606·3	298 12
DE	430·5	29 46

It is required to peg the point midway between A and E. Find the length and bearing of a line from station C to the required point.

(20) The coordinates of three points A, B and C are as follows:

	X	Y
A	5 091·4	5 902·0
B	19 284·3	5 523·7
C	24 425·1	14 392·9

and the following were the bearings observed from these points to a fourth point D:

AD	34° 31′ 37″
BD	82 03 59
CD	146 43 28

Find, with the help of a graphical construction, the most likely values of the coordinates of point D.

REFERENCES

BROWN H. C. 'Adjustment of traverse networks'. *Empire Survey Review*, No. 120, Apr. 1961.

DOWSON A. H. AND COLLINS M. O. 'The continuous revision of Ordnance Survey maps and plans'. *Chartered Surveyor*, Feb and Mar. 1956.

GOUSSINSKY B. 'Checking traverse computations'. *Empire Survey Review*, No. 49, July 1943.

— 'Some notes on the adjustment of traverses' *Empire Survey Review*, No. 60, Apr. 1946.

GWILLIAM R. 'A note on the adjustment of minor traverse networks'. *Empire Survey Review*, No. 114, Oct 1959.

THOMAS T. L. 'The correlation of local grid to national grid coordinates'. *R.I.C.S. Journal*, Dec. 1954 and Feb. 1955.

CHAPTER VI

ORDINARY LEVELLING

---

307. Levelling, or the determination of the relative altitudes of points on the Earth's surface, is an operation of prime importance to the engineer, both in acquiring data for the design of all classes of works, and during construction operations.

308. Definitions. *A Level Surface* is one which is at all points normal to the direction of gravity as indicated by a plumb line. Owing to the form of the Earth, a level surface is not a plane, nor has it a regular form, because of local deviations of the plumb line caused by irregular distribution of the mass of the Earth's crust. The surface of a still lake exemplifies a level surface.

A Level Line is a line lying throughout on one level surface, and is therefore normal to the direction of gravity at all points.

The *Horizontal Plane* passing through a point is the plane normal to the direction of gravity at the point. It is therefore tangential to the level surface at the point, and sensibly coincides with it within ordinary limits of sighting (see section 346).

A *Horizontal Line* passing through a point is one lying in the horizontal plane, and is tangent to a level line through the point and having the same direction.

A *Datum Surface* (*Line*) is an adopted level surface (line) to which the elevations of points may be referred.

The *Reduced Level* of a point is its elevation above the datum adopted.

PRINCIPLES

309. Difference of Level of Two Points. The simplest operation with the Level (section 310) is to determine the difference of level between two points so situated that, from one position of the instrument, readings can be taken on a staff held successively upon them. The precise situation of the instrument is immaterial but, to minimise the effects of possible instrumental error and other complications, the two sights should be of equal lengths or nearly so. After the instrument has been set up on firm ground, the levelling process appropriate to the instrument is carried out and the eyepiece is focused clearly on the diaphragm lines. The staffman is instructed to hold the staff vertically on the first point. The observer directs the telescope towards the staff and brings it into focus free from parallax. Any necessary further adjustment of the bubble is then made, and the staff reading is noted: no fine adjustment is required if an automatic Level is in use. The staff is then taken to the second point and the telescope is directed on to it. After any necessary fine adjustment of bubble, the observer takes the second reading on the staff.

310. Calculation of Difference of Level. Let the difference of level between A and B (Fig. 153) be required. The respective staff readings are 7·24 and 2·01 ft., so that A is 7·24 ft. below the first line of sight, and B is 2·01 ft. below the second. But the two lines of sight lie in the same horizontal plane, any want of coincidence between the two planes, due to relevelling the instrument for the second observation, being insignificant. The staff readings are therefore measurements made vertically downwards from a horizontal plane, and this horizontal plane practically coincides with the level surface through the telescope axis. The difference of 5·23 ft. between the readings is therefore the difference of level between A and B, the smaller reading being observed on the higher point.

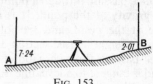

FIG. 153.

Now, let it be assumed that the reduced level of A is known to be 100 ft. above a particular datum, then that of B is obtained as 105·23 ft. above the same datum by application of the difference of level, adding a rise and subtracting a fall. Alternatively, the reduced level of B may be found by referring to that of the lines of sight. By adding the first reading to the reduced level of A, the instrument height, or height of collimation as it is termed, is found to be 107·24 ft. But B is 2·01 ft. below this level, and its reduced level is obtained as 105·23 ft. by subtraction.

311. Series Levelling. The more general case occurs when the two points to be compared are so situated, by reason of their distance apart, their difference of level, or the intervention of obstacles, that it is impossible from any one instrument station to read a staff held successively upon them. In these circumstances, the work is performed in a series of stages, to each of which the previous method is applied. Thus, in Fig. 154, the difference of level between A and D is determined by observing that from A to a convenient point B, and then proceeding similarly from B to C, and from C to D.

The instrument is therefore first set up in such a position as 1, from which a staff held on A can be read and a clear forward view obtained. When the sight on A has been taken, the staffman, proceeding up the slope, selects a firm point B on which to hold the staff. The first stage is completed by noting the reading on B, and the instrument is then transferred to position 2, the staff meantime being kept on B. When the instrument is levelled, the staff on B is again sighted, so that the level of B may be compared with that of a convenient point C in the same manner as before. A third step suffices to reach the point D.

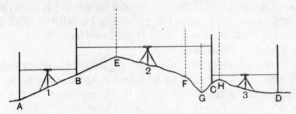

FIG. 154.

It will be evident that the essential feature of the system lies in the observation of two staff readings on each of the points B and C, one before, and the other after, moving the instrument. When the instrument is being shifted, the staff must not be moved, and, while the staff is being carried forward, the instrument must remain stationary.

Points such as B and C, on which the staff is held to permit the transfer of the instrument, are called *change points* or *turning points*. It is unnecessary that they should lie in the line AD: the essential requirement is that they should be on unyielding ground, to avoid any risk of settlement occurring between the two observations (section 343).

312. Backsights and Foresights. The word 'sight' is used to denote either an observation or the resulting reading.

A *Backsight* is the first sight taken after setting up the instrument in any position.

A *Foresight* is the last sight taken before moving the instrument.

The first sight on each change point is therefore a foresight, and the second a backsight: every line of levels must commence with a backsight and finish with a foresight. In Fig. 154, the sights 1–A, 2–B and 3–C are backsights, while 1–B, 2–C and 3–D are foresights. The two sets of sights are distinguished because they are differently applied in the calculation of the levels. A backsight is always taken on a point the reduced level of which is known or can be computed. By addition of the backsight to that level, the instrument height can be obtained. A foresight is always taken on a point of unknown level with the object of ascertaining its level by subtraction of the foresight from the known instrument height, and then using the result for continuing the line.

Having regard to the setting sensitivity of the bubble (or automatic device) of an ordinary Level, and to the ability of the observer to read the staff, the maximum sighting distance is of the order of 400 feet. However, in practice the sights will usually be much more closely restricted by the requirements of the survey and by the slope of the ground: if the instrument height is 5 ft. and the staff is 14 ft. long, and the slope of the ground is 1 in 10, the sight on the up-slope side must be less than 50 ft. and on the down-slope side it must be less than 90 ft. long.

The terms, backsight and foresight, are unfortunate because they imply direction. The sights may be taken in any direction, the terms having no other significance than that contained in the above definitions.

313. Intermediate Sights. When the backsight has been taken, the observer is in a position to determine the level not only of the next change point, but of any number of points within range. Thus, in the course of the operations of Fig. 154, it may be desired to obtain the elevations of certain points E, F, G and H. The staffman on his way from B to C holds the staff on E, F and G successively, and the leveller takes a single sight on each. The point H is observed from the next instrument station. Sights taken between a backsight and a foresight to ascertain the levels of points are called *Intermediate Sights.*

314. Datum. To define the relative altitudes of a series of points, it is sufficient to ascertain their elevations above any one datum surface. A reduced level for one of the points may therefore be assumed arbitrarily, and those of the other points are deduced from it. Many advantages, however, accrue from the adoption, throughout a country, of a standard datum of reduction. Mean Sea Level, as ascertained by prolonged observation, affords a universal datum, and is that most generally chosen.

The standard datum of Great Britain is that adopted by the Ordnance Survey. It was originally the mean level of the sea at Liverpool. Later, this site was considered to be unsuitable for sea-level determinations, and the datum now adopted is mean-sea-level at Newlyn, in Cornwall, as found from continuous observations during the period 1915–21.

315. Benchmarks. A benchmark is a fixed point of reference, the elevation of which is known. In mapped countries, benchmarks are established at intervals throughout the country by the appropriate national organisation, and their positions and their elevations above the standard datum are published. Surveyors have thus a ready means of expressing reduced levels in terms of standard datum by commencing a series of observations with a reading of the staff held on a benchmark.

Standard benchmarks take various forms. In Britain, Ordnance benchmarks are mostly of the type shown in Fig. 155, and these are chiselled on buildings and other permanent structures: the height is referred to the centre of the horizontal groove. Some benchmarks are on bronze tablets let into walls, etc.

FIG. 155.—ORDNANCE BENCH MARK.

Special brackets are obtainable to hold a staff at the correct height on an Ordnance benchmark.

The positions of benchmarks are shown on the six-inch and larger scale Ordnance Survey maps, and their heights can be obtained on application to the Director-General, Ordnance Survey, Romsey Road, Maybush, Southampton.

A surveyor may have occasion to establish benchmarks for his own use. Thus, in the construction of engineering works, repeated levelling between the nearest Ordnance benchmark and the site of the work may be obviated by carefully determining at the outset the elevations of a number of permanent points at the site, these being referred to as required. When levelling in un-mapped country, the surveyor should leave marks on rocks, etc., at intervals, for convenience in connecting on to his assumed datum at a future date.

316. Level Book. Levelling work must be recorded in a book ruled in suitable lines and columns, otherwise it can soon become quite untidy and unintelligible. A good arrangement is to have columns for readings and reductions on the left page, and keep the right page for notes and sketches. There should be separate columns for backsights, intermediate sights and foresights. Additional columns will normally be required for station identification, arithmetical working and, in many kinds of work, measured distances.

The object of the arithmetical work is to get heights of points from staff readings, and there are two recognised procedures, called *Height of Instrument Method* and *Rise and Fall Method:* the latter is probably the more generally used nowadays.

317. Height of Instrument Method. This method is illustrated by the working shown in Fig. 156. The staff readings are in the first three columns, and were taken as follows: staff held at A and reading 6·38 noted, staff moved to B and reading 1·17 noted, Level moved forward while staff remained at B and reading 5·97 noted, staff taken to C, D, E, F and readings noted, the 8·22 being the last from this instrument position, then Level moved, and so on.

Page .....   LEVEL BOOK

STAFF READINGS						
BACK	INTER	FORE	STAFF STATION	HEIGHT OF INST.	REDUCED LEVEL	DISTANCE
6·38			A	56·38	50·00	
5·97		1·17	B	61·18	55·21	
	2·10		C		59·08	
	6·35		D		54·83	
	10·20		E		50·98	
1·53		8·22	F	54·49	52·96	
	2·90		G		50·69	
		3·76	H		50·73	
13·88 13·15	13·15	13·15			0·73	
0·73						

FIG. 156.

It is to be noted that each horizontal line in the book refers to a particular staff position, thus, the readings 5·97 and 1·17 were both taken when the staff was at B. No distances are recorded, as the levelling was done solely for finding the heights of points.

The arithmetical working in the fifth and sixth columns follows exactly the geometry of the levelling process. Assuming the height of point A as 50·00, the reading 6·38 shows that the line of sight of the instrument was at height 56·38, then the reading 1·17 shows that B was 1·17 below the line of sight, that is the height of B was 56·38 − 1·17 or 55·21; and so on. The reduced heights are entered in the sixth column. Heights of intermediate points must of course be calculated from the height of instrument as determined from the previous backsight; thus, the height of D is 61·18 − 6·35 = 54·83.

318. Rise and Fall Method. The same staff readings are entered in Fig. 157, and reduced by the Rise and Fall method. The principle of

Page.....   **LEVEL BOOK**

STAFF READINGS							
BACK	INTER	FORE	STAFF STATION	RISE	FALL	REDUCED LEVEL	DISTANCE
6·38			A			50·00	
5·97		1·17	B	5·21		55·21	0
	2·10		C	3·87		59·08	50
	6·35		D		4·25	54·83	100
	10·20		E		3·85	50·98	150
1·53		8·22	F	1·98		52·96	180
	2·90		G		1·37	51·59	200
		3·76	H		0·86	50·73	245
13·88		13·15		11·06 10·33	10·33	0·73	
13·15 ·73				0·73			

FIG. 157.

this method is simply that two consecutive readings from the *same* instrument position give the difference of height of the two points where the staff was held. Thus, for instance, the difference of height between D and E is 10·20 − 6·35 = 3·85: and this is a *fall* from D to E because the second reading is greater than the first.

The rises and falls are entered in the appropriate columns, then they are successively added or subtracted, beginning with the initial point, the height of which must be taken as known.

The distances entered in the 8th column are information needed if the levelling is for a longitudinal section; they are cumulative distances along the route of the staff stations B to H and do not refer to positions of the instrument. In ordinary levelling, distances from instrument to staff are not normally of significance, but, in accordance with the principle mentioned in section 309, the surveyor should try to set up the instrument so that the distances to the *back* and *fore* staff positions are approximately equal. Thus, at the second set-up, the staff positions B and F should be about the same distances from the Level; the Surveyor may check this by taking stadia line readings.

319. Arithmetical Checks. Along a line of levelling, the height is carried forward on the principle that the change of height recorded at each set-up of the Level is equal to *backsight minus foresight*, as mentioned in section 310. Therefore the total change along the line is equal to *total of backsights minus total of foresights*, and this relation provides a simple check on the arithmetical work. These checks are shown on Figs. 156 and 157. An advantage of the Rise and Fall method is that three numbers should agree, as shown.

The arithmetic checks should always be applied, but it must be emphasised that *the checks only test the computer's arithmetic, they do not check the observer's accuracy.* If it is found that the checks are not satisfied, there must be an *arithmetical* mistake, and nothing further should be done until the mistake has been found and rectified.

The only check on observational accuracy is close agreement when levelling is done between two points where the heights are already known; otherwise, it is essential to run check levelling, preferably from different instrument stations and by another observer, if any confidence is to be had in the results.

320. Comparative Merits of Reduction Methods. The instrument height system is economical of figuring, and proves the more rapid method. It is well adapted for reduction in the field, particularly in setting out levels for constructional work. It is open to the objection that a mistake in an intermediate reduction may pass unnoticed. This disadvantage is entirely absent in the rise and fall method, which affords a complete check, only vitiated if mistakes balance. By reason of its greater certainty, the use of the latter method is sometimes insisted upon in important work. Note that the mistake in Fig. 156 would have been detected if the Rise and Fall method had been used.

To minimise the waste of time occasioned by an arithmetical slip, rises and falls should be checked before reduced levels are filled in. With the same object, instrument heights and levels of change points may be worked out and checked before reducing intermediate readings.

FIELD WORK

321. Ordinary levelling operations are generally for one or more of the following purposes:
(1) Determining the difference of level of two points.
(2) Running Sections.
(3) Locating Contours (see Chapter VIII).
(4) Setting out or giving levels for construction purposes (see Chapter X).

322. Determination of Differences of Height. This information may be required for the design of proposed works, or the levelling may be done to establish the heights of benchmarks for reference during construction. In such levelling, there will normally be no intermediate points; the shortest convenient route should be followed, and the sights may be as long as the nature of the ground will allow and the staff readings can be clearly seen. The surveyor should take some care to equalise back- and foresights.

Even if one point is vertically above the other, it will often be convenient to determine the difference of height by level and staff.

323. Sectioning. The running of a section is the most common levelling operation, and consists in obtaining a record of the undulations of the ground surface along a particular line, straight or curved, so that they may be represented to scale. This involves observing not only the elevations of a number of points on the

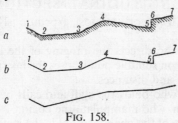

Fig. 158.

line, but also their distances along it. It is important that the points selected should be those at which the inclination of the ground surface sensibly changes, so that, having plotted them, it is justifiable to assume a uniform slope between each consecutive pair. Thus, in Fig. 158, *a* represents the actual undulations of a piece of ground: if the levels and chainages of the salient points numbered are observed and plotted, a good representation of the slopes is obtained even though the points are joined by straight lines as in *b*. Fig. *c* shows the serious misrepresentation produced by omission to observe points 3 and 5. On the other hand, if, as in running a section along a road or a railway, the points of change of gradient are not evident, it is best to observe equidistant points, say a chain length apart.

324. Longitudinal and Cross Sections. Sections are of two kinds: (*a*) Longitudinal Sections, or Profiles; (*b*) Cross Sections.

A longitudinal section is one which follows some predetermined line, usually the centre line of proposed work, *e.g.* a road, railway, canal, or pipe-line. By means of such sections the engineer is enabled to study the relationship between the existing ground surface and the levels of the new work in the direction of its length. In the design of works, it is frequently necessary to run longitudinal sections along various proposed centre lines and compare the costs of the several schemes.

Since a plotted profile cannot exhibit any particulars as to the character of the ground on either side of the centre line, it does not convey sufficient information for the complete design of works. In the case of those which occupy a very narrow strip of ground, *e.g.* a pipe-line, it will, however, be sufficient to assume that the ground within the limits of the width of the work is level in a direction normal to the centre line. Otherwise, additional information is required, and this is conveniently obtained by means of cross sections, which are sections taken at right angles to the centre line and of sufficient length to embrace the limits of the work on either side.

LONGITUDINAL SECTIONS

325. Field Party. The work is most expeditiously performed by a party of four, *viz.*:

The Leveller, who directs the ranging of the line, sees that the staff is being held on suitable points, uses the instrument, and books the staff readings and distances.

The Staffman, who holds the staff and calls out the chainages.

Two Chainmen, who manipulate the chain.

Commonly one of the chainmen holds the staff, the leveller having to wait while the chain is being pulled forward. If only one assistant is available, much time is wasted by the surveyor having to leave the instrument in order to take part in the chaining.

It may be found economical in some circumstances, such as where there are numerous intermediate points, to have two staffs.

326. Equipment. *Essential.* Level, Staff, Ranging Poles, Chain or Band, Arrows, Chalk, Level Book, and Pencil.

Optional. Tape and Plumb Line for stepping.

327. Running the Section. If necessary, the line of the section must first be set out by locating a sufficient number of points to define the straights and curves, intermediate poles being ranged by eye. In a long section only a sufficient length should be ranged in advance as will occupy, say, half a day to level.

It frequently happens that the section is run along the route of a traverse survey. If the traversing has been completed previously, it is necessary to recover the stations by reference to the notes describing their positions, and the lines must be ranged afresh. In mapped country it may not be necessary to run a traverse, and the line of the required section is drawn on a map, from which it has to be set out by the levelling party. Its position relatively to buildings, fences, and other definite features represented on the map is obtained by scaling, and the scaled dimensions are reproduced on the ground. In cases where a number of points so located are intended to lie on a straight line, it frequently happens that they fail to do so exactly on account of errors of scaling and of measurement, but it is usually quite sufficient to range a straight line by eye in such a position as will average out the irregularities. In districts where there is a scarcity of well-defined detail the transfer of the section line from the map to the ground is sometimes troublesome, and it becomes necessary to set it out by theodolite.

To refer the levels to a standard datum, levelling should start from an established national bench mark, and the position and value of one near the beginning of the section should be obtained before proceeding to the ground. Levelling is conducted from the bench mark, by observing backsights and foresights only, until the instrument can be placed in a position from which the first part of the section can be commanded.

Chaining is now commenced, and the staffman proceeds along the chain, and holds the staff at all points of change of slope. After each reading is taken, he calls out the chainage of the point to the leveller, or he may enter them in a notebook, from which they are transcribed into the 'distance' column of the level book at frequent intervals. When the length of sight reaches the maximum permissible for good reading (section 344), or when inequalities of the ground prevent the taking of further intermediate sights, the leveller signals the staffman that he wishes a foresight. The latter selects a firm change point, either on or off the chain line, marks it with chalk, and holds the staff upon it, taking due care as to verticality.

Having entered the reading, the leveller carries the instrument forward. He selects his next instrument station on firm ground, and in such a position that he can sight back on the staff, and also obtain an unobstructed forward view along the section. When the backsight observation is completed, the staffman proceeds to give intermediate sights as before, and this routine is followed to the end of the section.

As the various features lying on the section line are reached, such information regarding them as is likely to prove useful in the design of the proposed works should be acquired. Such items include the levels of the beds of streams, flood water levels, if indications of them can be traced, the chainage at which fences, etc., intersect the line, the names and levels of roads and railways crossed, the headroom of bridges, etc. When the section line passes below a bridge, readings are taken with the staff held inverted against the underside of the girders or arch, these being distinguished in booking by being marked with a plus sign or otherwise. The elevation of the road or rail surface on the top of the bridge is ascertained by levelling up the bank.

The chainage should be continuous from start to finish along the section itself, disregarding positions of the Level.

The use of salient points on the line of section as change points saves time, but accuracy should not be sacrificed by using an unstable place on the line in preference to a firm place on one side. In levelling along very soft ground, the staffman should carry a piece of wood or a flat stone on which to place the staff for a turning point.

Time should not be wasted by excessive refinement in reading at intermediate points; in most levelling work, it suffices to take intermediate readings to the nearest tenth of a foot.

For notes on speed in levelling, see section 349.

328. Checking Levelling. It is impracticable to check every result of a levelling except by a second levelling over all the points. Short of this, a good check is obtained if the levelling line closes on itself or on a benchmark of known reliable height: however, such closure does not check any reading on an intermediate point, but an error at an intermediate point is isolated and cannot be carried forward.

Errors in readings at change points are carried forward, and certain steps can be taken to avoid such errors.

One method is to use double change points: in this way, each instrument height is obtained from two sources, and the two values should agree. The booking of the readings on the subsidiary change points can be done on the 'notes' page of the field book.

Almost all Levels (and theodolites) are provided with stadia lines on the diaphragm, and a check can be obtained by reading all three lines at change points: a common source of error is to read a stadia line instead of the centre line.

On a very long line, it is advisable to mark some of the change points permanently, thus dividing the line into sections so that each section can be checked before the next is levelled.

Another possible check method is to have the back of the staff graduated in a different unit, say metres and decimetres. At change points the readings are taken on both sides and the difference back-fore are compared at each station before the instrument is moved on.

329. Plotting the Profile. When the heights of all the points along the section have been calculated, a profile can be drawn. The usual procedure is to draw a straight line to represent a convenient level well below the heights of all the points on the section; this line is marked off, at a suitable scale, with the chainages measured, and vertical lines are drawn at these points representing the differences of ground height above the reference level. See Fig. 159. A line joining the top points of these ordinates then represents the ground section.

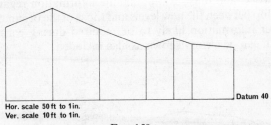

Hor. scale 50 ft to 1 in.
Ver. scale 10 ft to 1 in.
Datum 40

FIG. 159.

Since the horizontal distances involved are in general very much greater than the variations in level, it is usual to plot vertical dimensions to a larger scale than horizontal distances. In this way the irregularities of the ground are made more apparent. The steepness of slopes is exaggerated, and artificial features intersected suffer corresponding distortion. Exaggeration is of value in enabling the relationship between the original surface and the proposed levels of new work to be clearly and accurately shown. The ratio of exaggeration adopted runs from 5 to 15 times and upwards, depending upon:

(a) *The Character of the Ground.* A greater exaggeration is required to exhibit the irregularities of flat ground than those of rough country.

(*b*) *The Horizontal Scale*. The horizontal scale may be chosen arbitrarily, but it is commonly that of the plan upon which the section is drawn. The smaller the horizontal scale, the greater should be the ratio of exaggeration.

(*c*) *The Purpose of the Section*. The vertical scale should be increased in cases where a highly accurate representation of vertical dimensions is required, but there is no point in exaggerating the vertical scale out of relation to the accuracy of the observations. If readings are taken to 1/10 ft., a scale of $\frac{1}{2}$ foot to 1 inch would be silly.

To finish the drawing, the profile and the datum line should be inked up; the ordinate lines should be left in thin pencil or inked in a light colour. Scales should be stated. Other information, as described in the next section, may be added to the drawing.

Printed profile paper can be obtained, suitably ruled for drawing at various horizontal and vertical scales.

330. Working Profile. When the location of an engineering scheme has been decided, and the design made, a working section is prepared for the use of the resident engineer. This profile incorporates the features of the original ground surface, as well as the levels of the new work, and must exhibit definite information regarding the relationship between the new levels and those of the original ground. All further information likely to be required during construction, and which can be clearly shown, is also included.

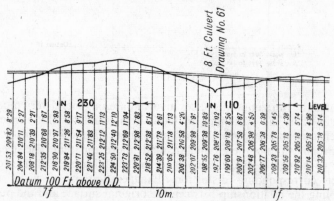

FIG. 160.—WORKING PROFILE OF RAILWAY.

The character of this information depends upon the nature of the work. Part of a working profile for a railway is shown in Fig. 160.

The new work is represented by two parallel lines, the lower, in red, denoting formation or sub-grade level, *i.e.* the surface level of the earthwork, and the upper, usually in blue, representing rail level. Ordinates are drawn, in this case at 1 chain intervals, and the datum line is figured to show distances from the commencement of the railway in miles, furlongs, and chains. The figures written against the ordinates represent original ground level, formation levels, and depth of cutting or bank. Original levels are written in black, formation levels in red, and depths of earthwork in red or blue, according as they refer to excavation or embankment. The gradients of the new work are figured boldly, and the limits of each clearly shown by arrows against ordinates drawn in red. The positions of bridges, culverts, level crossings, etc., as well as brief particulars of existing features crossed by the line of section, are also entered.

CROSS SECTIONS

331. These are usually taken during the progress of the longitudinal section. If the best results are desired, the observations may be made with level and staff, particularly if the cross sections are long. More rapid methods by clinometer or hand level are sufficiently accurate for many purposes, and are especially suitable for short sections, the length of which precludes the accumulation of serious error.

332. Interval between Sections. The purpose of cross sectioning is to furnish the engineer with sufficient information regarding the levels of the ground on either side of the longitudinal section to enable him to design the intended works and compute the quantities of earthwork, etc., involved. To facilitate estimation of the character of the ground between cross sections, they should be taken at every marked change of slope along the longitudinal section, so that it is valid to assume that the ground surface changes uniformly from one section to the next. This desideratum is often neglected in practice, cross sections being commonly taken at constant intervals, either 1 chain or 100 feet.

333. Setting Out. The lines of cross sections are in general perpendicular to the longitudinal section line and radial on curves. When for any reason a cross section is run in another direction, the angle it makes with the longitudinal section must be measured, so that its position may be shown in plan. In sectioning a wide area, *e.g.* for a reservoir or a dock, the cross sections may have a considerable length, and their lines should be set out by theodolite, box-sextant, or optical square. In the case of a narrow strip of ground, *e.g.* for a road or a railway, their perpendicularity is usually judged by eye. Since cross sections must be long enough to include the width of the new work, a sufficient margin must be given in cases where the latter is not known exactly.

334. Cross Sectioning by Level and Staff. The procedure is similar to that followed in longitudinal sectioning, but in the case of short sections the distances are taped. A chainman remains with the tape box at the point on the centre line from which the section is projected, while the staffman takes the ring and proceeds along the section. The chainman guides the staff holder into the perpendicular, and calls out the measurements of the points observed. If the leveller has only one assistant, he should first mark by arrows or twigs the salient points to be levelled, and note the distances on sketches of the sections.

The levelling of cross sections on flat ground is performed from the instrument stations used for the longitudinal section, but on sidelong ground it may be impossible to complete any cross section from one instrument station. In these circumstances, the repeated shifting of the instrument renders the progress very slow, particularly if each cross section is finished before the next is begun. The number of instrument stations may be greatly reduced by sighting from each the several points within range on a number of sections.

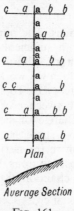

Plan

Average Section

FIG. 161

Thus, in Fig. 161, the points *a* may be observed from the instrument station selected for the sighting of the points *a* of the longitudinal section. On shifting the instrument up the slope, points *b* are levelled, their distances out being measured from marks left on the centre line. Finally, points *c* are observed from a lower position of the instrument.

In booking cross-sections, the staff readings are entered in the appropriate columns in the usual manner. Distances along cross-sections are measured from the main section line and distinguished as 'left' or 'right' in relation to the direction of measurement along the longitudinal section. These cross-section distances should not be entered in the main 'distance' column, and another column may be provided for them, but they may be entered in the 'notes' space of the book. Some typical entries for a cross-section at point E of Fig. 157 are shown in Fig. 162; all these readings were taken from the instrument position after turning-point B.

6·35		D		54·83	100	
10·20		E		50·98	150	
11·8		E₂		49·4	"	20 L
10·7		E₁		50·5	"	40 L
9·5		E₃		51·7	"	20 R
7·9		E₄		53·3	"	50 R
1·53		8·22	F	54·49	52·96	180

FIG. 162.

The plotting of cross sections observed as above is similar to that of profiles, except that, for the purposes of showing and measuring new work, it is more useful in this case to have vertical and horizontal measurements plotted to the same scale. The scale is commonly that used for the vertical dimensions of the profile. It is usual to arrange the cross sections on a sheet in rows on a series of vertical centre lines. The elevations represented by their datum lines may be frequently altered to keep the ordinates reasonably short. To economise room, datum lines and ordinates are sometimes erased, and the reduced level at the centre is written horizontally on the ground line.

335. Cross-sections by Hand Level. It may be quicker to observe cross-sections independently by using a hand-level. The surveyor stands on the main section line and notes the height of the hand-level above the ground; then he takes readings on a staff placed at suitable points on the cross-section, and the readings and measured distances may be recorded on the 'notes' pages of the field book.

The simplest way to plot such sections is to draw a line to represent the line of sight of the hand level and drop perpendiculars from it of lengths representing, to scale, the staff readings.

336. Cross Sectioning by Theodolite. On steep ground the labour of frequently shifting the instrument is obviated by the use of an inclined line of collimation roughly parallel to the ground and of known inclination. This is most accurately given by the theodolite.

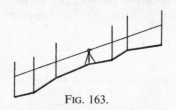

FIG. 163.

In Fig. 163, the theodolite is set over the centre line mark, the level of which is known. Having levelled up, the height of the horizontal axis above the ground is taped and noted. The line of collimation is set roughly parallel to the ground, and the vertical circle is clamped at the nearest whole degree, the angle being booked. The inclinations of the line of collimation on either side of the instrument need not be the same, and may differ considerably on rough ground. Where the general slope of the ground is fairly uniform, however, the value of the angle of elevation on the uphill side is reproduced on the circle as an angle of depression on the down-hill side in order to simplify the plotting a little. The readings of the staff held vertically at the various salient points are recorded. Distances are taped along the line of collimation, the staffman holding the ring of the tape against the staff as nearly at the reading as he can judge, while the measurements are read by the chainman at the horizontal axis of the instrument.

The results of the observations are usually booked in columns, but may be noted on sketches. In the tabular arrangement, the middle column should show the chainage at which the cross section is taken, the height of the telescope axis above the ground, and the elevation of the instrument station, if it is already known. The adjacent columns to right and left are reserved for the vertical angles observed on the corresponding sides of the section, the entries being marked plus or minus, or *e* or *d*, to distinguish elevations and depressions. In the remaining columns on either side are booked the staff readings, with the corresponding distances written below. If sketches are preferred, they may be made on the lines of Fig. 163, with the dimensions entered in place.

To plot the section, the reduced level of the centre point is set up from the datum, and from it is scaled the height of the instrument axis above the ground. From a horizontal line through the point obtained, the line of sight is plotted by protractor. The several distances are then marked off along the line of sight, and from those points verticals equivalent to the staff readings are measured down. A continuous line through their lower ends represents the ground line.

337. Cross Sectioning by Clinometer. The Abney or other type of clinometer may be used in place of the theodolite. This is the most rapid means of cross sectioning, and the results are sufficiently accurate for many purposes. The clinometer is much used in location surveys.

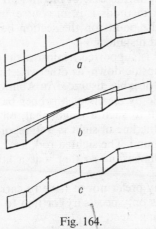

Fig. 164.

Three systems of observation are shown in Fig. 164.

(*a*) The observer, standing over the centre point, holds the clinometer at a known height above the ground with the line of sight roughly parallel to the ground. The slope angle on each side is noted, readings are taken on a levelling staff or ranging pole, and slope distances measured as in using the theodolite.

(*b*) The observer at the centre sights a mark on a ranging pole at the same height as the instrument is held above the ground. The various slope angles and distances are noted.

(*c*) The angle and length of each slope are separately measured, the observer using method (*b*) and proceeding along the section.

Of these methods, the first is the most rapid in observation, booking, and plotting, and is to be preferred unless the ground is so irregular that the required staff readings cannot be obtained from two lines of sight. In this case the other methods are preferable. Method (c) is likely to prove slower than (b); but it is useful if the sections are long, as it avoids lengthy tape measurements. If, in the course of using method (a) or (b), a hump occurs on a section, so that the sighting cannot be completed from the centre point, it becomes necessary to move the instrument to a point on which an observation has already been taken in order to continue the section beyond the summit. Circumstances may warrant the combination of two or more methods on one section.

The booking is performed in columns or with the aid of sketches, as described for theodolite sectioning. Sketches are useful in cases where the instrument has to be moved from the centre station, since the points from which the angles are observed can be indicated clearly. The plotting in the case of method (a) has been described above. In the other methods it is, of course, unnecessary to draw the lines of sight, the points on the section being plotted directly from the angles and distances.

If the clinometer is supported against a ranging pole, it is well to hold the latter upside down, as otherwise the pointed shoe will enter the ground a variable distance. An Abney clinometer may be attached to the pole by means of a rubber band. Alternatively, a wooden rod is sawn to such a length that, on placing the instrument on top of it, the line of sight is 5 ft. or other convenient distance above the ground. The sighted rod, used in methods (b) and (c), is cut to this dimension, so that it is a little longer than the other.

The observer may prefer not to have to carry a support for the clinometer, and it is only necessary for him to know the height of his eye when standing erect.

If a levelling staff is used in method (a), it should have an open, bold graduation. If readings are estimated on a long ranging pole, care is required to avoid mistakes of a whole foot.

338. Setting Out Levels. The levels to which work is to be built may be shown by driving a peg or making a mark either at the desired level or at a stated distance above or below it. In the former case, the required staff reading when the foot of the staff is at the correct level must first be deduced, and the staff is raised or lowered until this reading is obtained. Instrument height booking will be found the more convenient for such work. In the latter method, an arbitrary point is established, and the staff is read upon it, the difference between its level and that of the construction being communicated to the foreman. There is less likelihood of mistakes on the part of the workmen if this distance is an exact number of feet.

The leveller must be constantly on his guard against mistakes, as these may have very serious consequences in setting out. The work must always be well checked against mistakes.

In setting pegs to a required level, time is wasted by driving a peg too far, as it may have to be removed and driven afresh. The latter stages of the driving must be performed with caution, and the reading observed at frequent intervals. The staffman should be told after each reading by how much the peg still requires to be lowered. If, however, a peg is only a small fraction of an inch too low, a nail may be driven in it to the correct level to save time.

Example. An engineer on works is required to give a number of levels. He observes a backsight on a temporary bench mark (TBM) of elevation 297·34, and reads 4·06. He then takes a reading of 5·62 at a change point, and transfers the instrument to a position from which he can see the work. His new backsight is 3·81. He is required (*a*) to give a mark at elevation 295·60, (*b*) to correct a peg which is roughly indicating a height of 5 ft. above the bottom of an excavation to be taken out to level 291·00, (*c*) to check the finished level of different points of a concrete foundation which is intended to be at 293·75. In the second case he reads 3·45 when the staff is held on the peg, and in the last his readings are 5·88, 5·86, 5·85, and 5·89.

The height of instrument from which the required observations are taken is $297·34 + 4·06 - 5·62 + 3·81 = 299·59$.

(*a*) To set out a level of 295·60, the staff must be adjusted until the reading obtained is $299·59 - 295·60 = 3·99$, the mark being made at the bottom of the staff.

(*b*) The elevation of the peg is $299·59 - 3·45 = 296·14$. As the peg is intended to be at 296·00, it must be lowered by 0·14 ft., or until a reading of 3·59 is obtained with the staff held on it.

(*c*) If the foundation were at the correct level of 293·75, each staff reading would be $299·59 - 293·75 = 5·84$. The work is therefore too low by $\frac{1}{2}$ in., $\frac{1}{4}$ in., $\frac{1}{8}$ in. and $\frac{5}{8}$ in. at the respective points.

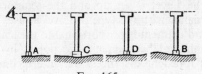

FIG. 165.

Levels given in this way can be transferred within a limited distance by the workmen by the use of a straight edge and a small spirit level. This is also accomplished by means of T-shaped crossheads or boning rods, which furnish a line of sight whereby, from two given pegs, points at the same level or on the same gradient may be established in their line. In Fig. 165, A and B are pegs set on a particular gradient and, say, 100 ft. apart. The foreman, holding a boning rod on A and looking along the top of it at the top of a similar rod held on B, can direct the adjustment of points such as C and D into the gradient by judging when the upper surfaces of the boning rods held upon them are in the line of sight.

339. Setting Slope Stakes. See section 463.

SOURCES OF ERROR IN
ORDINARY LEVELLING

340. Numerous sources of error may affect the accuracy of a line of levels, but the precautions against them are of a simple nature, so that it is not difficult to obtain good results without delaying progress. It is to be understood that the precautions detailed below refer more particularly to the observation of backsights and foresights, since an error introduced at a change point is carried forward through the subsequent work. An error in an intermediate sight affects the recorded level of that point only, and may not prove of much consequence; although it should be recognised that some intermediate sights may be of great importance, especially in engineering work. The various errors and mistakes may be classified as:

    (1) Instrumental Errors.
    (2) Errors and Mistakes in Manipulation.
    (3) Errors due to Displacement of Level and Staff.
    (4) Errors and Mistakes in Reading.
    (5) Mistakes in Booking.
    (6) Errors due to Natural Causes.

341. Instrumental Errors. *The Level.* The testing and adjustment of the instrument have been discussed in Chapter I. The important desideratum is that the line of collimation should be exactly horizontal when the Level is set up and the temporary adjustments made in accordance with the type of Level in use. The error introduced by non-adjustment is proportional to the length of sight, and is entirely eliminated between change points by equality of the backsight and foresight distances, but intermediate sights, being usually of various lengths, will be thrown into error by different amounts. Errors arising from imperfect estimation of the equality of backsight and foresight distances are generally compensating, but if the backsights are consistently longer or shorter than the foresights — a tendency to be guarded against on steep slopes — the error becomes cumulative.

A defective bubble tube may have a considerable influence. If under-sensitive, the bubble may apparently come to rest in the central position although the tube axis is not horizontal, thus giving rise to a compensating type of error. On the other hand, over-sensitiveness in an instrument for ordinary use leads to waste of time in levelling up. Irregularity of curvature of the tube is a serious defect, and a bubble found to be in such condition should be replaced as soon as possible.

The tripod should be examined and loose joints tightened, as instability of the instrument causes waste of time, and leads to erroneous readings.

The Staff. It is advisable to test the graduation of a new staff by a steel tape or a good graduated scale, but the error is likely to be negligible in ordinary work. There is greater probability of error through wear of the staff at the joints or by dirt adhering to hinges or sockets. A telescopic staff should be let down gently to minimise wear, and the ends of the separate pieces of a socketed staff should be kept clean to enable them to be pushed firmly home in the sockets. It is also important that the separate parts should all belong to one staff, otherwise very serious errors of graduation may occur at the joints.

It may be noticed that wear at the bottom of a staff is of no consequence since it is unnecessary that the zero of the graduation should be placed at the foot of the staff in order that differences of staff readings may represent differences of level. An exception occurs, however, in obtaining the difference of level between a point below the plane of sight and one above, the staff being held inverted on the latter point. In this case, the difference of level is the sum of the two staff readings, and an error would be produced equal to twice the distance between the zero of the graduation and the foot of the staff.

342. Errors and Mistakes in Manipulation. *The Level.* When the instrument is not of the automatic type, the most serious and common mistake in observing is the omission to have the bubble central at the instant of sighting. In the instruments such as that illustrated in Fig. 65, the observer obtains, by reflection, a view of the bubble while he is reading the staff. In using a level without a reflector, he should examine the bubble before sighting, and bring it central if necessary. After reading, he should again glance at the bubble, and, if correct, it may be assumed that it remained so during the observation. After the instrument is levelled, it should not be handled unnecessarily. The tripod should not be grasped, and, in turning the focusing screw or orienting the telescope, the application of vertical pressure should be avoided.

The Staff. The staff should be held quite vertical. If held off the plumb, it will be intersected by the line of collimation farther from the foot than it should be, and the reading will be too great (Fig. 166). As the errors caused by a given deviation from the vertical are proportional to the readings, special care must be taken with large readings. Errors are avoided, (*a*) by having a spirit level or a pendulum plumb bob attached to the staff, to facilitate holding it plumb, or (*b*) by swinging or waving the staff.

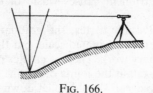

FIG. 166.

The latter is a useful and simple method. The staffman, holding the staff on the point in the ordinary manner, inclines it slowly towards and away from the instrument, on both sides of the vertical. The observer sees the reading vary against the horizontal hair, but the smallest reading corresponds to the vertical position, and that is noted. Waving should be performed in the direction of the line of sight only, as the leveller can detect lateral non-verticality by means of the vertical line or lines on the diaphragm. It is unnecessary and inadvisable to swing the staff if the reading is below about 3 ft., since the bottom of the graduated face is raised appreciably off the ground when the staff is leaning away from the instrument. This difficulty could be overcome by fitting the bottom of the staff with a knife edge or pin placed in the plane of the graduated face, but such a design, although convenient for a precise levelling staff supported on a peg or a plate, is unsuitable for use in ordinary levelling.

Errors due to non-verticality of the staff tend to compensate at change points, but, if the backsight readings are consistently greater or smaller than the foresights, the error becomes cumulative. Thus, in levelling up a slope the observer will read well up the staff in taking backsights and near the bottom for foresights. Careless staff holding increases the former without appreciably affecting the latter, and, in consequence, too great a rise is recorded between change points, and the slope appears steeper than it really is. In levelling downhill, the foresights are the larger readings, and their increases makes the fall between change points appear too great, so that the slope is again exaggerated. In levelling over a hill, therefore, the error accumulated in working up one side is more or less completely neutralised in descending the other side, and the levels may check at the finish, but the deduced elevation of the hill is too great.

A further precaution to be observed by the staffman is to see that dirt does not accumulate on the foot of the staff, as this would cause a variable relationship between the zero of the graduation and the foot of the staff.

343. Errors due to Displacement of Level and Staff. *The Level.* If the instrument is set up on soft ground it may gradually settle from the moment of the backsight observation to that of the fore-sight. This will always make the foresight reading smaller than it should be, giving too great a rise or too small a fall between change points. The error is cumulative, as every settlement of the instrument increases the reduced level of all subsequently observed points by the amount of the sinkage. It follows that the level should, as far as possible, be placed upon solid ground with the legs thrust firmly into the ground, and that time should not be wasted between the backsight and foresight observations. If the engineer must plant the instrument upon staging, he should avoid treading on the planks which support it.

If the level is disturbed by the tripod being accidentally kicked, the mishap will be noticed, and no error need result. If the positions of change points have been marked, it is only necessary to relevel the instrument and again backsight on the last change point, substituting the new reading for the previous one. Any intermediate points taken prior to the dislevelment must have their readings correspondingly altered, or may be observed anew. If change points are not marked, it is necessary to return to the start or to the first change point which can be identified with certainty.

FIG. 167.

The Staff. A serious and common error is that occasioned by change of level of the staff at a change point while the instrument is being carried forward. It is commonly caused by choosing unsuitable turning points. Soft ground should be avoided owing to the probability of the staff sinking between the foresight and backsight observations. A flat stone embedded firmly in the ground makes a good support. If only irregular or rounded boulders are available, the staff should be held on the highest point as at *a* (Fig. 167): if held as at *b*, it is difficult to maintain the foot on the point while turning the staff to face the new instrument station. The use of a peg or a foot-plate as a support in soft ground prevents sinkage. Having selected a suitable point, the staffman should first mark the spot, and should keep holding the staff upon it until the backsight observation is completed.

Since any change of level of the staff will nearly always be in the direction of sinkage, the error is cumulative. The backsight reading on the settled staff will be too great, and the reduced levels of all subsequently observed points made too high.

344. Errors and Mistakes in Reading. Small compensating errors occur in the estimation of the decimal part of the readings. The increased size of the image makes estimation easier with short sights than with long ones, and it is desirable that important sights should not exceed about 300 ft., but this limit depends upon the quality of the telescope as regards resolving power, and also upon the character of the staff graduation and the clearness of the atmosphere (see section 312). Focusing must be carefully performed to eliminate parallax (section 28). The observer should keep moving his head up and down while sighting, and should adjust the focusing screw until no apparent movement of the horizontal line relatively to the staff can be detected.

In sighting an openly graduated staff, it is sometimes difficult to choose between two possible readings differing by 0·01 ft. The smaller should be preferred owing to the possibility of non-verticality, especially if the staff is not swung.

Common mistakes made by beginners are: (*a*) reading upwards, instead of downwards; (*b*) reading against a stadia line; (*c*) concentrating the attention on the decimal part of the reading, and noting the whole feet wrongly; (*d*) omitting the zero from decimals under 0·10; (*e*) reading downwards, instead of upwards, when the staff is inverted.

345. Mistakes in Booking. These include: (*a*) entering a reading in the wrong column; (*b*) omitting an entry; (*c*) noting a reading with the digits interchanged; (*d*) entering the wrong distance or remarks opposite a reading.

A fruitful source of erroneous booking occurs when the end of a line is reached and check levelling is to be carried back along it, as the last change point may be used twice. Thus, in Fig. 168, from A, the last position of the instrument for the forward levelling, the reading on the change point B is first observed, and entered as a backsight. When the intermediate readings on C, D, and E have been noted, the reading on B is again observed, and must be entered as a foresight. Uncertainty is avoided by keeping in mind the definitions of the terms, backsight and foresight (section 312).

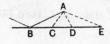

FIG. 168.

To detect mistakes in writing down readings, the best method is to read the staff, book the reading, and then sight the staff again to see that the figure booked is the correct reading.

346. Errors due to Natural Causes. *Wind and Sun.* (See section 237.) If levelling must be performed in a high wind, an endeavour should be made to shelter the instrument, and high readings should be avoided owing to the difficulty of holding a long staff sufficiently steady and plumb.

On hot sunny days the apparent vibration of the staff caused by irregular refraction makes close reading impossible, and, as a partial remedy, the lengths of important sights should be reduced. Distortion of the instrument by unequal heating and expansion or contraction of the staff produce errors which are negligible in ordinary work.

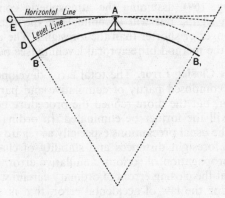

FIG. 169.

Curvature and Refraction. In consequence of the curvature of the Earth, the point read on the staff is not strictly at the same level as the horizontal line of the reticule, since the line of sight is not a level line. In the observation from A to a staff BC (Fig. 169), a difference CD is developed between the horizontal and level lines through the instrument. If the line of sight coincides with AC, the graduation C observed is the distance CD above the instrument height, and the reduced level of B is made out to be lower than it really is, by this amount.

Denoting the length of sight by L, then by geometry,

$$CD = \frac{L^2}{\text{Diameter of Earth.}}$$

If L is in miles, CD = 0·667 L^2 ft. (practically $\frac{2}{3}L^2$)

Actually, the line of sight is not straight, but, in consequence of the refraction of light in passing through layers of air of different densities, is, in general, a curve concave towards the ground. It is represented by AE, so that the graduation at E is that actually read. Under normal atmospheric conditions, arc AE may be taken as circular and of radius seven times that of the Earth. The effect of refraction is therefore $\frac{1}{7}$ that of curvature, but is of opposite sign, so that the combined error,

$$ED = 0·57 \ L^2 \text{ ft.,}$$

by which amount the point occupied by the staff is made out lower than it really is.

For ordinary lengths of sights the error is very small. It is eliminated by equality of backsights and foresights, *e.g.* between B and B_1 (Fig. 169), assuming the atmospheric conditions the same for both sights, but will accumulate if the backsights are consistently longer or shorter than the foresights. The error is also eliminated by the method of reciprocal levelling (see below).

347. Allowable Closing Error. The total error developed in a line of levelling is composed partly of cumulative and partly of compensating error, but the more refined the procedure becomes, the more nearly will the former be eliminated. In ordinary levelling, adherence to the usual precautions especially as regards equality of backsight and foresight distances and stability of change points, prevents the propagation of serious cumulative error, and experience shows that the closing errors of ordinary careful work may be taken as obeying the law of accidental error, that is, the liability to error is proportional to the square root of the number of instrument stations. As the number of stations per mile will not vary greatly for a particular kind of country, the estimate of error E in feet, developed in a distance of M miles may be represented as

$$E = C\sqrt{M}$$

where C is a factor depending on such circumstances as the observer's experience, the quality of the instruments, the character of the ground, and the atmospheric conditions.

For ordinary levelling on moderately flat ground, $E = 0.05\sqrt{M}$ represents good work and is not difficult to attain in reasonable conditions, while $E = 0.10\sqrt{M}$ is satisfactory on rough ground and is quite sufficient for most purposes.

348. Reciprocal Levelling. When the difference of level between two points has to be determined under conditions necessitating considerable inequality between the sights, the effects of collimation error, as well as of curvature and refraction, may be eliminated by reciprocal levelling. The routine involves two sets of observations yielding two erroneous differences of level, the mean of which is the true result.

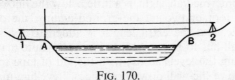

FIG. 170.

Thus, to ascertain the difference of level between A and B (Fig. 170), on opposite banks of a wide river and remote from a bridge, readings on a staff held on each point are taken from instrument station 1. The instrument is then transferred to position 2, so that 2B is equal to 1A, and the staff is again observed on A and B. If, due to the combined effect of instrumental error, curvature, and refraction on the long and short sights, the difference of level between A and B, as determined from one of the instrument stations, is made too great, the other determination evidently makes it too small by the same amount. The mean of the two differences of level so obtained is therefore the required difference.

349. Expedition in Levelling. The rate of progress possible depends greatly upon the character of the ground. If a line of levels intersects deep wooded gorges, the necessity for numerous instrument stations on the slopes, combined with difficulty of sighting, makes progress much slower than in open country. Attention to the following items prevents unnecessary delay.

(1) The surveyor should endeavour to select instrument stations from which he will be able to command as much ground as possible, and in particular should avoid setting the instrument too high or too low to read the backsight.

(2) To avoid delays arising from mistakes on the part of an inexperienced staffman, he should be warned of the importance of his share of the work, particularly with regard to change points,

and should be instructed not to remove the staff from a point until signalled to do so.

(3) Misunderstandings are largely obviated by using a code of signals, such as:

A quick upward movement of the right hand – Observation completed.

Both hands above the head – Hold on a change point.

Right hand waved up and down – Swing staff.

Right (left) hand up and moved to left (right), and left (right) hand down and moved to right (left) – Plumb staff as indicated.

Right (left) arm extended – Move staff to right (left).

(4) It is sometimes permissible to adopt expedients which are not allowable when the best results are required. Thus, when the staff, held for an intermediate sight, is a little below the line of sight, it can be raised a foot or two off the ground against the divisions of a ranging pole, and the extra length is added to the staff reading. If a high wall has to be crossed, a circuitous route may be avoided by continuing the levels across with the aid of tape measurements from the top of the wall down both sides. Levelling may be carried across a sheet of water by taking advantage of the fact that the surface of still water is level. The results are quite good if the obstruction is a still pond or lake, but appreciable error may be introduced in the case of a river, unless the points on the water surface are directly opposite and on a straight reach with a symmetrical channel.

EXAMPLES

(1) A reading taken on a levelling staff is 9·18, but the staff was held at 5° off vertical. What would be the true reading if the staff were held vertically?

(2) An account of the observations taken by a Leveller runs: Staff held on benchmark of height 273·17, staff reading 4·08: staff moved to picket (1), reading 9·67: instrument moved: back reading on staff at picket (1) 5·36: reading on staff placed at intermediate position 7·04: reading on staff held on TBM 8·45. Write out these observations in level-book form and calculate the heights of the intermediate position and the TBM.

(3) Write out the level readings given below, calculate and check the results, and find the height of the TBM:

Backsight	Foresight	Notes
7·61		on benchmark 86·55
9·62	2·07	
9·09	5·55	
6·37	4·13	on TBM
4·29	6·18	
	7·58	on benchmark 98·06

(4) Readings on a staff placed at two points A and B, 3 chains apart, are 6·97 and 3·22 respectively. What should the staff readings be at points on the line AB on a uniform gradient, at 1 chain and 2 chains from A?

(5) Two points are 56·23 metres apart, and readings on a staff placed on them are 1·55 m. and 3·89 m. Express the slope of the line between the points, as a fractional gradient, and as an angle.

(6) Staff readings taken at 50-foot intervals along the line for a trench to be dug were 11·6, 9·2, 7·2, 5·9, 4·9 and 3·7. The bottom of the trench is to be at uniform gradient, depth 3 ft. at the first point and 4 ft. at the other end. Find the depth of digging at the other four points levelled.

(7) A surveyor sets up a tilting Level midway between two pickets that are 180 ft. apart, and obtains staff readings of 5·816 at picket A and 4·224 at picket B. He then moves the Level to a point 20 ft. beyond picket B and thus 200 ft. from picket A. The staff readings are then 6·440 and 4·866. What should the surveyor now do in order to make the line of sight truly horizontal?

(8) A surveyor sets up a Level midway between two pickets A and B which are 170 ft. apart and obtains staff readings on them 4·334 and 5·108 respectively. He then moves the Level to a point 20 ft. from A and 150 ft. from B, and gets readings 4·962 and 5·790 What is the inclination of the line of sight of the telescope?

(9) The following readings were taken on a staff placed on a benchmark and then at three points A, B, C down a nearly uniform slope:

Backsight	Inter. Sight	Distance (ft.)	Notes
2·07			benchmark 286·53
	2·02	0	point A
	6·94	150	point B
	11·85	300	point C

Estimate the positions of the 285-ft. and 280-ft. contours on the line ABC.

(10) A surveyor is using a Level in which the line of sight is depressed 58 seconds below horizontal when the bubble is central. In levelling up a slope, he makes his backsights 30 m. long but his foresights average only 20 m. What error will be accumulated in a kilometre of levelling?

REFERENCES

HALLIDAY E. X. 'An Experiment in Hydrostatic Levelling'. *Survey Review,* Vol. XX, No. 152, Apr. 1969.

HERBERT H. R. 'The design and application of a new instrument for levelling'. *R.I.C.S. Journal,* Apr. 1955.

HOGG F. B. R. AND ARMSTRONG J. A. 'Two new self-aligning Levels'. *Survey Review,* Vol. XV, No. III, Jan. 1959.

KARREN R. J. 'Recent studies of leveling instrumentation and procedures'. *Surveying & Mapping,* Vol. XXIV, No. 3, 1964.

SCHELLENS D. F. 'Design and application of automatic Levels'. *Canadian Surveyor*, June 1965.

STEYNBERG C. A. 'Levelling from island to island in the South China Sea'. *Survey Review*, No. 127, Jan. 1963.

WAALEWIJN Ir A. 'Hydrostatic Levelling in the Netherlands'. *Survey Review*, Nos. 131 and 132, Jan. and Apr. 1964.

WILLIAMS J. W. 'Level transfers across water gaps by trigonometrical methods'. *Conference of Commonwealth Survey Officers*, 1967.

CHAPTER VII

PLANE-TABLE SURVEYING

350. The distinctive feature of Plane-tabling is that it is a method in which the field observations and plotting proceed simultaneously. It has been extensively employed for recording topography in national mapping and in engineering surveys. For topographical mapping, this method has now been largely superseded by air-photography.

In certain circumstances, however, the plane-table can be a very convenient instrument. The preparation of a sketch map, the rapid survey of a small area, the amendment of existing plans, are some jobs that can be done very expeditiously with its aid. The surveyor who knows how to use a plane-table does not have to wait for elaborate arrangements to be made for other kinds of survey: he produces the result at the time, on the spot, and in most cases without needing any assistance. He has only to wait for fine weather.

THE PLANE-TABLE AND
ACCESSORY INSTRUMENTS

351. The Plane-table (Figs. 171 and 172) is a drawing-board mounted on a tripod so that it can be (*a*) levelled, and (*b*) rotated about a vertical axis and clamped in any position. An essential item of equipment is a Sight-rule or Alidade, which is a straight-edge carrying sighting vanes, so that it can be aligned on visible objects that are marked or are to be fixed on the plan. The plane-tabling equipment is described in detail below.

352. The Board. Ranging in size from about 15 ins. × 15 ins. to perhaps 30 ins. × 24 ins., this is made of thoroughly seasoned wood and built up from several pieces so as to counteract any tendency to warping. Stiffening strips of wood or metal may be fixed to its underside. In the centre underneath there is a fitting with a screw-threaded hole or some other arrangement for attaching the board to the tripod. The upper surface of the board must be plane. The board may be fitted with a circular bubble or two short linear bubbles at right-angles, set flush with the upper surface in one corner, for levelling purposes.

353. The Tripod and Movement. The tripod is of open-frame type, combining lightness with rigidity, and it may be of telescopic construction, as in Fig. 172. It is essential that the hinges at the tops of the legs should be as wide as possible, to ensure rigidity. On the simplest and lightest types of plane-table, the underside fitting of the board bears directly on the flat top of the tripod, and levelling is effected by manipulation of the tripod legs.

More elaborate designs, incorporating levelling-screws, ball-and-socket joints, and other devices, are available, but their advantages, if they have any, are not likely to be appreciated by the surveyor who has to carry all the equipment about on his own person. What is important is that the bearing surface of the board on the tripod should be as wide as possible so as to give firm clamping; and the rotation of the board should not put it appreciably out of level.

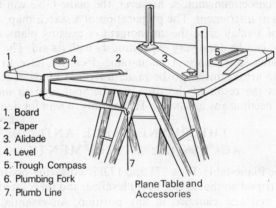

1. Board
2. Paper
3. Alidade
4. Level
5. Trough Compass
6. Plumbing Fork
7. Plumb Line

Plane Table and Accessories

FIG. 171.

354. The Alidade. In its simplest form the alidade consists of a wooden or metal ruler, of length about equal to the smaller dimension of the board, and furnished with a pair of sight-vanes (Fig. 171). One of the vanes has a narrow slit, and the other is an open frame with a stretched vertical hair or wire. The slit and wire must be perpendicular to the under surface of the ruler, but they need not be in the same vertical plane as the ruling edge or even in a plane parallel to it. For convenience of transport the sight-vanes may be hinged so that they can be folded down on to the ruler. It is also convenient if the ruler has a movable edge on a parallelogram linkage.

If the tops of the sight-vanes, directly above the slit and the wire, are connected by a light thread, it is possible to set the alidade on very steep sights. Note that if a steep sighting is made, the verticality of the vanes, and hence the horizontality of the board, is

important. A bubble may be fitted on the alidade, and then it can
be used for levelling the table.

355. Telescopic Alidade. In plane-table mapping on smaller scales,
when marks at distances of several miles have to be sighted, the
ease and accuracy of the work are improved by use of an alidade

FIG. 172.—PLANE TABLE EQUIPMENT WITH TELESCOPIC ALIDADE.
(By courtesy of Hilger Watts Ltd.)

fitted with a sighting telescope. Such an instrument is illustrated in
Fig. 172. If the telescope has a vertical circle and stadia lines, the
scope and utility of the plane-table method are greatly enhanced,
because it can then be used for making a complete large-scale

survey of a small area, including spot heights or contours, by tachymetric methods (see Chapter XI). The telescope is usually equivalent to that of a small theodolite. Vertical collimation and bubble adjustment should be carefully attended to, so that only a single reading of the slope need be taken. Tachymetric observations are much facilitated by use of special Tables, such as those prepared by Prof. F. A. Redmond, or by graphical-mechanical aids like Cox's Stadia-Computer.

356. Trough Compass. It is very convenient, though not strictly necessary, to have a method for approximately orienting the plane-table when an exact orientation is not immediately obtainable. This can be done by using a magnetic compass. A suitable instrument is a magnetic needle some 5 inches long pivoted in a narrow wooden rectangular box with a glass cover and a removable protective lid. Placing the lid on the box automatically raises and clamps the compass needle. Another type of compass is shown in Fig. 172.

To make use of compass orientation the plane-table must be correctly oriented as soon as possible, such as by setting it up at a point marked on the drawing and orienting it on another mark at a good distance away, then putting the trough compass, without lid, on the drawing at a suitable place near the edge, setting the box so that the compass needle points to the meridian mark, and drawing lines along the sides of the box with the hard pencil. The table can then be oriented at any other point by reversing the process, and the setting will be correct to a degree or two provided there is no local disturbance of the magnetic meridian.

357. Plumbing Fork. This is illustrated in Figs. 171 and 172. Its use is necessary only in surveying on very large scales, where the size of the board is appreciable in relation to the lengths of sights to be taken. The upper point of the fork is placed on the point representing the plane-table station, and the table must then be set up so that the plummet hanging from the lower point of the fork is over the ground mark.

358. Other Equipment. Pencils of two hardnesses should be used. The soft pencil is for construction lines which must be easily erased, and the hard pencil should make a permanent mark.

Good rubber erasers are essential. A plane-tabler does a lot of erasing; if he loses his eraser he is effectively put out of action, so he should carry one or two spares.

A piece of fine-grain sandpaper is excellent for keeping pencils sharp. It is common practice to tack a piece of sandpaper to a tripod leg or to the underside of the board.

Paper of good quality without a prepared surface should be used for plane-table work, except the most rough and simple jobs. Special cloth-mounted papers are obtainable, but good drawing paper is quite satisfactory if properly fixed to the board.

The alidade may have a graduated scale along one edge, but it is advisable to use a scale graduated to suit the scale of the map, so that distances can be read off directly.

ATTACHMENT OF PAPER TO THE BOARD

359. Drawing-pins should not be put in the upper surface of the board, and in any case they would interfere with the use of the alidade. Paper may be held to the board with adhesive tape, or with strong metal clips round the edges. Some boards have slits near the edges so that the paper can be put through them and fixed on the underside.

However, paper fixed to the board in the dry state is sure to buckle when atmospheric conditions change. The best way to fix the paper is to cut it larger than the board, make it thoroughly wet, place it on the board, and paste the overlaps on the underside. Fixed like this, the paper becomes very taut when dry and is not likely to buckle during use.

SETTING UP THE PLANE-TABLE

360. Centering. If the plane-table is to be set up at a point of known position on the drawing, it should be centered sufficiently closely to the ground-mark, considering the scale of the map. This means that in plane-tabling on scales like 1/25,000 and smaller, it is quite good enough to put the table within a foot or two of the ground position. Otherwise, centering by eye will be adequate unless the scale is so large that the centering-fork ought to be used. However, it is always important to centre fairly carefully when running a plane-table traverse, as described below.

361. Levelling. Experienced plane-tablers consider that they can level the board sufficiently well by eye, but for work in hilly country or at very large scale it is advisable to use a small bubble.

362. Orientation. When the table is set up at a predetermined point it can of course be oriented at once by setting with the alidade on another recognisable mark, preferably at a considerable distance. Otherwise, approximate orientation only can be done with the compass. The procedure for getting accurate orientation at new positions (by *resection*) is perhaps the most distinctive feature of the art of plane-tabling.

TECHNIQUES IN PLANE-TABLING
363. Plane-tabling may be either:

  (*a*) controlled by surveys done by other methods, or

  (*b*) independent.

Systematic topographical mapping formerly done on plane-tables was mostly controlled by points fixed by a network of conventional triangulation. Coordinates of the ground marks were calculated on a chosen map-projection and plotted at the chosen scale on the plane-table sheets. The surveyor's job was to get the detail drawn on the paper in correct positions relatively to the control points, and, as a rule, to add contours too.

Another type of controlled survey is that done when an existing map or plan is placed on the board, and plane-tabling methods are used to add new detail such as roads, housing estates and engineering works, to bring the plan up to date.

An independent survey is one done without previous control, and obviously such a survey cannot be very extensive, because the inevitable errors of graphical construction will begin to have their effects as the drawing proceeds further from the starting point. To provide scale for an independent survey, some form of linear measurement must be made on the ground.

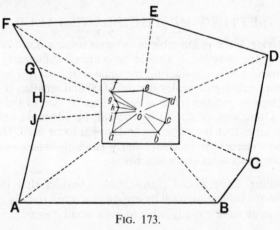

Fig. 173.

Methods of surveying with the plane-table may be classed under four distinct types: Radiation, Intersection, Traversing and Resection. In the diagrams which follow, to illustrate these methods, the ground points are indicated by capital letters, and the corresponding points on the drawing are indicated by the corresponding small letters, the size of the plane-table itself being greatly exaggerated.

364. Radiation. This method is suitable only for the survey of a small area such as a courtyard or a small field. Select an instrument station O from which all the points to be surveyed are visible (Fig. 173). Set up and level the plane-table and clamp it in a suitable position having regard to the shape of the area to be surveyed. Mark the point *o* on the paper to represent the instrument station, and with the alidade edge close to *o*, direct the alidade successively to A, B . . . and draw lines through *o*. Measure the distances to A, B . . . from O and plot them at the desired scale: join up *abc . . . ja*. Obviously this method will be inconvenient if the distances to be measured are more than a few hundred feet. A telescopic tachymetric alidade can speed up this kind of work and enable somewhat larger areas to be covered. In any case, an assistant or two will be needed for making the tape measurements or holding the staff.

365. Intersection and Graphical Triangulation. This method has more extended application. Lay out and measure a base line AB (Fig. 174). Plot *ab* on the paper at a suitable scale, drawing the line through *a* and *b* at least as long as the alidade. Set up the plane-table with point *a* centred over A and orient by laying the alidade

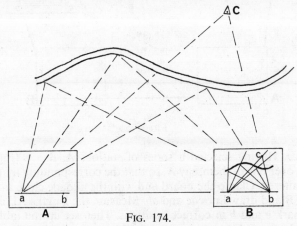

FIG. 174.

along *ab* and turning the board until the sight is on B. Clamp, and with the alidade edge close to *a* draw lines to points defining the surrounding detail. Lines may also be drawn in the directions of points like C which have been selected and marked as possible future instrument stations. Then proceed to B and orient by backsighting on to A, and draw rays to intersect the points of detail, to points such as C, and to any new points that are conveniently situated and were not intersected from A.

The survey of a small area can be completed by this intersection method. Otherwise, by fixing points like C it is possible to extend the survey by a procedure of graphical triangulation to provide a framework of instrument stations from which details can be surveyed by intersections or radiation. It is important that lines drawn through instrument stations should be drawn the full length of the alidade, so as to give maximum control of orientation.

366. Traversing. This method can be used in surveying round a large area, particularly where the detail to be surveyed is an irregular boundary; and it may be the only practicable method in conditions of restricted visibility such as wooded country or a built-

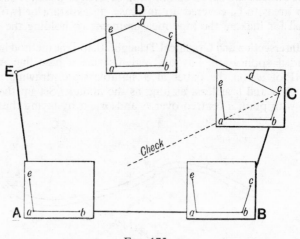

FIG. 175.

up area. Having selected a series of stations A, B . . . (Fig. 175), set up over one of them, say A, so that the corresponding map point *a* is centered. Clamp the board and with the alidade close to *a* sight E and B and draw lines *ae* and *ab*. Measure the lengths AE and AB and mark *e* and *b* in correct positions. Then set up the table at B, orient back on A, measure and draw *bc*, and so on. It is essential in this method to draw all the lines along the full length of the alidade.

A traverse survey should always return to the starting point A, or to some other previously fixed point: there will often be a visible misclosure, due to drawing, centering and orienting errors, hence it is advisable to complete the traverse and adjust any misclosure before surveying any details from it.

As described above, the orientation is done by back-sighting on to instrument stations, which must be marked with poles or some other suitable signals. Alternatively, it is possible to traverse by orienting the board with the magnetic compass: in this case, there is no point in drawing the traverse lines beyond the stations.

Once the traverse framework is fixed, details can be surveyed by any of the methods previously described.

367. Resection. In extensive topographical surveys on the plane-table, this is the most commonly used method for fixing the position of a station occupied by the surveyor. The typical situation is illustrated in Fig. 176. Three points, A, B, C are visible, and may be trigonometrical stations or points of detail already mapped. The table is set up at station R whose position r on the board is not known: thus, the surveyor is unable to orient the board by use of the alidade.

A △

△B

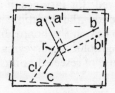

C △

FIG. 176.

Suppose he approximately orients it as shown in broken lines. The points a, b, c will be located at a', b', c', and the three rays drawn through these points by sighting on A, B, C, respectively will fall as shown by the dotted lines. That is, they will appear on

the drawing as shown in Fig. 177, each being rotated about the
control point *a* or *b* or *c* through equal angles from the true positions *ar*, *br* and *cr*. Thus, the practical method is as follows: set up
and level the plane-table, orient as accurately as possible (by compass usually) and draw lines back through the three control points.
If the three lines do not intersect in one point, and usually they will
not, the true position of *r* is such that each line must rotate through
the same angle about its control point *a*, *b* or *c* in order to pass
through *r*: and the rotations must be all clockwise or all anti-
clockwise. The angle of rotation is of course the error in the initial
orientation of the board.

FIG. 177.

When the correct position of *r* is determined, the board can be
truly oriented, preferably on the most distant visible control point.

The valuable feature of the resection method is that the plane-
table can be set up at suitable points from which a maximum
amount of detail can easily be surveyed: for instance, if a resected
position is established at a road junction, the directions as well as
the positions of the roads can be marked on the board. Indeed, the
art of rapid plane-tabling in topographical surveying depends
largely on the intelligent choice and use of resected points.

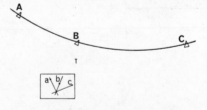

FIG. 178.

There are, however, dangers inherent in the method. The surveyor
must avoid occupying a resection point on or near to the circumcircle of A, B and C. If he is on this circle, the three rays he draws
will be concurrent, whatever the error of orientation of the board,
and the point of concurrence will be somewhere else on the circle.

This follows directly from the well-known property of angles sub-tended at the circumference of a circle by points also on the circle.

There is no objection to being well outside the circle, as indicated in Fig. 178. With incorrect orientation, a construction like that shown in Fig. 179 might be obtained: again, the rule about equal angles of rotation in the same direction will lead to the correct position for r. In practice, the rule may also be put in the form that r must be at distances from the lines proportional to the distances of the control points from the surveyor's position.

However, if a choice is possible, it is best to fix from control points situated so that the resection station is inside the triangle ABC.

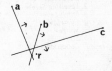

FIG. 179.

It is important to remember that a resection from only three points contains no check; an apparently good fixation is obtained by the procedure described above, even if one or more of the con-trol points is wrongly identified on the ground or mis-plotted on the map. A check ray or two from other control points or previously fixed details is therefore always desirable.

Other methods of fixing by resection are available: for instance, the rays may be drawn on a separate piece of tracing paper which is then fitted over the control points, and the resected point is pricked through. This means carrying more materials about, and most experienced surveyors are quite happy with the trial-and-error method.

368. Contouring: Vertical Control. Plane-table surveying for the production of topographical maps normally includes the third dimension, shown by contours at vertical intervals to suit the nature of the country and the scale of the map. Heights of the trigonometrical stations determined in the course of the establish-ment of the control framework will usually be the basis of a contour survey. What the plane-tabler has to do is to find the heights of many other points, so that he has a close network of heights on the board and can accurately interpolate the contours, aided by his experienced eye for the lie of the land and the way the contours run.

The instrument most commonly used is the Indian Clinometer

(Fig. 180) which is specially adapted to plane-tabling. The clinometer is placed on the board and, when it has been directed
towards the point to be observed, it is adjusted by turning the levelling screw until the bubble is central. If the bubble has been correctly set, the line of sight from the eye-hole to the zero mark on the
graduated vane will then be horizontal.

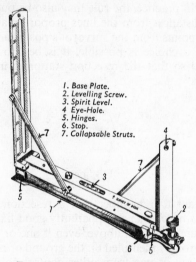

1. Base Plate.
2. Levelling Screw.
3. Spirit Level.
4 Eye-Hole.
5. Hinges.
6. Stop.
7. Collapsable Struts.

FIG. 180.—INDIAN CLINOMETER.
(Reproduced by kind permission of the
Surveyor General of India.)

A clinometer can be tested by measuring the slope of a line between two points whose heights are already known, or by measuring the slope of any line in both directions.

The folding vanes of the clinometer are about 8 inches apart.
The front vane is graduated in degrees on one side of the slot and
in natural tangents on the other side. Tangents can be read to
0·001 easily. Suppose a signal at a trigonometrical station has the
height 876 ft. and the reading of the clinometer is 0·022 elevation:
the distance measured on the plan is 2780 ft. Then the difference
of height is 0·022 × 2780 = 61 ft. Allowing 3 ft. for height of table
above ground, this makes the surveyor's station height to be
876 − 61 − 3 = 812 ft. Many plane-tablers take clinometer readings
on to a point estimated to be 3 ft. above the reference level at the
observed point, and so obviate the correction for table height.

Points on contours can be located by inverting the above calculation. Suppose that the surveyor with the aid of the clinometer,

estimates the maximum slope of the ground as *arctan* 0·037: then the 800 ft. contour will be at a distance of 12/0·037 = 324 ft. from the plane table station, down the slope.

In open country, a good method for surveying a contour is to get the plane-table at a point actually on the contour and use the clinometer as a level to locate other points on the contour.

Heights of stations should always be determined from two or more points of known heights.

ACCURACY OF PLANE-TABLING

369. Provided the control is adequate, and the surveyor can start work from at least a dozen or fifteen points well scattered over the board, plane-tabling at any scale can be as accurate as the thickness of a pencil line, except that the contours, which are located with the aid of some measure of personal judgement and estimation, cannot be expected to be everywhere in precise position. The plane-table, with its auxiliary equipment, is a very inexpensive and handy instrument of surveying, requiring little more than common-sense in use, and instantly available for surveying small areas without previous instrumental work or computation.

CHAPTER VIII

CONTOURS AND CONTOURING

---

370. Representation of Three Dimensions. The value of a plan or map, whether of large or small scale, is greatly enhanced if the altitudes of the surface features are represented, as well as their relative positions in plan. One way in which the conformation of the ground may be presented on the plan is by delineation of the surface slopes on some conventional system of shading, intended merely to convey an impression of relative relief, but without indicating actual elevations. These systems of representations are sometimes used in geographical mapping on small scales, but they do not impart sufficiently definite information to be of service for engineering purposes.

In plans for location or construction work, vertical dimensions should be represented at least as accurately and clearly as horizontal distances, and this points to two requirements in delineation:

(*a*) The reduced levels of numerous points in terms of a known datum should be shown.

(*b*) These should be arranged in such a manner that the form of the surface can readily be interpreted.

The first requirement may be met by the plotting of spot levels, the elevation of each being written over a dot representing the position of the point. The utility of spot levels is, however, limited, as they can convey only a vague idea of the form of the ground. Both conditions are fulfilled by contour lines, a method of representation of great value, as it is equally suitable in flat as in mountainous country, and for large as well as small scale plotting.

371. Contour Lines. A contour line, or contour, may be defined as the line in which the surface of the ground is intersected by a level surface. It follows that every point on a contour has the same elevation—that of the assumed intersecting surface. If the contour lines determined in this way by several equidistant level surfaces are imagined to be traced out on the surface of the ground and surveyed, the resulting plan will exhibit the contours in their proper relative positions, and will portray the character of the ground. Thus, Fig. 181 (a) represents a hill which is shown intersected by a series of level surfaces at elevations of 100, 150, etc., ft. above datum. The contours may run as shown on the plan (b), which serves to depict the shape of the hill.

It follows from the definition that the water edge of still water is a contour line, and the student may at first find it helpful to conceive contours as shore lines of a water surface which can be adjusted to any desired level. Regarded in this way, Fig. 181 (b) shows the varying outline of an island when the level of the water stands at 100, 150, etc., ft. In the case of a valley or other depression, the contours may similarly be considered as indicating shore lines of a lake.

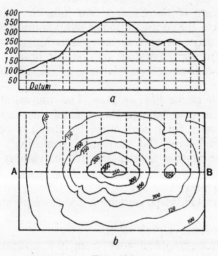

FIG. 181.

372. The Contour Interval. On the assumption that the contours of Fig. 181 (b) are correctly located and plotted, full information is available regarding the surface along each, but not about the intervening ground. Considerable irregularities may occur between adjacent contours, but, having escaped intersection, they are not represented on the plan. The number of available level surfaces is, however, infinite, and the representation of the ground can be carried to any degree of refinement by sufficiently increasing the number of contours. The constant vertical distance between the contours in any case is called the *Contour Interval.* The most suitable interval to adopt on a survey is a matter to be decided at the outset from consideration of the following items:

(*a*) *Time and Expense of Field and Office Work.* The smaller the interval, the greater is the amount of field work, reduction, and plotting required in the preparation of the plan.

(*b*) *The Purpose and Extent of the Survey.* Close contouring is required in cases where the plan is intended to be utilised for the detailed design of works or for the measurement of earthwork quantities. In general, the area included in such a plan will be comparatively small, so that it may be quite practicable to locate contours with an interval as small as one foot. In location surveys for lines of communication, reservoirs, and their drainage areas, etc., since a rather large area may be involved, a wider interval must be made to suffice. The requirements of geographical mapping are met by the use of still greater intervals.

(c) *The Nature of the Country.* An interval which would be sufficient to show the configuration of mountainous country would be quite inadequate to portray the undulations of comparatively flat ground. The representation by contour lines of a stretch of very flat ground is made possible only by the adoption of a very small interval.

(d) *The Scale of the Plan or Map.* The interval should, *ceteris paribus,* be in inverse ratio to the scale of the map. By using a close interval on a small scale map, the detail is obscured, and, unless the ground is very flat, the result is confusing.

Values of the contour interval adopted for various purposes are as follows:

For building sites: 1 or 2 ft.

For reservoirs, landscape gardening, and town-planning schemes: 2 to 5 ft.

For location surveys: 5 to 10 ft.

For general topographical work: from 10 ft. upwards, depending upon the scale and the character of the ground; for average country a common rule is

Contour interval = 50 ft. ÷ number of inches per mile.

Whatever interval is adopted, it should be constant within the limits of a map. If the regularity of the intervals is interrupted by the interpolation of additional contours for the better definition of flat ground, these should be distinguished in drawing by dotting or otherwise. Unless this is done, the general appearance of the map is misleading, the closely contoured ground appearing steeper than it really is.

373. Characteristics of Contours. By virtue of the fundamental property of a contour, that every point on it has the same elevation, a contour map with constant interval portrays elements of configuration in a characteristic manner. A knowledge of the more important attributes of contours facilitates the interpretation of a map, and, in plotting, enables the draughtsman to avoid possible misrepresentations false to nature. The following items should be kept in view in map reading and plotting.

(1) Since the rise or fall from a point on one contour to any point on the next above or below is the constant contour interval, the direction of steepest slope is that of the shortest distance between the contours. The direction of steepest slope at a point on a contour is therefore at right angles to the contour.

(2) Steep ground is indicated where the contours run close together; flat ground where they are widely separated. Uniform distance between contours indicates a uniform slope, and a series of straight, parallel, and equally spaced contours represent a plane surface.

(3) Contour lines of different elevations can unite to form one line, only in the case of a vertical cliff.

(4) Two contour lines having the same elevation cannot unite and continue as one line, nor can a single contour split into two lines. This is evident from the conception of contours as water marks. The single line would indicate an impossible knife-edge ridge or hollow.

(5) There are conceivable circumstances in which two contours of the same elevation may touch at a point, giving them the appearance of crossing, but this occurs very rarely. Contours of different elevations can cross each other only in the case of an overhanging cliff or a cave penetrating a hillside. At the intersections, two or more points of different levels on the ground are represented by one point in plan.

(6) A contour cannot end anywhere, but must ultimately close on itself, although not necessarily within the limits of the map. This property is obvious from the idea of water marks. A series of closed contours indicates either a hill, or a depression without an outlet, according as their elevations increase or decrease towards the centre of the series.

(7) The same contour appears on either side of a ridge or valley, for the highest level surface which intersects the ridge and the lowest one which intersects the valley do so on both sides. It is impossible for a single higher or lower contour to intervene between two of equal value.

(8) Contour lines cross a watershed or ridge line at right angles. They curve round it with the concave side of the curve towards the higher ground.

(9) In valleys and ravines, the contour lines run up the valley and turn at the stream so that the convexity is next the higher ground. They are intersected at right angles by the stream. If the scale allows of both banks being shown, the contours should be stopped

at the edge of the water unless they have been located across the bed of the stream.

METHODS OF CONTOURING

374. Since in the location of contours both vertical and horizontal measurements are involved, the field work may be executed in various ways according to the instruments used. Field methods may, however, be divided into two classes—*Direct*, and *Indirect*.

Direct methods comprise those in which the contours to be plotted are actually traced out in the field by the location and marking of a series of points on each. These points are surveyed and plotted, and the appropriate contours are drawn through them.

Indirect methods are those in which the points located as regards position and elevation are not necessarily situated on the contours to be shown, but serve, on being plotted, as a basis for the interpolation of the required contours. This system is used in all kinds of survey, and proves less laborious than the first.

In both methods, but particularly in the case of the latter, the accuracy of the resulting map will greatly depend upon the number and disposition of the selected points. The careful location of numerous points may be out of the question owing to the time required in the field, and for some purposes sufficient accuracy is attained by interpolating between widely spaced points. In such work, horizontal positions may be determined by intersection, and elevations by barometric or trigonometric levelling, but this class of field work belongs more properly to the subject of topographical and geographical surveying, and is dealt with in Vol. II. The field methods to be considered here are those directed to the production of a map suitable for location or construction work. Variation in procedure will depend upon which of the following instruments are used:

Vertical Control: Dumpy, Tilting or Automatic Level, Hand Level or Clinometer.

Horizontal Control: Chain and Tape, Theodolite or Tachymeter, Compass, Plane Table.

Combined Control: Tachymeter, Plane Table with Tachymetric Alidade or Clinometer.

Contours may be sketched on air-photographs under a stereoscope, or they may be plotted accurately from air-photographs by use of photogrammetric apparatus. These methods of contouring are described in books on photogrammetry.

375. Direct Methods. Direct contouring is necessarily very slow, and on this account it is comparatively seldom adopted on large surveys.

It has the merit of superior accuracy, and is suitable for the close contouring of small areas where considerable precision is required. The field work consists of two steps: (1) the location of points on the contours; (2) the survey of those points. These operations may be conducted nearly simultaneously if performed by two parties, one levelling and the other surveying, but in dealing with a small area the pegging out of the contours may be completed before their survey is commenced.

Although the field routine is largely dependent upon the instruments employed, two principles of general applicability are to be observed.

(1) The degree of accuracy required in the location of contours should depend upon the scale to which they are to be plotted, and, for a given scale, upon the use to which the plan is to be put.

(2) Since the contours are to be drawn through the plotted points, only such points should be located that the contours are nearly straight between them. At places of sharp curvature more locations are necessary than elsewhere. Salient points on ridge and valley lines are of special importance, and should never be omitted.

376. Direct Vertical Control. *By Level and Staff.* Having levelled from the nearest bench mark to the site of the survey, set the instrument to command as much ground as possible. From the known instrument height deduce the readings to be observed when the staff is held on the various contours within range of the level. Taking one contour at a time, direct the staffman up or down hill until the required reading is sighted. On receiving the signal to mark, the staffman should insert a lath, or peg down a piece of paper on which is noted the reduced level of the contour. Having located one contour over the length visible from the instrument, proceed in the same manner for the others with the new staff reading, until a fresh instrument station is required.

FIG. 182.

Fig. 182 shows a possible arrangement of the points fixed from one set-up, the arrows showing the route followed by the staffman. If the instrument height were 86·37, the staff readings would be 1·4, 6·4, and 11·4 in locating points on the 85, 80, and 75 contours respectively. It is unnecessary to read to two places of decimals.

It is advisable that the levelling from the B.M. to the first control station should be checked before contouring is commenced. A temporary B.M. should therefore be established on the site of the survey, and checked first of all.

In walking forward, the staffman should try to avoid going up or down hill. It is an advantage to have the assistance of an experienced staffman, who, when remote from the instrument, can be relied upon to select suitable points, but the surveyor should nevertheless be on the lookout for possible omissions at salient points. A special signal should be used to indicate that the staffman has to move to a different contour.

Fixing adjacent points on several contours occasions loss of time and confusion with the marking tags; the method of working along each contour as described, is much preferable.

By Hand Level. The hand level may be of any type, an Abney clinometer clamped to zero proving suitable, and the instrument may be used in conjunction with a levelling staff or simply a ranging pole marked off in feet. By levelling from a B.M., first locate a point on one of the contours, preferably one having an elevation about the mean of those to be traced. This contour is to be pegged out first in order to provide points of known elevation from which the remaining contours can be located. To trace the contour, stand on the initial point, and direct the staffman until the point on the staff or pole corresponding to the height of the instrument above the ground is in the line of sight. The level should be held against a pole, so that the reading is a whole number, say 5 ft. Locate as many points on the contour as can be conveniently sighted from the instrument station, and then move forward to the last point marked.

When a sufficient length has been set out to form the basis of the day's work, the fixing of points on the contours on either side can be commenced. Let it be assumed that the contour interval is 2 ft., and that the instrument is held 5 ft. above the ground. Send the staffman uphill until a reading of $(5-2) = 3$ ft. is obtained, and fix a few suitable points with that reading. The next series gives a reading of 1 ft., but to locate the higher contours the surveyor must take up a position over one of these points and start anew. In setting stakes on the downhill side of the reference contour, the readings from it will be 7, 9, etc., ft. If a sufficiently long pole is not available, the staffman can hold his pole on the reference contour, while the surveyor places himself approximately on the lower contour and shifts about until he sights 3 ft. as before. Having in this manner located portions of the several contours with respect to the initial point, the other points on the reference contour are utilised in the same manner.

Greater accuracy is attained if the preliminary work of levelling from the B.M. and setting out the reference contour is performed by level. A reference contour may be dispensed with by establishing a number of temporary B.M.'s over the area by means of ordinary levelling.

If the contour interval exceeds the height of the instrument above the ground, the contour above the reference contour cannot be located directly from it, but the levelling must be carried up from the reference points in a series of stages.

377. Direct Horizontal Control. The survey of the positions of the stakes defining the contours is a somewhat tedious operation. The system to be adopted must be decided upon from consideration of the size and shape of the area involved and of the accuracy required. For small areas, chain surveying may prove suitable, but in general it is necessary to execute a traverse, either by theodolite, compass, or plane table. If contouring is confined to a narrow strip of ground, a single traverse can be run approximately along the centre line, and offsets taken to the stakes on either side. Some of these offsets may be longer than is generally desirable, but this is not a serious objection, and one or two 100 ft. tapes should be carried. When the width is too great for effective control by a single traverse, a network, or framework, must be laid out.

The survey should be conducted as soon as possible after the location to avoid possible errors due to displacement of the stakes. If only one surveyor is in the field, he should therefore locate and survey on alternate days, or employ the forenoons for locating and the afternoons for surveying. A considerable saving of time may be effected by the use of the tachymeter, which, although best adapted for indirect contouring, can be utilised in the direct system for horizontal control only. As soon as the staffman has been placed on a contour by the leveller, an observation is taken of the bearing and distance of the staff from the tachymeter set over a traverse station. In this way the need for marking located points is eliminated.

However the points are surveyed, a note must be taken of the contours on which they are situated, this information being obtained by reading the tag on the stake when offsetting, or from the leveller in the above tachymetric system. When the points are plotted, the contours are drawn through them as curved lines, due attention being paid that the fundamental characteristics of contours are nowhere violated.

378. Indirect Methods. In these methods, the points to be completely located may be either (1) situated along a series of straight lines set out over the area, or (2) scattered spot heights at representative points. In (1), the straight lines simply constitute lines of sections, and in general no cognizance is taken of the ground between these sections. In (2), more particular attention can be paid to the salient features upon which the topography depends.

379. *By Cross Sections.* Suitably spaced sections are projected from traverse lines, the observations being made in the usual manner by level, clinometer, or theodolite. The sections should be closer than the normal spacing at places where the contours curve abruptly, as on spurs and in ravines, and in the latter case it is expedient to run a section approximately along the line of the stream, taking its direction by compass. The configuration may in places suggest the running of a number of sections radiating from a point.

The sections need not be plotted if an ordinary level has been used, so that the reduced levels of the various points are known. To draw the contours, the points levelled are first marked off along the section lines, and the elevation of each point is written against it. On the assumption of uniform slope between adjacent points, the contours are then interpolated by estimation as shown in Fig. 183.

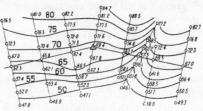

FIG. 183.

If the sections are plotted, the interpolation may be performed more quickly and mechanically by the following method, which must be used when the sectioning is performed by clinometer. Rule on a sheet of tracing paper a series of equidistant parallel straight lines, the distance between which represents the contour interval to the scale of the sections. Place this sheet over a section in such a position relatively to its datum or its ground line that the ruled lines coincide with the intersecting planes defining the required contours. In Fig. 184 the lines on the tracing paper are shown dotted, the interval between them being 5 ft. The tracing has been placed so that one of the lines coincides with the datum line of the section, and, reckoning upwards, it will be observed that the ground surface is intersected by lines of elevations 65, 70, 75, and 80. The distances from the centre line to those intersections on right and left are transferred to the section line on the plan, and the contours are then drawn as in the direct method.

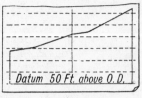

Datum 50 Ft. above O. D.

FIG. 184.

If the base lines of the cross sections have not been drawn, it is necessary to subdivide a portion of the vertical centre line on the tracing paper into feet. The tracing paper can then be placed with the horizontal lines in their proper position relatively to the centre line level marked on the cross section.

The transfer of the right and left distances from the section to the plan is expeditiously performed if the work is shared by two men, one using the sections and calling out the distances and elevations of the various contours, the other plotting the points on the plan and drawing the contours through them.

The cross section method of contouring does not fall far short of the direct method in point of accuracy, provided always that additional sections are run where called for, and that the ground is fairly uniform in slope between the points located. It has the merit that the sections can be employed in connection with estimates and construction.

380. *By Squares.* If the area to be contoured is not very extensive, it may be divided up into a series of squares, the corners of which define the points to be levelled. Bounding lines at right angles to each other are set out to enclose the area, and ranging poles are placed at regular intervals along them. If the ground is not too rough, the pegging out of each separate square may be obviated by providing sufficient poles (as in Fig. 185) to enable the staffman to place himself at each point. To prevent misunderstanding, every fifth line may be flagged, and, for referencing, one set of lines should be lettered and the other numbered. If the site is one where it is proposed to carry out earthwork, the enclosing lines should be permanently pegged for convenience in running levels for progress measurements.

The interpolation is similar to that of the previous case, and is exhibited in Fig. 186. It is to be observed that the assumption of uniform slope between the located points is not so justifiable here as in the cross section method, but the smaller the squares the more valid will be this hypothesis. The sizes employed range from 10 ft. to 100 ft. side, small squares being used for close contouring or on

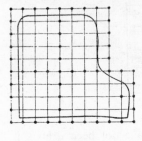

FIG. 185.

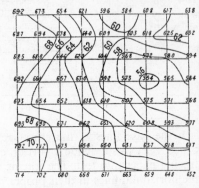

FIG. 186.

rough ground. In ground of varying character, the squares need not be of the same dimensions throughout, but can be reduced on rugged parts, or, alternatively, a few additional observations on points within the squares can be located on the diagonals or by other measurements from the corners.

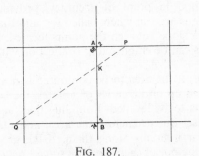

FIG. 187.

A method of interpolating between spot-heights on a square system is illustrated in Fig. 187. Between the spot-heights 68·2 and 74·3, the 70-ft. contour is to be located, assuming uniform slope along the line AB. This line must obviously be divided in the ratio 1·8 to 4·3. Along the lines perpendicular to AB and in opposite directions, measure off AP and BQ as shown, proportional to 1·8 and 4·3 in any convenient units, e.g., these lengths could be made 18 and 43 millimetres. Place a ruler with its edge on P and Q, and mark K on the line AB; then the 70-ft. contour passes through K.

381. *By Spot Heights* or *Representative Points.* If spot heights only are observed, the number required for contour location is considerably reduced, but the facility of horizontal control possessed by the previous two methods does not obtain. Notwithstanding, this system is the most popular, particularly on large surveys, because of its suitability for tachymetric methods, which, since they are designed to furnish both horizontal and vertical control, are specially adapted for contouring. This branch of surveying is described in Chapter XI.

382. Contour Drawing. The contour lines, having been drawn in pencil, are inked, either in black or brown, the latter being preferred as being less likely to lead to confusion where roads, buildings, etc., appear on the plan. A drawing pen gives a better line than a writing pen, and French curves should be used as much as possible. The elevations of the contours must be written against them in a uniform manner, either on the higher side or preferably in a gap left in the line. The figures should be normal to the contours, and several rows of figures are required on a large sheet. In small scale mapping it is sufficient to figure every fifth contour, which should be distinguished by a bolder line, but the values of intermediate contours must be shown where there is a possibility of misinterpretation.

383. Interpolation of Contours. When contours have been directly located with a rather wide interval, intermediate contours can be interpolated on the plan. This practice should, however, be avoided; it is unsound in principle since it adds nothing to the genuine information given by the surveyed contours, and it uses up draughtsman's time. If such contours are interpolated without reference to field observations, they should be distinguished by being drawn in broken lines.

384. Contour Gradients. A line lying throughout on the surface of the ground, and preserving a constant inclination to the horizontal, is called a contour gradient. The path of such a line can be traced in the field or on a map (section 387), if its inclination is known, and it is required to pass through a given point.

In the field, the location is most quickly performed by means of a clinometer, theodolite, or a level with gradient drum (section 107). The instrument is first set up at the given point, and the line of sight is laid at the given inclination. A man, carrying a level staff, or simply a graduated ranging pole, proceeds in the estimated direction of the contour gradient to a greater or less distance from the instrument, according as the ground is of uniform slope or undulating. He holds up the pole, and is directed by the instrument man to move up or down hill until the pole reading equals the height of the instrument above the ground. The line from the instrument station to the point on which the pole is held is then parallel to the line of sight, and forms a part of the required contour gradient. The point so obtained is pegged, and is used as the instrument station for the location of the next point in the same manner.

Alternatively, a level may be used, and in this case the instrument need not necessarily be set up over the contour gradient. Since the line of sight is horizontal, it is necessary to measure out from the starting-point, or from the last point pegged, the distance, say 100 ft., to the point to be fixed. The required staff reading is computed from this distance, the gradient, and the elevation of the instrument. Thus, if a down gradient of 1 in 75 is being traced, and the line of sight is at 4·2 ft. above a peg on it, the reading on a staff 100 ft. from that peg should be 5·5 ft. The staffman, holding the end of a 100-ft. tape or chain stretched from the peg, is moved up or down hill until the required reading is observed. Indeed, in this case it would be simpler to have the staffman move 75 ft. each time.

In the clinometric method, the instrument has to be moved forward after each observation, unless the ground surface is plane. A hand clinometer is therefore the most convenient of the instruments available. In the method of levelling, the same instrument station may be used for the pegging of several points.

From any point there will usually be two or more directions in which a given gradient may proceed. There is, however, less difficulty than might be supposed in deciding which of the directions should be followed, since the gradient required in practical cases is that which is least winding.

If the gradient as set out has to be surveyed, it is usually sufficient to take compass bearings from peg to peg and note the distances. Otherwise, offsets may be taken to the pegs from traverse lines or other existing horizontal survey.

The tracing of contour gradients is a common field operation in the location of road and railway routes. When the locating engineer has to find the best route for, say, a road over a range of hills, it is generally impossible to adhere, even roughly, to the straight line route between the controlling points on either side because of the steepness of the ground. By laying off a contour gradient equal to the steepest slope which is considered allowable on this part of the road–the limiting or ruling gradient–he discovers a route over the difficult ground which, if it were followed, would enable the road to be constructed without excavation or embankment along its centre line.

The contour gradient located is generally that passing through the lowest point on the ridge near the straight line route, and it is best set out from the ridge down both sides of the hill. It affords a most useful guide to the location of the road, but its adoption as the actual centre line would result in a very tortuous route except where the ground is of uniform slope. In crossing a gorge, for example, the contour gradient runs up the gorge, and crosses the stream in a manner similar to a contour line. The road, when constructed, will necessarily be carried across the gorge at a point downstream from the contour gradient, as it must be above the stream. The engineer, instead of locating the true contour gradient at such a place, therefore examines the ground for the most suitable site for the road crossing, and projects the contour gradient across in the air. It is likewise unnecessary for him to trace the true gradient round small hollows and ridges, at which it would have a sharp curvature. At every place, however, where the finally adopted route for the road deviates from the contour gradient of the same value as the road gradient, earthwork will be required on the centre line. Deviations downhill necessitate embankments, while those to the uphill side involve excavations.

USES OF CONTOUR PLANS AND MAPS

385. Drawing of Sections. A section along any line, straight or curved, on the map can be drawn by marking on a datum line the distances along the line of section at which the various contours are intersected. The respective contour elevations are then set up as ordinates from these points. Thus, in Fig. 181, a is the section AB of the hill represented in the plan b.

386. Determination of Intervisibility. Whether two points of given position and elevation are intervisible or not can be ascertained from the map without having recourse to drawing the section between them. Let it be required to test the points A and B (Fig. 188) in this respect. Join AB. From its length and the known levels of A and B, determine by calculation, or other method of interpolation, the position of the points on it which have the same elevation as the contours. These are marked with sloping figures. Determine by inspection whether these points are above or below the ground. Examining the point of 90 ft. elevation on AB, it is seen to fall on the lower side of the 90 contour, and therefore is above ground, since a vertical line through it would cut the ground at an elevation of less than 90 ft. Similarly, the points at 80 and 50 ft. are clear, but those at 70 and 60 are below the surface, and in consequence A and B are not mutually visible. C and D, the points at which AB and the surface are estimated to have the

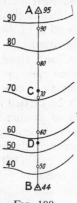

FIG. 188

same levels, about 72 and 57 respectively, show the limits of the obstruction. The determination of intervisibility, of importance in military work, is sometimes of service to the surveyor. The above method of solution does not apply to very long sights, on account of the effects of curvature and refraction.

387. Tracing of Contour Gradients. The method of locating on a map the route of a given contour gradient through a given point

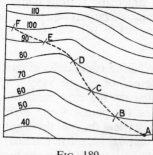

FIG. 189.

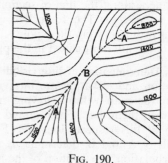

FIG. 190.

is illustrated in Fig. 189, in which it is required to trace the upward course from the point A of a contour gradient of 1 in 30. It will be sufficient to locate the points at which the gradient intersects the given contours, and, since the contour interval is 10 ft., the horizontal distance between the successive required points must be 300 ft. With centre A on the 50 contour, describe an arc of this radius to cut the 60 contour at B, and, with centre B, describe a similar arc to cut the next contour, and so on. Join these points with a curve. This line represents the path of the contour gradient with sufficient accuracy for most practical purposes notwithstanding that the lengths intercepted between adjacent contours are not now 300 ft. precisely. The actual line would be more tortuous than is shown.

Each of the arcs described in Fig. 189 will cut its appropriate contour at two points, so that the contour gradient drawn is only one of those which can be marked out from A. The others which could be shown within the limits of the diagram have a zigzag course.

388. Measurement of Drainage Areas. For water supply and irrigation purposes the extent of these can be estimated by measurement of the area contained within the watershed line separating the basin from those adjoining. This line is to be traced on the map in such position that the ground slopes down on either side of it, *i.e.* it lies along ridges, as at A (Fig. 190), and on cols or passes, B. Care must be observed to ensure that the whole area draining into the valley under consideration is included.

389. Intersection of Surfaces. If two intersecting surfaces are each represented in plan by a system of contours, then the points in which the contours of one surface cut those of equal elevation belonging to the other are situated on the line of intersection of the two surfaces. Such points are common to both surfaces, and the line of intersection can be drawn by joining up successive points. This construction is of practical utility in affording a rapid method of determining on a plan the boundary of proposed earthwork, which is simply the line in which the earth slopes cut the original surface of the ground.

Thus, in Fig. 191, let it be required to draw the plan of an earth dam, of the given dimensions, to be built across a valley. Having drawn the top surface in the proposed position, set off the contours of the sloping sides for the same elevations as the contours representing the original ground. The new contours are shown dotted, their positions being obtained from the given side slopes and the top level of the dam. The line joining the points of intersection of the equivalent contours of the new and original surfaces shows the extent of the earthwork in plan. In the above case, since the top of the dam is level, the contour lines along the side slopes are parallel to the top edges.

Fig. 192 illustrates the method of finding the limits of an excavation for a road which has an upward gradient of 1 in 22 towards the right. From a known formation level the positions of contours 150, 155, etc., 110 ft. apart, are obtained on the formation

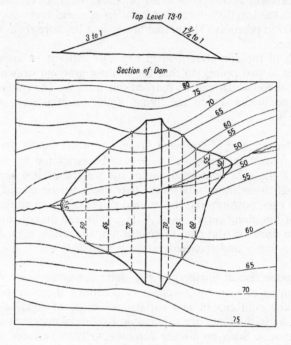

Top Level 78·0

Section of Dam

FIG. 191.

surface, and, on the assumption that this surface is horizontal transversely, the contour lines are drawn straight across it. Along the side slopes, which are to be $1\frac{1}{2}$ horizontal to 1 vertical, the contours will deviate from the sides of the formation surface at the rate of $7\frac{1}{2}$ ft. in 110 ft. The intersection points of the two sets of contours are obtained and joined up as before.

A similar application of the principle to earthwork occurs when it is desired to show where excavation and filling respectively will be necessitated in grading a piece of ground. This problem is illustrated in Fig. 218, where the solid lines represent the contours of the original ground, and the dotted lines the altered surface. The line *aaa*, obtained as above, marks the boundary between cutting and filling; the ground within it must be excavated, and that outside filled within the limits of the diagram.

In tracing these intersection lines, there may at times be some doubt as to the direction of the line between the points located on it. In such cases, the interpolation of a few intermediate contours on both series will yield additional intersections and fix the line

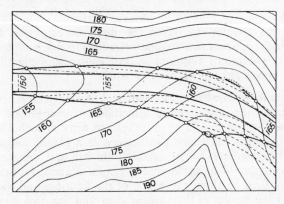

FIG. 192.

more definitely. Thus, in Fig. 192, two points are obtained on the upper intersection line by the interpolation of small parts of the $172\frac{1}{2}$ contour.

390. Measurement of Earthwork. See sections 420–422.

REFERENCES

CRONE D. R. 'Notes on terrestrial altimetry'. *Empire Survey Review*, No. 122, Oct. 1961.

RANGAN V. 'Reduction of aneroid readings'. *Empire Survey Review*, No. 103, Jan. 1957.

SHEWELL H. A. 'Accuracy of contours'. *R.I.C.S. Journal*, Sept. 1951.

WILSON G. U. 'Barometric determination of elevation'. *Empire Survey Review*, No. 118, Oct. 1960.

CHAPTER IX

MEASUREMENT OF
AREAS AND VOLUMES

391. In this chapter are given the methods employed for the measurement of areas and volumes. Needless to say, such measurements are commonly required in civil engineering practice, and form an important branch of office work.

392. Necessity for System in Calculation. At the outset it is necessary to emphasise the advantage of a methodical arrangement of calculations. The aim should be to set out the work in such a manner that the result of each step can be seen at a glance, and so that all the intermediate work can be followed and checked. The calculations should be performed in books of about foolscap size, one being used to show the whole of the figuring, and the other to record the results of the various steps and the final quantities. As in all other operations in connection with surveying, verification of the results is necessary before they can be employed for any purpose. If the checking is performed by the person who made the original calculations, the most reliable check is afforded by working out the result by a different method, when possible. If the method first adopted is considered preferable to any other, care should be taken to ensure that the checking embraces every part of the work involved in the first determination. In office work, verification by another person is preferred, and the checker should adopt the same routine, and should work quite independently from beginning to end.

393. Accuracy of Results. Arithmetical slips being eliminated, the degree of accuracy of the final result will depend upon (a) the accuracy of the field work, (b) the accuracy of plotting, when the calculations are made from scaled measurements, (c) the method of calculation adopted. It must be realised that the data employed in computation are subject to error and that the accuracy of the final result cannot be increased by needless refinement of figuring.

As there is considerable difficulty in assessing the precision of the data, it is not easy to judge the degree of refinement justifiable in calculation. One must, to a large extent, estimate the probable accuracy of the field and office work, and the final result should be expressed with no more significant figures than can be relied upon.

The beginner is apt to overlook this point, and, in computing areas for example, would find it helpful to form a mental picture of the extent of ground corresponding to the last significant figure in his result. Keeping in view the probable accuracy of the field methods and the total area under consideration, he will acquire some conception of how to express results consistently with the value of the data.

MEASUREMENT OF AREA

394. By the area of a piece of ground is meant its area in *plan*. The British unit of land measurement is the imperial acre, 1/640th of a square mile, subdivided as follows:

$$
\begin{aligned}
1 \text{ acre} &= 4 \text{ roods} &&= 160 \text{ poles (or perches)},\\
&= 10 \text{ sq. chains} &&= 100{,}000 \text{ sq. links},\\
&= 4{,}840 \text{ sq. yards} &&= 43{,}560 \text{ sq. feet}.
\end{aligned}
$$

Fractional parts of an acre are expressed either in decimals or in roods and poles.

The decimal connection between the acre and the square chain facilitates reduction when the Gunter chain has been used in the field. For example, let the area of a piece of ground be calculated as 349,435 sq. links. By moving the decimal point five places to the left we have at once 3·49435 acres, which, having regard to the circumstances of the measurement, might be stated as 3·494 acres. Conversion of the fractional part to roods and poles can be done as shown: at each multiplication, only the decimal part is multiplied. The area is 3 acres, 1 rood and 39 perches.

$$
\begin{array}{r}
3\text{·}494 \\
4 \\
\hline
1\text{·}976 \\
40 \\
\hline
39\text{·}04
\end{array}
$$

In the metric system, the customary unit of area is the *hectare* of 10,000 sq. metres, equivalent to 2·471 acres. Areas given on Ordnance Survey large scale maps will in future be expressed in both systems.

395. General Methods of Measurement. Areas may be obtained (*a*) from the plotted plan, (*b*) by direct use of field notes. The former is the more common and less troublesome method, but the latter is susceptible of greater accuracy since errors introduced in plotting and scaling are eliminated.

396. Measurement from Plan. Areas of plans may be found by several methods:

   (*i*) division into simple geometrical figures,
   (*ii*) division into strips with addition by ordinates or computing scale,
   (*iii*) division into squares which are counted,
   (*iv*) mechanically by planimeter.

397. Division into Simple Geometrical Figures. The most convenient method is to divide up the survey into triangles, either by pencil lines on the plan or on superimposed tracing paper. The base length and perpendicular height of each triangle are scaled and noted, care being taken not to omit any, nor to record the same one twice. When one or more of the boundaries of the plot of ground consist of irregular curves, these lines are replaced by straight ones such that the area contained within them is equal to that of the original figure. This 'equalising' of the boundaries is easily accomplished by the use of a piece of black thread or a transparent set-square, the straights being drawn so that, as nearly as can be judged, the areas contained between them and the irregular boundary are equally disposed on either side of the straight (Fig. 193). No attempt should be made to equalise large irregularities, but small triangles should be introduced in such cases.

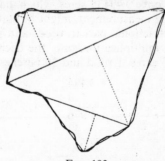

FIG. 193.

398. Division into Strips. A piece of tracing paper or thin transparent plastic is covered with equally-spaced parallel lines at a convenient distance apart according to the scale of the plan: for example, for use on a plan at the scale of two chains to one inch the lines could be $\frac{1}{4}$ inch, representing $\frac{1}{2}$ chain, apart. This tracing is placed over the plan, as shown in Fig. 194 so that the drawing fits a whole number of line widths. The length of plan covered by each strip is measured, the operator having to make estimates of 'give and take' at the ends of the strips as shown on the figure. The lengths of the strips are added up and, with the dimensions as given above, each inch of strip will represent a square chain, and ten inches of strip will therefore represent an acre.

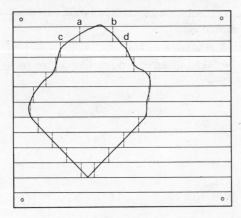

FIG. 194.

A very valuable aid to the strip method is the Computing Scale illustrated in Fig. 195: the cursor is set to zero and the scale is placed on the ruled tracing so that the cursor line is on the line a in Fig. 194. The scale is held down and the cursor is moved to line b. The scale is then lifted up without disturbing the cursor setting and placed so that the cursor line is on line c; then the cursor is moved to line d: and so on. Thus the computing scale accumulates the area. The graduations of the scale can be spaced so as to read directly the acreage with suitable subdivisions.

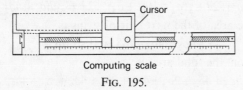

Computing scale

FIG. 195.

A very similar method, particularly useful in finding the area of a long strip of land, is operated by measuring ordinates in relation to a base-line drawn through the strip, as shown in Fig. 196. At equal distances along the base-line, perpendiculars $A_1M_1B_1$, $A_2M_2B_2$, . . . are drawn and their lengths measured. Let the ordinates be at distances d apart, and let their successive lengths be $o_1 = A_1B_1$, $o_2 = A_2B_2$ . . .

There are two formulas commonly used for calculating the area:

The Trapezoidal Rule. It is assumed that each strip like $A_1A_2B_2B_1$ is a trapezoid, then the areas of the successive trapezoids are

$$\tfrac{1}{2}(o_1+o_2)d, \quad \tfrac{1}{2}(o_2+o_3)d, \ldots$$

Thus the total area is:

$$A = d(\tfrac{1}{2}o_1+o_2+o_3+ \ldots +o_{n-1}+\tfrac{1}{2}o_n).$$

Care must be taken at an end point like M_n to include this last ordinate as $o_n = 0$.

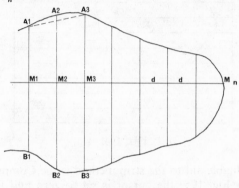

FIG. 196.

Simpson's Rule. Some allowance for the curved shape of the boundaries can be made by assuming that portions like $A_1A_2A_3$, covering two strips, are parabolic. From the geometrical properties of parabolas it follows that the area of a double section like $A_1A_3B_3B_1$ is

$$\tfrac{1}{2}(o_1+o_3) . 2d+\tfrac{2}{3}\{o_2-\tfrac{1}{2}(o_1+o_3)\} . 2d = \tfrac{1}{3}d(o_1+4o_2+o_3).$$

Similarly, the area of the next pair of strips is

$$\tfrac{1}{3}d(o_3+4o_4+o_5), \text{ and so on.}$$

By summation, the total area is:

$$A = \tfrac{1}{3}d(o_1+4o_2+2o_3+4o_4+2o_5+4o_6+ \ldots +4o_{n-1}+o_n)$$

Clearly, it is necessary to have an even number of strips in order to make use of Simpson's Rule, but if this is inconvenient it is a simple matter to add an extra fictitious zero ordinate at one end.

399. Division into Squares. The plan is covered with a transparent sheet having a pattern of equal squares drawn on it. The size of the squares should be simply related to the scale of the plan, as with the parallel lines method.

The number of squares inside the boundary is counted and the result converted to true area.

Incomplete squares can easily be estimated to one tenth of a square; otherwise, it may be considered accurate enough to count portions greater than half as whole squares, and ignore portions less than half.

400. Planimeter. The planimeter is a mechanical integrator used for the measurement of plotted areas. The various forms which have been devised possess the common feature that a point of the instrument is guided round the boundary of the area, and the resulting displacement of another part of the mechanism is such as to enable the area to be recorded. The most commonly used form is Amsler's polar planimeter, and it only will be considered.

The operation of the instrument is based on a consideration of the area swept out by a moving line of fixed length In Fig. 197 (a), the line AB, of length L, moves to the position A'B', and the movement is made up of a translation from AB to A"B' plus the rotation from A"B' to A'B'. The translation distance is m perpendicular to the direction of the line, and the rotation angle is θ.

To the first order of infinitesimals, the area swept out is thus $Lm + \frac{1}{2}L^2\theta$: ($\theta$ in radian).

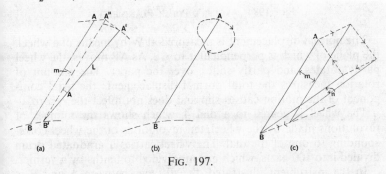

FIG. 197.

Suppose the end A travels round a closed curve, while B moves along another line and returns to its original position, as indicated in Fig. 197 (b). With a convention that areas swept by a left to right movement are positive, and areas swept by a right to left movement are negative, the total area swept out is $L\int m + \frac{1}{2}L^2 \int \theta$. and this is the area of the closed curved, because the rest of the area swept by the line is covered both ways and cancels out.

If the line L returns to its original position without having done a complete rotation, the $\int \theta = 0$, but $\int m$ is not zero: this can be understood by reference to Fig. 197 (c) where the total left to right translation is $m-n$. Therefore, if the operation is carried out as described above, the area is $L\int m$, and can be determined if L is known and m is measured.

401. Amsler's Polar Planimeter. This instrument (Fig. 198) consists of two bars AB and BC hinged at B. The first carries at A a tracing point, which is guided round the boundary of the area to be measured. C is a stationary point or pole, the bar BC terminating at C with a needle point, which is fixed in the paper and held down by a weight. B is therefore constrained to move along the circumference of a circle of radius CB while the tracing point travels round the area under measurement.

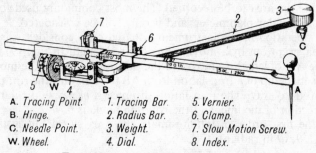

A. Tracing Point.	1. Tracing Bar.	5. Vernier.
B. Hinge.	2. Radius Bar.	6. Clamp.
C. Needle Point.	3. Weight.	7. Slow Motion Screw.
W. Wheel.	4. Dial.	8. Index.

FIG. 198.—AMSLER'S POLAR PLANIMETER.

The normal displacement is measured at W by means of a wheel, the plane of which is perpendicular to AB. As AB moves, the wheel partly rotates, and partly slides, over the paper. The amount of rotation measures the total normal displacement: the axial component of the motion causes slip and does not affect the record.

The wheel is geared to a dial 4, which shows the number of revolutions made by the wheel, ten revolutions of the wheel corresponding to one of the dial. The wheel carries a graduated drum divided into 100 parts, which are subdivided to tenths by a vernier.

In the instrument illustrated, the distance between A and B is adjustable, and, on setting the appropriate mark on the tracing arm opposite the index 8 by means of the clamp and tangent screw, readings are obtained in square inches, square centimetres, acres, etc.

402. Use of the Planimeter. To measure an area, first adjust the tracing arm so that the result may be given in the desired units. Fix the needle point in the paper outside the area and in a position which will enable the tracing point to reach all parts of the boundary. Mark a point on the boundary from which to start, and place the tracing point there. Read the dial and wheel, or alternatively, set these to zero. Guide the tracing point in a clockwise direction along the boundary, and, on returning it exactly to the starting-point, again read the dial and wheel. The final reading less the initial reading gives the required area.

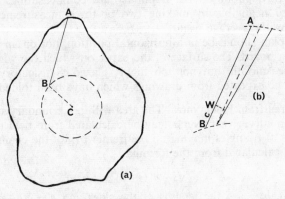

FIG. 199.

If the area is large, it is possible to place the planimeter needle inside and take A round as shown in Fig. 199 (a). The point B traces out a complete circle and the arm BA makes a complete revolution, so that $\int \theta = 2\pi$. If the wheel is at W as indicated on Fig. 199 (b), at distance c from B, it will record not only the translations of the arm BA but an additional $2\pi c$ due to the rotation of the arm: in the planimeter illustrated, the wheel is on the other side of B and it will then record a total movement of $\int m = 2\pi c$.

The total area inside the curve is

$$\pi b^2 + L\int m + \tfrac{1}{2}L^2 \cdot 2\pi,$$

where b is the length of the arm BC, and the area can be written as

$$L(\int m - 2\pi c) + \pi b^2 + \pi L^2 + 2\pi Lc ;$$

the quantity in brackets is recorded by the wheel; the values of the other terms must be known from the constructional dimensions of the planimeter.

Usually, it is preferable to divide a large area into small pieces by arbitrary internal lines and compute the areas of the pieces separately. Another possibility is to draw a large rectangle, triangle or circle inside the area and use the planimeter to find the rest of the area, divided into suitable pieces.

The accuracy of setting of the tracing arm for any particular unit may be verified by drawing a trial square of known size and measuring its area.

In moving the tracing point round the area under measurement, it should be guided as far as possible by a straight-edge or a French curve. Small errors arising from the tracing point not being maintained throughout on the boundary are compensating, and are rendered negligible by making two or three measurements and adopting the average result.

The plan should be in a horizontal position, not on an inclined drawing board. The surface of the paper on which the wheel rolls must be smooth, but not too highly glazed. Satisfactory results cannot be expected from drawings which have been folded.

403. Area from Field Notes. The area within a boundary surveyed by chain survey or traverse may be calculated from field notes.

Areas of triangles making up the framework of the chain survey may be calculated from the formula

$$A = \sqrt{s(s-a)(s-b)(s-c)},$$

where a, b, c are the lengths of the sides and $s = \frac{1}{2}(a+b+c)$. A traversed polygon can be divided into triangles the areas of which are calculated from the above formula, or from $\frac{1}{2}$(base)(height). See Fig. 200.

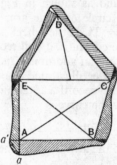

FIG. 200.

If the plan is very large, a rectangle of convenient dimensions can be drawn inside it and the rest of the area found by methods given above.

Areas between survey lines and the plan boundary may be calculated from the lengths of the offsets and the distances between them. To include corner bits, the survey lines can be extended as indicated in the figure; in many cases, offsets from extended lines would be available in the notes.

404. Area from Coordinates. The area contained within the survey lines of a closed traverse may be computed directly from coordinates. Let the traverse be illustrated in Fig. 201 by the polygon ABCDEA having coordinates as shown. Consider the figure M_1ABM_2 in which $M_1A = y_1$, $M_2B = y_2$, and $M_1M_2 = x_1 - x_2$.

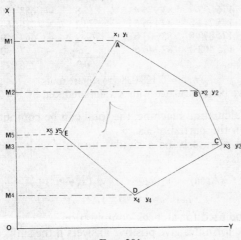

FIG. 201.

The area of this figure is obviously equal to M_1M_2 multiplied by the mean of M_1A and M_2B, that is

$$(x_1 - x_2) \cdot \tfrac{1}{2}(y_1 + y_2).$$

To avoid continual use of the $\tfrac{1}{2}$, let us consider twice the area. It is clear that the area of the whole polygon is equal to the areas M_1ABM_2, M_2BCM_3 and M_3CDM_4 *less* the areas M_4DEM_5 and M_5EAM_1. So we get:

$$2(\text{Area}) = (x_1 - x_2)(y_1 + y_2) + (x_2 - x_3)(y_2 + y_3) +$$
$$+ (x_3 - x_4)(y_3 + y_4) - (x_5 - x_4)(y_5 + y_4) - (x_1 - x_5)(y_1 + y_5).$$

After some cancellation and re-arrangement, this expression can be written as:

$$2(\text{Area}) = x_1(y_2 - y_5) + x_2(y_3 - y_1) + x_3(y_4 - y_2) +$$
$$+ x_4(y_5 - y_3) + x_5(y_1 - y_4).$$

Evidently, if there are n sides of the polygon the formula will be:

$$2(\text{Area}) = x_1(y_2 - y_n) + x_2(y_3 - y_1) + \dots$$
$$\dots x_{n-1}(y_n - y_{n-2}) + x_n(y_1 - y_{n-1}).$$

310   PLANE AND GEODETIC SURVEYING

To avoid computing with very large numbers, a temporary local origin for the X coordinate may be adopted. Suppose the co-ordinates are as given in the table:

| | Co-ordinates | | (y_2-y_5) | Double |
	X	Y	etc.	Areas
A	176 509·4	83 297·7	+1629·8	+4 089 820·12
B	176 027·9	83 976·4	+ 817·8	+1 658 416·62
C	175 213·2	84 115·5	− 673·4	− 816 968·88
D	174 377·5	83 303·0	−1768·9	− 667 759·75
E	175 402·0	82 346·6	− 5·3	− 7 430·60
			0·0	4 256 077·51

Area = 2 128 038·755 square units.
The areas have been calculated with X values reduced by 174 000.

Using a calculating machine, the total can be computed without writing down the partial areas.

There will obviously be a similar formula with x and y interchanged; it is:

$$2(\text{Area}) = y_1(x_n-x_2)+y_2(x_1-x_3)+ \dots$$
$$+ y_{n-1}(x_{n-2}-x_n)+y_n(x_{n-1}-x_1).$$

This can be used for a check computation.

The above formulas give positive answers if the suffix numbering of the co-ordinates goes clockwise round the perimeter.

405. Subdivision of an Area into Given Parts from a Point on the Boundary. Let ABCDEFA (Fig. 202) be a plot of land, and let it be required to cut off a definite area by a line drawn from the point H on the boundary.

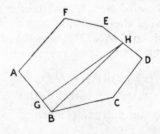

Fig. 202.

Calculate the area of the figure ABCDEFA from the coordinates and also plot the figure on a fairly large scale. By inspection, or by trial and error on the plotted plan, find the station B so that the area bounded on one side by the line HB is nearer in value to the given area than that bounded by a line from H to any other station. Compute the length and bearing of HB and the area of the figure HBCDH. Let A be the area required and A' the area of the figure HBCDH. Then, if HG is the line needed for the subdivision, the point G is found from the relation:

$$A - A' = \tfrac{1}{2} . BG . BH . \sin HBG$$

or,

$$BG = \frac{2(A - A')}{BH . \sin HBG}.$$

The length and bearing of the line BH have been computed from the coordinates of H and B and, since the bearing of BG is known, the angle HBG is known. Consequently, BG can be computed and the coordinates of G found.

406. Subdivision of an Area into Given Parts by a Line of Given Bearing. Let it be required to divide the area ABCDEFGA (Fig. 203) into two parts by a line whose bearing is given.

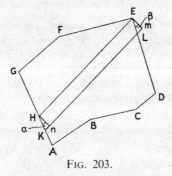

FIG. 203.

Calculate the area of the figure from the coordinates and plot it on a fairly large scale. Draw the line EH from one station E, and in the given direction, so that it cuts off an area HEFGH approximately equal to A, the area required. Calculate the bearing and distance of GE. Then, since the bearings of the lines GE, EH, and HG are known, the three angles of the triangle GEH are known and, from these and the computed distance GE, the lengths HE and GH can be calculated. Hence, the coordinates of H can be found. Using these coordinates, and those of the points E, F, and G, calculate the area of the figure HEFGH. Let A' be this area. Then, if LK is the line needed to cut off the area A, we must have:

$$A - A' = \text{Area of figure HKLEH.}$$

From E draw Em perpendicular to EH to meet KL in m, and from H draw Hn perpendicular to KL. Let Em = Hn = x and angle

$KHn = \alpha$, and $ELm = \beta$, these angles being known since the bearings of the different lines are known. Then, length $KL = HE - x \cdot \tan \beta + x \cdot \tan a$. Hence,

$$\text{Area of figure HKLEH} = \frac{x}{2}(HE + KL)$$
$$= \frac{x}{2}\left(2HE + x(\tan a - \tan \beta)\right)$$
$$= x \cdot HE + \frac{x^2}{2}(\tan a - \tan \beta).$$

Hence,
$$A - A' = x \cdot HE + \frac{x^2}{2}(\tan a - \tan \beta).$$

This is a quadratic equation which can be solved for x. Then, having found x, we have:

$$EL = x \cdot \sec \beta; \qquad HK = x \cdot \sec a.$$

Hence, the coordinates of K and L can be found.

407. Areas of Cross-sections. A typical simple cross-section for determining earthwork in railway or road construction is shown in Fig. 204. The line XCY represents the original ground surface, with C on the centre-line of the proposed route. Slopes shown as 1 to n and 1 to n_1 are obtained from levelling observations or clinometer readings. In many cases, n and n_1 will be equal.

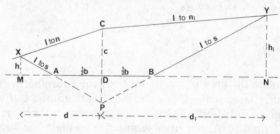

Fig. 204.

The side slope, 1 to s, the formation width b, and the depth of cutting c on the centre-line, will be available after the proposed work has been designed.

The quantities mentioned above are sufficient to determine the geometry of the cross-section.

The side widths, or half-breadths', are d and d_1, and their sum $(d + d_1)$ is the total horizontal width of the excavation on original ground surface. The side heights h and h_1 are the depths of formation level below original ground at the extremities.

It is evident from the figure that $d = (c-h)n = \frac{1}{2}b + hs$. From the second equality we get

$$h = \frac{nc - \frac{1}{2}b}{n+s},$$

and then

$$d = (\tfrac{1}{2}b + cs)\left(\frac{n}{n+s}\right)$$

If the triangle ABP is completed as shown, then $DP = \dfrac{\frac{1}{2}b}{s}$, and the area of the triangle CXP is $\frac{1}{2}d\left(c + \dfrac{\frac{1}{2}b}{s}\right)$

Thus, the required area CXAD is $\frac{1}{2}d\left(c + \dfrac{\frac{1}{2}b}{s}\right) - \dfrac{\frac{1}{8}b^2}{s}$.

For the geometry of the other side of the centre-line, it is easy to show that

$$h_1 = \frac{n_1 c + \frac{1}{2}b}{n_1 - s}, \quad d_1 = (\tfrac{1}{2}b + cs)\left(\frac{n_1}{n_1 - s}\right),$$

and then the area of the whole section is

$$\tfrac{1}{2}(d + d_1)\left(c + \dfrac{\frac{1}{2}b}{s}\right) - \dfrac{\frac{1}{4}b^2}{s}.$$

A partially embanked section is shown in Fig. 205.

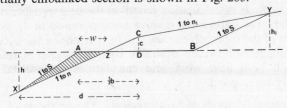

FIG. 205.

The width of the embanked portion at formation level is:

$$w = \tfrac{1}{2}b - cn.$$

We have: $d = \tfrac{1}{2}b + hs = (c + h)n,$

whence, $h = \dfrac{\frac{1}{2}b - cn}{n - s}, \quad d = (\tfrac{1}{2}b - cs)\left(\dfrac{n}{n - s}\right).$

The area CZD in cutting is $\frac{1}{2}c^2 n$, and the area in embankment $\frac{1}{2}h^2(n-s)$.

For the two-level section, Fig. 206, $n = n_1$.

For the one-level section, Fig. 207, the formulae are very simple:
$h = c, \qquad d = \tfrac{1}{2}b + cs, \qquad \text{area} = c(b + cs).$

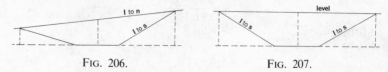

FIG. 206.   FIG. 207.

The derivation of formulae for the case where the centre-line is in embankment is left as an exercise for the student.

MEASUREMENT OF VOLUME

408. The reference made at the beginning of the chapter to the sources of error affecting the accuracy of calculated quantities applies with special force to the measurement of volume. The calculations are based upon the results of levelling, and the field work is subject to the usual errors of observation, but an additional and more important source of inaccuracy consists in the deliberate omission to record many of the surface irregularities. Although it would in general be a mere waste of time to measure all the minor undulations, the value of the field work is greatly increased by so setting out the lines of levels that the most faithful record of the surface features may be secured.

The methods of calculation based on the field measurements are not of a precise nature, but involve assumptions as to the geometrical form of the solid, which may still further impair the accuracy of the result. It is, however, desirable that the degree of refinement to which both field work and calculations are carried should mutually correspond.

409. General Methods of Measurement. The methods of calculation may be classed according as the form of the solid is defined by (*a*) cross sections, (*b*) spot levels, (*c*) contour lines.

Because of its general application, the measurement of earthwork is more particularly referred to below. The unit employed for such quantities is the cubic yard. When the methods are directed to the estimation of the capacity of a reservoir, the results are expressed in millions of gallons.

410. Measurement from Cross Sections. In this universally applicable method, the total volume is divided into a series of solids by the planes of cross sections. The spacing of the sections should depend upon the character of the ground and the accuracy required in the measurement. They are generally run at 66 or 100 ft. centres, but sections should also be taken at points of change from excavation to embankment, if these are known, and at places where a marked change of slope occurs either longitudinally or transversely.

Except when the calculations are made directly from the field notes, the sections are plotted, without vertical exaggeration; and, in the case of earthwork measurement, the new surface is represented on each, its level being obtained from the profile. On drawing the side slopes, the sectional form of the earth to be handled is represented on each section, and considering two adjacent cross sections, these areas form the plane ends of a solid of length L equal to the distance between the sections (Fig. 208).

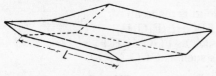

FIG. 208.

The cubic content of each of these solids is to be determined separately, the addition of those in cutting giving the total volume of excavation, and similarly for banking.

In water supply work, measurement of the cubic capacity of a proposed reservoir has to be made for various water levels to determine the surface level and area of the reservoir when impounding a given volume. In this case, therefore, water levels are drawn across the sections, and the volume up to any of them is computed in the same way as for earthwork.

411. The Prismoid. In general, a straight length of excavation may be described as a solid bounded by two parallel plane ends having the form of polygons, joined by longitudinal faces which will usually, though not necessarily, be plane surfaces.

Fig. 209 illustrates a prismoidal volume with quadrilateral end faces ABCD and EFGH, and longitudinal faces ABEF, etc. If the longitudinal faces are plane surfaces then adjacent edges must intersect in pairs as indicated by the thin lines and the dots; in fact, the longitudinal faces must be trapezia.

Project the end EFGH on to the other end as $E_1F_1G_1H_1$; each of the perpendiculars EE_1, etc., is of length L. It is easily seen that the sides of $E_1F_1G_1H_1$ must be parallel to the corresponding sides of ABCD.

Drop perpendiculars F_1F_2 FF_2 G_1G_2 GG_2 on to BC; G_1G_3

GG_3 on to CD, etc., as shown on the figure; not all these perpendiculars are shown as that would make the figure even more confused.

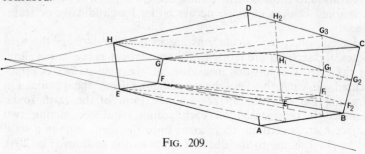

FIG. 209.

The whole solid is made up of:
(i) the prism $EFGHE_1F_1G_1H_1$:
(ii) 4 wedges such as $FGG_1G_2F_2F_1$;
(iii) 4 pyramids such as $GCG_2G_1G_3$.

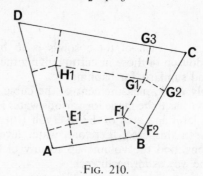

FIG. 210.

The complete geometry on the face ABCD is shown in Fig. 210.

Let the areas of the end faces be: A_1 = area of ABCD, A_2 = area of EFGH.

The whole solid can be seen to be made up of nine pieces:
(i) a prism standing on base $E_1F_1G_1H_1$ which has area A_2;
(ii) 4 wedges on bases $G_1G_2F_2F_1$, etc., which have total area W, say; and
(iii) 4 pyramids on bases $CG_2G_1G_3$, etc., which have total area P, say.

The bases of the nine portions make up ABCD, hence:

$$P+W+A_2 = A_1.$$

All the nine solids have 'height' L.

Now the volume of a prism is: height × base,
    the volume of a wedge is: $\frac{1}{2}$ height × base, and
    the volume of a pyramid is: $\frac{1}{3}$ height × base.

Thus, the volume of the whole solid is:
$$V = LA_2 + \tfrac{1}{2}LW + \tfrac{1}{3}LP.$$

The cross-section parallel to the end faces and midway between them will be a quadrilateral similarly divided into nine portions: the area of the central portion will again be A_2 but the areas of the mid-sections of the wedge-like portions will be half the areas of their bases, and their total area will therefore be $\frac{1}{2}W$, while the areas of the mid-sections of the pyramidal portions will be quarter of their bases and their total area will therefore be $\frac{1}{4}P$.

Thus, the total area of the median section will be:

$$M = A_2 + \tfrac{1}{2}W + \tfrac{1}{4}P.$$

It can now be easily verified from the above equations, that the volume $V = LA_2 + \frac{1}{2}LW + \frac{1}{3}LP$ is also equal to:

$$\tfrac{1}{6}L(A_1 + 4M + A_2).$$

This formula is exact for the type of solid described above.

Indeed, the simplest way to prove this formula is by showing that it applies equally to prisms, wedges, and pyramids.

For a prism, $M = A_2 = A_1$ and $V = \frac{1}{6}(6A_1)L = LA_1$,

for a wedge, $M = \frac{1}{2}A_1$, $A_2 = 0$ and $V = \frac{1}{6}(3A_1)L = \frac{1}{2}LA$,

and for a pyramid, $M = \frac{1}{4}A_1$, $A_2 = 0$ and $V = \frac{1}{6}(2A_1)L = \frac{1}{3}LA_1$.

Thus, the formula as given applies to any solid made up of prisms, wedges and pyramids all of the same length.

412. Prismoidal Correction. The volume formula just derived is a kind of 'Simpson's Rule' for volumes. However, its application requires a knowledge of the area of the mid-section of the solid, and, in practice, information for calculating this area will not usually be available. It is not $\frac{1}{2}(A_1 + A_2)$.

It is obvious, however, that in many cases a reasonable approximation of the volume will be obtained by taking the mean area of the ends and using the simple formula $\frac{1}{2}L(A_1 + A_2)$ for the volume, and it would be convenient to have a correction term to be applied to this result, provided that the correction could be calculated from quantities already known from the dimensions of the end sections.

Referring to Fig. 209, it will be seen that the simple mean-area formula would give the correct answer for the prism and the four wedges, but if $\frac{1}{2}LP$ were taken as the volume of the pyramids, instead of the correct $\frac{1}{3}$LP, a positive error of $\frac{1}{6}$LP would be made in the volume.

The *Prismoidal Correction* is therefore a subtraction of $\frac{1}{6}$LP from the volume calculated from the simple mean-area formula. To obtain the prismoidal correction, it is only necessary to obtain the area of those portions of the larger end-section which would be the bases of pyramidal volumes. The correction will usually be small and the required areas can be found by simple graphical estimation from the section drawings.

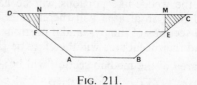

Fig. 211.

A very simple case is illustrated in Fig. 211, where ABCD and ABEF are the dimensions of two consecutive cross-sections of a cutting. The shaded portions will be bases of pyramidal volumes, and the prismoidal correction will be calculated from the total areas of these portions.

413. Skew Surface. If two corresponding edges of the end faces are not parallel, the longitudinal face joining them cannot be a plane, and it must be some form of three-dimensional or skew curved surface. Such a solid is illustrated in Fig. 212.

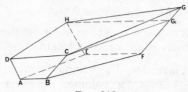

Fig. 212.

If the line HG_1 is drawn parallel to DC, the solid $ABCDHG_1FE$ is a prismoid as already described. The extra volume DCG_1GH has a skew upper surface: a little consideration will show that its general form is something wedge-like, its midway section will be a triangle with the short side half of CG_1 but its height will probably be somewhat less than that of triangle HGG_1. Thus, the extra volume will be less than that of a wedge on base HGG_1 but it will be greater than that of the pyramid on the same base, because that is the volume it would have if the upper surface were formed of the two plane triangles HDC and HCG.

However, it will generally be quite accurate enough to take the volume of the extra piece as slightly less than that of the wedge. In any case, if the face DCGH is the ground surface, as it usually will be, and if it is so irregular that the simple procedures already discussed will not give the volume as accurately as it is needed, then the proper solution is to make a more detailed survey of the irregular surface by spot-levels or contours.

414. Cutting and Embankment. If, between two levelled sections there is a change from embankment to cutting, it is best to take an extra section there and deal with the two portions separately. If no survey is done, the dimensions of the intermediate section can be obtained fairly accurately by a proportional calculation based on the dimensions of the end sections.

415. Curvature Correction. Where the centre-line of a cutting is a circular curve, the cross-sections are usually run on radial lines, and the solids to be considered in calculating volumes do not have parallel end surfaces. In computing the volumes of such solids, the common practice is to employ the usual methods, treating the ends as parallel, and then, if necessary, to apply corrections for curvature.

In principle, the corrections are based on Pappus' Theorem which states that the volume of the solid formed by rotating a plane figure about an axis in its plane is equal to the area of the figure multiplied by the length of the path traced out by the centre of gravity (centroid) of the figure.

The quantities available in practice are the radius R of the curve of the centre-line, and the length L between sections, measured along the centre-line.

Thus, referring to Fig. 213, the angle θ between two cross-sections is L/R radian.

Let the centroid, G, of the cross-section be at distance e from the centre-line. The radius of the path of the centroid is therefore $R-e$ and its length is $(R-e)L/R$. Thus if A is the area of the section, the volume between ends is:

$$A(R-e)\frac{L}{R}, \text{ or } AL - Ae\frac{L}{R},$$

and the second term is the curvature correction.

On the three-level section ABYCX, draw the line CY_1 symmetrically opposite to CX. Then the 'moment' of the extra triangle CYY_1 about the centre-line will determine the value of e.

In the notation used in section 407, the area of the extra triangle is the difference of the areas of the two portions on either side of the centre-line, that is

$$\tfrac{1}{2}(d_1 - d)\left(c + \tfrac{1}{2}\frac{b}{s}\right).$$

The distance of M, the mid point of YY_1, from the centreline is $\tfrac{1}{2}(d_1 + d)$, so the distance of the centroid of the triangle is $\tfrac{2}{3}$ of this, that is $\tfrac{1}{3}(d_1 + d)$. Thus, the moment of the triangle about the centre-line is:

$$\tfrac{1}{6}(d_1^2 - d^2)\left(c + \tfrac{1}{2}\frac{b}{s}\right)$$

and this is also equal to Ae, so the curvature correction Ae(L/R) can be calculated.

If the centroid of the cross-section is further from the centre of curvature than the centre-line, then the correction will be positive.

In practice, the cross section will probably vary from place to place. In such cases, it will generally be sufficiently accurate to find the values of A and e for each cross-section and use the mean value $\tfrac{1}{2}(A_1e_1 + A_2e_2)$ in calculating the correction.

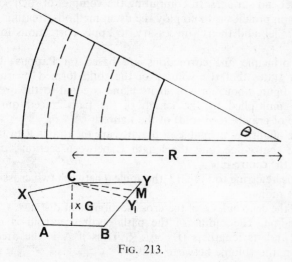

FIG. 213.

In the case of a side-hill two-level section, where the sections of embankment and cutting are triangular, it will be found that the curvature correction can in either case be calculated from the formula:

$$\frac{Lwh}{6R}(d + b - w)$$

provided that the appropriate values of w, h and d (Section 407) are used for each triangle.

416. Measurement from Spot Levels. This method of measurement is sometimes applied to large excavations. The field work consists in dividing up the site of the work into a number of equal triangles, squares, or rectangles and observing the original surface level at each corner of those figures. After completion of the earthwork, or during its progress, the lines dividing up the area are again set out, and the levelling is repeated on the new surface, the differences of level thus determined representing the depths of earthwork at a number of points of known position. In estimating the volume of proposed earthwork, the levels of the new surface are obtained from the drawings. The differences of level are regarded as the lengths of the sides of a number of vertical truncated prisms, *i.e.* prisms having non-parallel end-planes, the horizontal area of each of which is known. The volume of each prism is obtained as the product of the area of horizontal section by the average length of the vertical edges.

417. Division of Ground. The size of the unit areas into which the site should be divided will depend upon the degree of accuracy required in the measurement and upon the character of the ground. The aim should be to have the partial areas of such dimensions that the assumption that the ground surface within each is a plane is not greatly in error; it is therefore desirable to reduce their size on irregular ground.

Good results are obtained by dividing up the site into a series of squares or rectangles and regarding each as divided by a diagonal into two triangles, the diagonal which more nearly lies along the surface of the ground being selected and noted (Fig. 214). The surface within each rectangular area is then treated as consisting of two triangular planes, and the volume is computed from triangular prisms. This system involves the same amount of levelling as the method of rectangles, but the freedom of choice in each rectangle as to the pair of planes which more nearly coincides with the ground surface necessarily tends to greater accuracy.

It most frequently happens that the boundary of the work is irregular in plan, in which case there are a number of unequal partial areas adjoining it. These figures approximate in shape to triangles or trapezoids, and may with sufficient accuracy be treated as such. Their areas are obtained from measurements in the field or by scaling from the plan, and, since they are unequal, the volume of each prism must be separately computed.

418. Method of Calculation. Calculations of volumes within the limits of the grid pattern are based on the fact that the volume of a triangular prism with plane ends (not necessarily parallel) is equal to

$\frac{1}{3}A(h_1 + h_2 + h_3)$, where A is the area of a normal section and h_1, h_2, h_3 are the lengths of the three parallel edges. See Fig. 215.

In the case of a gridded area as shown in Fig. 214, the volume under each triangle may be computed separately, but the arithmetical work can be reduced by taking advantage of the fact that all the triangles have the same plan areas.

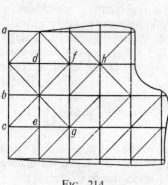

FIG. 214.

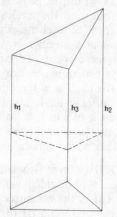

FIG. 215.

It will be seen that corner a belongs to one triangle only, and the difference of level between the new and the original surfaces at this point is therefore employed only once in the calculations. The difference at b is used twice, that at c is used three times, and so on to the difference of level at h which belongs to eight prisms, and this is the maximum number of times a particular height can be employed.

Thus, if A is the area of one triangle, and

H_1 is the sum of all heights used once,

H_2 is the sum of all the heights used twice, and so on, then the total volume will be

$$\tfrac{1}{3}A(H_1 + 2H_2 + 3H_3 + 4H_4 + 5H_5 + 6H_6 + 7H_7 + 8H_8).$$

To this total must be added the volumes contained under any irregular boundary strips.

If certain parts of the ground are much rougher than the rest, one or more of the grid squares can be subdivided into a grid of smaller triangles.

419. Simple Volumes. In some cases, the division of a volume into truncated prisms can be done in more than one way, and the calculation of the volume by two methods provides a very good check.

Consider the case shown in Fig. 216, which might represent an excavation for a storage reservoir. The dimensions of the bottom of the excavation are as shown, the original ground surface slopes at 1 to 15, and the side slopes of cutting are at 1 to $1\frac{1}{2}$. Depth of excavation above A and B is 8 ft. and above C and D is 18 ft.

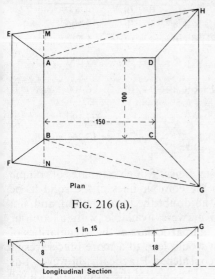

Plan

FIG. 216 (a).

FIG. 216 (c).

Longitudinal Section

FIG. 216 (b).

(In calculating the areas of the last two triangles on the list, use is made of the geometry shown in Fig. 216 (c). By the well-known theorem about the areas of triangles between parallels, the area of AEH is equal to that of AE′H′, i.e. $\frac{1}{2}AM \times E′H′$.)

It is easily calculated that the height of GH above formation level is 20 ft. and the plan width between CD and GH is 30 ft. Similarly, EF is 7·37 above formation and 10·91 ft. horizontally from AB. Further, AM = BN = 12 ft.

The volume may be divided into triangular prisms with *vertical* edges as indicated in the calculations on next page.

The same volume can be divided into only two triangular prisms, having horizontal edges, by the plane ABGH, as indicated in Fig. 216 (a).

The volume ABCDHG has edges 100, 100 and 160, and its normal section is a triangle with base 150 and height 20. The rest of the volume has edges 100, 160 and 121·82, and its normal section is a triangle which, by making use of the theorem mentioned above, has area $\frac{1}{2} \times 8 \times (190\cdot91)$. Thus the total volume is:

$$\frac{1}{2} \times 150 \times 20 \times \frac{1}{3}(100+100+160)+\frac{1}{2} \times 8 \times (190\cdot91) \times$$
$$\times \frac{1}{3}(100+160+121\cdot82) = 180{,}000+97{,}190 = 277{,}190 \text{ cu. ft.,}$$

as before.

Triangle	Dimensions of Prism	Volume
ABD	$\frac{1}{2} \times 100 \times 150$ $\times \frac{1}{3}(8 + 8 + 18)$	85,000 cu. ft.
BDC	$\frac{1}{2} \times 100 \times 150$ $\times \frac{1}{3}(8 + 18 + 18)$	110,000
DCH	$\frac{1}{2} \times 100 \times 30$ $\times \frac{1}{3}(18 + 18 + 0)$	18,000
CGH	$\frac{1}{2} \times 160 \times 30$ $\times \frac{1}{3}(18 + 0 + 0)$	14,400
ABE	$\frac{1}{2} \times 150 \times 10 \cdot 91)$ $\times \frac{1}{3}(8 + 8 + 0)$	2,909
BEF	$\frac{1}{2} \times (121 \cdot 82)(10 \cdot 91) \times \frac{1}{3}(0 + 0 + 0)$	1,772
ADH ⎫ BCG ⎭	150×130 $\times \frac{1}{3}(8 + 18 + 0)$	39,000
AEH ⎫ BFG ⎭	$(190 \cdot 91) \times 12$ $\times \frac{1}{3}(8 + 0 + 0)$	6,109

$$\overline{277,190}$$
$$= 10,266 \text{ cu. yd.}$$

420. Measurement from Contour Lines. Rough estimates of volume may be made by reference to the contour lines of the solid to be measured. In principle, this is no doubt the ideal method, and, if a highly accurate contoured plan were available with a contour interval sufficiently small that full particulars of the irregularities of the solid could be obtained, it would lead to a more precise result than the methods previously considered. Practically, however, this degree of accuracy is not realised because of the trouble of locating contours with an interval small enough to record the minor features of the ground. In dealing with contour lines, one must assume that the surface of the ground slopes uniformly from one contour to the next, and in most cases this assumption will be incorrect, the resulting error depending upon the vertical interval used as well as upon the character of the ground. To make an estimate of any practical value, the contours should not have a greater vertical interval than 5 ft. on ground of average character, but, if the surface is very irregular, an interval as small as 2 ft. might be required. The method therefore cannot compare with cross sectioning in point of convenience, and is not much used in practice except in the determination of the capacity of reservoirs or the measurement of subaqueous excavation, but even in these cases the method of cross sections is often preferred.

When contours are located by photogrammetric methods, they can be drawn at small vertical intervals, and the calculation of volumes is greatly facilitated.

421. First Case. The problem is a simple one in reservoir work and in earthwork when the made surface is level, for the new surface is then parallel to the equidistant level surfaces which define the contour lines.

To consider this case, let it be desired to determine the capacity of the reservoir shown in Fig. 217 when the water level is 165·0. The outline of the water surface is first obtained by interpolating the 165 contour. The volume is measured by regarding it as being divided up into a number of horizontal slices by the contour planes.

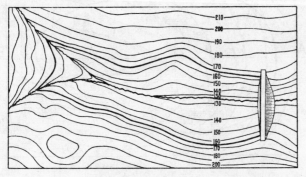

FIG. 217.

The depths of these are known, and their end areas are obtained from the plan by any of the usual methods. The area measured in each case is the whole area lying within a contour line and not that of the strip between two adjacent contour lines.

The nature of the data is precisely the same as in cross section work; the contour interval corresponds to the distance between cross sections, and the volume may be calculated either by the prismoidal formula or by the end area method. In using the prismoidal formula, every second area may be treated as a mid-area, or the mid-areas may be measured from contour lines interpolated midway between each original pair of contours. The latter method is recommended when the slopes are flat, so that there is a considerable difference between the areas within successive contour lines. For most purposes, however, the method of end areas yields sufficiently good results.

In connection with the location, design, and operation of reservoirs, it is not sufficient to know the storage capacity for one particular surface level. The volumes in the reservoir for various water levels, such as the contour elevations, should be determined. If these computed volumes are plotted against the corresponding surface levels, the resulting curve is available for reading off the volume up to any water level or for finding the level corresponding to any given volume.

422. Second Case. The more general case in earthwork is that in which the ground is not brought to a level surface.

In Fig. 218, let the full lines denote the contours of the original surface, and the dotted lines those of the proposed new surface.

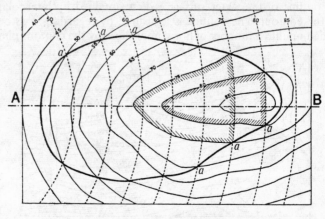

FIG. 218.

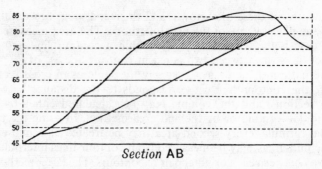

Section **AB**

FIG. 219.

By joining up the intersections *a* of original and new contours of equal value, the line in which the new surface cuts the original is obtained: within this line excavation is necessary, the surrounding parts shown being in embankment. The methods of measurement may be considered with reference to the excavation, the same methods being applicable to embankments.

First Method. It will be seen from the section (Fig. 219) that the contour planes divide the solid to be measured into a series of horizontal layers. The end areas of these can be obtained by planimeter from the plan, those for the strip between the 75 and

80 contours, which is hatched in section, being shown shaded in plan. It frequently happens that an original contour line closes on itself without intersecting the equivalent contour of the new surface, as is the case with the 85 contour in the diagram, and the area enclosed by it is that to be measured. If necessary, a sketch section will usually clear away any doubt as to the areas required. The required volume is then obtained by applying the method of end areas to each layer.

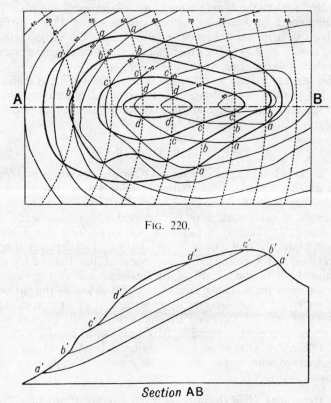

FIG. 220.

Section AB

FIG. 221.

Second Method. Instead of dividing the solid by the contour surfaces, it may be divided into layers parallel to the new surface (Figs. 220 and 221).

Just as *aaa* is the line along which the depth of excavation is zero, the depth will be exactly 5 ft. along the line *bbb* got by joining up the intersections of the original contours with new ones of 5 ft.

lower elevation. The total areas contained within *aaa* and *bbb* are the horizontal projections of the sloping surfaces which appear in section as the lines *a'a'* and *b'b'*, and the mean of those two areas times the *vertical* distance between them gives approximately the volume of the lowest strip. By joining the intersections of the original contours with the new ones 10 ft. lower, the line *ccc* of 10 ft. excavation is obtained, *ddd* being similarly derived as the line along which the cut is 15 ft., and the areas within these lines are used in the same manner. By inspection of the plan, the greatest depth of excavation appears to be about 17 ft., so that the highest strip has a vertical depth of 2 ft., the same method being applied to it as to the others, since, in this case, it more nearly resembles a wedge than a pyramid.

This method usually involves less measurement of areas than the first, but has the disadvantage that the areas to be measured have to be specially traced out, and this may necessitate the interpolation of intermediate contours.

THE MASS DIAGRAM

423. In the planning and execution of earthwork certain problems arise which are most simply studied by reference to what is termed a mass diagram.

Labour in earthwork may be analysed under four heads: (*a*) Loosening; (*b*) Loading; (*c*) Hauling; (*d*) Depositing. For a given class of plant, the unit cost of items (*a*), (*b*), and (*d*) depends almost entirely upon the character of the material, but that of (*c*) is a function both of the weight of the material and of the varying distance from the working face of the excavation to the tip end. Haulage is the most variable item in the cost of earthwork, and in cases where more than one scheme of distribution from excavation to embankment is possible, it is desirable to be able to compare different projects as regards economy of haulage. This may be accomplished with ample accuracy by means of the mass or haul diagram.

424. Definitions. *Haul Distance* (*d*) is the distance at any time from the working face of an excavation to the tip end of the embankment formed from it.

Average Haul Distance (*D*) is the distance from the centre of gravity of a cutting to that of the tipped material.

Haul is the sum of the products of each load by its haul distance $= \Sigma vd = VD$, where V is the total volume of an excavation.

Change of Volume. Since excavating involves the loosening and breaking up of the material, the volume available for the formation of an embankment is greater than that measured *in situ* in the unworked excavation. On a rough average, this increase may be estimated at 20 per cent for earth. When banked, however, the material commences to shrink, and after the lapse of a year or two will be found to occupy less space than it did originally in the excavation. The amount of shrinkage depends on several factors, but for the present purpose may be taken as 10 per cent for all earths. On the other hand, excavated rock is permanently swelled by about 50 per cent. Change of volume does not influence ordinary measurements, which should always be based upon original volumes of excavation.

425. Construction of the Mass Diagram. The mass diagram is a curve plotted on a distance base, the ordinate at any point of which represents the algebraic sum up to that point of the volumes of cuttings and banks from the start of the earthwork or from any

Distance (ft.)	Volume (cu. yd.)		Total volume (cu. yd.)
	Cut+	Bank−	
0			0
	490		
100			+ 490
	927		
200			+ 1,417
	982		
300			+ 2,399
	279		
380			+ 2,678
		31	
400			+ 2,647
		226	
500			+ 2,421
		654	
600			+ 1,767
		1,160	
700			+ 607
		933	
800			− 326
		92	
831			− 418
	220		
900			− 198
	428		
1,000			+ 230

arbitrary point. In obtaining the algebraic sum, the usual convention is to consider cuttings plus and embankments minus. The
mass diagram is therefore simply an integral or sum curve of the
volumes of the several cuttings and banks. When excavation and
banking occur on the same section, as in side-hill work, their difference only is used in the summation, the sign being that of the
greater volume.

The construction and properties of the diagram are best followed
from a simple example. In Fig. 222 let 1 represent a longitudinal
section of, say, a railway siding. The quantities required for plotting
the mass curve are tabulated above.

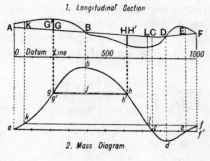

FIG. 222.

In the example it is assumed that the nature of the material is
unknown, and no allowance is made for change of volume. In
plotting, the curve should be placed directly above or below the
longitudinal section, the same horizontal scale being used for each.
Positive total volumes are plotted above, and negative quantities
below, the base line, and the ends of the ordinates are joined by a
smooth curve, the resulting diagram being as in 2, Fig. 222.

426. Characteristics of the Mass Diagram. With the sign convention adopted:

(1) Upward slope of the curve in the direction of the algebraic
summation indicates excavation. Downward slope indicates
embankment.

(2) A maximum point occurs at the end of an excavation; a
minimum point, at the end of an embankment.

(3) The vertical distance between a maximum point and the
next forward minimum point represents the whole volume of an
embankment; that between a minimum and the next forward
maximum point, the whole volume of a cutting.

(4) The vertical distance between two points on the curve which

have no maximum or minimum point between them represents the volume of earthwork between their chainages.

(5) Between any two points in which the curve cuts the base line the volume of excavation equals that of embankment, since the algebraic sum of the quantities between such points is zero. The points a and c, for example, show, on being projected to A and C, that the earthwork is balanced between A and C, *i.e.* the material excavated from AB would suffice to form the embankment up to the point C. There is also balance from E to C.

(6) Any horizontal line intersecting the mass curve similarly serves to exhibit lengths over which cutting and banking are equalised. Thus, gh is a balancing line, the cut from G to B just filling from B to H, the volume moved being represented by bj.

(7) When the loop of the mass curve cut off by a balancing line lies above that line, the excavated material must be hauled forward, *i.e.* in the direction of summation of the volumes. When the loop lies below, the direction of haul is backward.

(8) The length of balancing line intercepted by a loop of the curve represents the maximum haul distance in that section. Thus, taking the base line as the balancing line, the greatest haul distance involved in disposing of excavation AB is $ac = $ AC, so that no material should be hauled past C. In general, the haul distance increases from zero at B to this maximum, and its value at any stage of the work is given by the length of a horizontal line intercepted within the loop. Thus, gh is the haul distance when the face of the excavation as at G.

(9) The area bounded by a loop of the mass curve and a balancing line measures the haul in that section. To take the case of area $abca$, since haul $= \Sigma vd$, consider a small volume or load v, initially situated at GG′, and whose final position in embankment is HH′ obtained from the horizontals gh and $g'h'$. From (4) above, the vertical distance between gh and $g'h' = v$, and therefore area $ghh'g' = $ the haul of v. The summation of all such small areas $= $ the area of the loop $= $ the haul involved in transferring cut AB to bank BC. Similarly, area $cde = $ the haul from E to C. Regard must be paid to the scale. If the horizontal scale is 1 in. $= x$ ft., and the vertical scale is 1 in. $= y$ cub. yd., an area of n square inches represents a haul of nxy cub. yd. ft.

(10) The haul over any length is a minimum when the balancing line is so situated that the sum of all areas cut off by it, without regard to sign, is a minimum.

427. Use of the Mass Diagram. The exact interpretation of a mass diagram is entirely dependent upon the balancing line, each position of which exhibits a possible method of distributing the excavated material, and the selection of the most economical scheme is made by comparing those shown by various balancing lines. Par. (10) above is an important guide, but it most frequently happens that the condition of minimum haul necessitates the wasting or spoiling of material at one place and borrowing at another, the advisability or possibility of which depends upon circumstances. The haul involved in proposed wasting and borrowing is difficult of estimation, but it may be comparatively small if wasting is effected by widening an embankment, and borrowing by widening a cutting. The limit of profitable haul distance, beyond which it is economical to waste and borrow, is, of course, reached when the cost of excavating and hauling one cubic yard equals the cost of excavating and hauling to waste one cubic yard plus that of excavating and hauling one cubic yard from the borrow pit.

Considering first the previous example of Fig. 222, it will be seen that it is impossible to secure balance over the whole length, as there is an excess ff' of excavation. Trying the base line as a balancing line, there is shown balance from A to E and wasting of the material between E and F. But the haul ($abc + cde$), and the maximum haul distance, ac, can evidently be reduced by raising the balancing line, so that, if the haul required in wasting be assumed constant, the balancing line klf shows a preferable scheme, the material from A to K being wasted.

Fig. 223 illustrates a second example. In this case, excavation and embankment are equal over the length shown, and borrowing

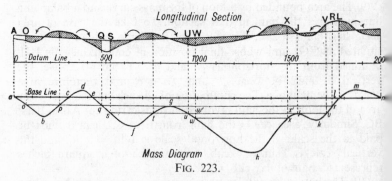

Mass Diagram
FIG. 223.

and wasting would be unnecessary provided the excavated material were distributed by the method indicated by the base line as the balancing line. But, on account of the large area below the base line, this arrangement involves considerable haul, with a maximum haul distance from l to e of about 1,350 ft., and it may be more economical to reduce the haul by wasting and borrowing. A possible method is that shown by balancing lines *opqgr* and *ln*, in which there is borrowing from A to O and spoiling of the same volume from R to L. The maximum haul distance is reduced to that from *r* to *g*, or about 880 ft. Another possible arrangement is given by balancing lines *opq*, *stujv*, and *ln*, with a maximum haul distance *ju* of about 590 ft. This scheme necessitates borrowing from A to O and Q to S and wasting from V to L, and is shown in longitudinal section. By increasing the amount of borrowing and wasting, haul and maximum haul distance may be still further reduced. In particular, if the haul distance is not to exceed 500 ft., the line *wx*, scaling that length, will be used as a balancing line. In selecting the most economical of the various methods of disposal, one must be guided by the relationship between the cost of haul and that of borrowing and wasting in so far as it can be estimated for the particular case.

428. Allowance for Change of Volume. In comparing schemes for the disposal of proposed excavations, the available information regarding the nature of the material is not usually sufficiently reliable to warrant making refined allowances for the change of bulk caused by excavating, and the common practice in such cases is to neglect these and proceed as above. If, however, sufficient data have been obtained, allowance should be made for the fact that rock increases considerably in bulk on being excavated. Soft earths also swell when loosened, but, chiefly owing to loss of material, they ultimately shrink to a smaller volume than they occupied in their original position in the excavation. The initial swell of earth is largely compensated for by the necessity for making embankments higher than they are intended to be, to allow for subsidence, and the ultimate change of volume is in the direction of shrinkage.

Allowance for change of volume is made, before summing the volumes, in either of two ways: (*a*) by multiplying each computed volume of excavation by a factor which will convert it to the volume of embankment ultimately formed from the material; (*b*) by multiplying each volume of embankment by a factor to give the volume of excavation from which it can be made. If we suppose that, on a rough average, 100 cub. yd. of solid rock will suffice for forming 150 cub. yd. of embankment, and that 100 cub. yd. of earth in an excavation will ultimately form 90 cub. yd. of embankment, the respective factors would be 1·5 and 0·9 in the first method, and 0·67 and 1·1 in the second. Such average values are usually sufficient, since difficulties are encountered in estimating the best allowances for change of volume in a particular case. Shrinkage factors for soft earth depend upon the material, and are greater for low than for high embankments. In addition, mixtures of different earths, and of earth and rock, may have to be dealt with, and, until the mass diagram is prepared and studied, it is not known from which cuttings the material for a given embankment will be derived.

429. Overhaul. The terms of contracts for earthwork may stipulate either that the price per cubic yard of excavation is inclusive of the cost of haul regardless of the haul distance involved or that the price includes the cost of haul within a specified distance only. In the latter case, extra payment is made for the haulage of each cubic yard which has to be moved a distance exceeding this specified distance, termed the free haul distance. The excess of haul distance above this amount is called overhaul distance, and the sum of the products of volumes by their respective overhaul distances is termed overhaul, and is the quantity for which extra payment is made. The unit of measurement of overhaul distance is commonly 100 ft., and that of overhaul is therefore cub. yd. × 100 ft.

The mass diagram affords a convenient aid to estimating overhaul. In the case of Fig. 223, let the scheme with balancing lines *opq*, *stujv*, and *ln* be that adopted, and let the free haul distance be 500 ft. Overhaul is required on the section JU. The horizontal line *wx* of length 500 ft. having been drawn on the mass curve and projected to *w'x'* upon the balancing line, the area *w'whxx'* represents free haul. The overhaul required is given by the product of the volume between U and W or X and J multiplied by the excess beyond 500 ft. of the shift of its centre of gravity in being transferred from cutting to bank. The positions of the centroids may be computed or estimated, but, since the total haul on the section JU is represented by the area *whj*, and the free haul by *w'whxx'*, the difference, or overhaul, is given by the sum of the two areas *uww'* and *x'xj*. These may be measured by planimeter, and are expressed in terms of cub. yds. × 100 ft.

EXAMPLES

(1) A surveyor sets up a theodolite in a five-sided field and reads directions and measures distances, as given below, to the five corners. Find the area of the field, in acres.

Direction	Distance (links)
00° 00′	317
69° 27′	387
156° 48′	346
197° 49′	267
271° 31′	290

(2) Distances and directions read on a tachymeter from a position in a private car-park to six points on the boundary were:

Direction	Distance (metres)
00° 00′	53·8
118° 16′	77·6
172° 51′	83·8
204° 36′	46·1
238° 41′	85·0
302° 17′	53·5

Assuming the boundary to be made up of straight lines joining the surveyed points, find the area of the car-park, in hectares.

(3) A chain line 200 ft. long joins two points on an irregular boundary which lies wholly on one side of the line: the successive offsets at 20 ft. intervals including the ends of the line are:

0, 11·8, 18·1, 21·0, 16·3, 15·2, 25·8, 40·1, 46·9, 36·2, 0

Estimate the area between the chain-line and the irregular boundary.

(4) The adjusted coordinates of the successive points of a closed, five-sided, traverse, are:

X (ft.)	Y (ft.)
1000	1000
1212	927
1380	1238
1175	1461
709	1052

Find the area enclosed within the survey lines.

(5) At formation level, a cutting for a new road is 48 ft. wide with a cross-slope of 1 in 12. It is cut into ground that slopes at 1 in 15 in the direction perpendicular to the centre-line. These two slopes are in opposite senses. The side slopes of the cutting are at 1 vertical to $1\frac{1}{2}$ horizontal. A certain cross-section shows a depth of 8 ft. at the centre-line. Find the area of this cross-section.

(6) Staff readings taken at various distances along a section of a dry stream bed as follows:

Distance	0	30	66	85	104	130	146	163	184
Reading	3·8	6·2	6·9	9·1	9·4	8·6	7·0	5·5	3·8

Find the cross section area of the stream bed, up to the level of the first and last staff positions.

(7) If areas are measured by planimeter as square inches of map surface, what are the factors for converting the results to acres in the cases of the '6-inch' and 1/1250 Ordnance Survey maps?

(8) A rectangular field, as shown on the '6-inch' Ordnance map, measures 2·36 × 1·77 inches. What is its area in acres? If the dimensions could be in error by 1/100 inch, what could be the error in the calculated area?

(9) Staff readings taken on a cross-section of a small embankment were:

Staff Reading (ft.)	Position
10·8	23 ft. left: original ground
4·5	7 ft. left
4·2	centre line
4·4	7 ft. right
6·7	12 ft. right: original ground

Find the area of the cross section, assuming the original ground to be of uniform slope.

(10) A road is being made by cut and fill, on ground with transverse slope of N :1. It is level at formation, the cut surface on the upper side slopes at n_1 :1, the slope of the fill is n_2 :1, and the volumes of cut and fill are equal. Show that the line between cut and fill, at formation level, divides the formation width in the ratio

$$\sqrt{\frac{N-n_1}{N-n_2}}.$$

(11) An excavation is made in ground having a uniform surface slope of 1 vertical to N horizontal. The bottom of the cutting is level, of rectangular shape, a ft. $\times$ b ft., with one of the a-length sides coinciding with a contour of the original surface. The sides of the cutting slope at 1 vertical to n horizontal. Show that the total volume of excavation is

$$\frac{b^2}{N-n}\left(\frac{a}{2}+\frac{bn}{3(N-n)}\right) \text{ cu. ft.}$$

(12) An excavation for an irrigation channel is 12 ft. wide at the bottom and has side slopes of 1 in 3. It is cut into ground with a cross-slope of 1 in 10. The centre line is curved, with radius 400 ft. and the centre of curvature is on the high side of the channel. Find the cross section area where the depth of cutting at centre-line is 6 ft., and find the curvature correction for 100 ft. measured along the centre line.

(13) An embankment of 30 ft. formation width with side slopes of 2 to 1 is to be formed on a curve of 1000 ft. radius. If the original ground surface slopes at 5 to 1 downward towards the inside of the curve, calculate the curvature correction per 100 ft. section when the centre height of the bank is 8 ft.

(14) The contoured plan of a proposed new dam and reservoir show that the areas enclosed within certain contours will be as given below:

Contour	Enclosed area (millions of sq. ft.)
420	11·95
400	9·10
380	6·32
360	3·49
340	0·75

Estimate the capacity of the reservoir up to contour 420, ignoring the volume below 340.

REFERENCES

GOUSSINSKY B. 'Some notes on the Elling method of computing areas by machine'. *Empire Survey Review,* No. 26, Oct. 1937.

RAY P. N. 'Earthwork volumes–the prismoidal correction'. *Empire Survey Review,* No. 96, Apr. 1955.

CHAPTER X

SETTING OUT WORKS

430. This part of the duties of an engineer on location or construction involves the placing of pegs or marks to define the lines and levels of the work, so that construction may proceed with reference to them. Setting out is, in a sense, the reverse of surveying in that data are transferred from the drawings to the ground.

In the case of works which may be completely defined by a series of straight lines, setting out is a simple operation, and requires little explanation. The greater part of this chapter is therefore devoted to a description of the setting out of curves.

The final plans, from which a construction work is set out on the ground, are based on the results of various processes of surveying which can be broadly classified into two stages—Reconnaissance Survey and Location Survey.

431. Reconnaissance Survey. The function of this part of the surveying is to provide information from which it is possible to understand the problems that will have to be faced, and to find suitable locations for the proposed works, and to estimate relative costs of any alternative solutions. The reconnaissance may be anything from a cursory examination and sketch to an extensive survey operation involving the establishment of a triangulation or traverse system over the area, heights by levelling or aneroid barometer, and the preparation of a contoured topographical map.

If topographical maps are available, it will probably be necessary for a surveyor to go over the ground with them and bring them quite up to date by adding new buildings and amending details no longer correctly shown, and so on. For this purpose, recent air photographs of the area may be very helpful. And indeed, especially in poorly mapped country, the reconnaissance stage may be carried out under contract by an air-survey company which will prepare maps and indicate on them certain special information that the engineers may require in connection with the proposed work. It is desirable that engineers concerned with works covering great areas and distances should have some knowledge of air-survey techniques, and be able to make satisfactory specifications for any survey work given out to specialist contractors, and be able to assess the information that can be obtained from air photographs. (See Chapter XIII.)

As a rule, reconnaissance surveys should be done by the most economical method applicable in the circumstances, as it is likely that a considerable part of the work will have no further use after decisions as to the final location have been taken. In many cases, traversing by prismatic compass and heights by barometer will give quickest results for reconnaissance survey. However, the preliminary survey work should not be regarded as entirely ephemeral; for instance, if trigonometrical observations are taken, it is advisable to leave a ground mark at each theodolite station so that it could be incorporated in later definitive work if found suitable.

432. Location Survey. This is the stage of survey when all the information required for final design is obtained: the work may be quite straightforward, or still somewhat exploratory, according to circumstances.

For instance, if the reconnaissance survey has produced a contoured topographical map, and the proposed work is a motorway, the route can probably be set out fairly precisely with the aid of the map. The location survey will then consist of the establishment of permanent control points along the route, perhaps by traverse, and the levelling of longitudinal and cross sections. During this work, the surveyor will probably measure additional details such as property boundaries and local geological features, and collect other information that might influence the final design.

On the other hand, if the work is an irrigation scheme in featureless country, the location survey may mean covering a large area with accurate levelling and survey of details, before the layout of channels, sluices, etc., can be considered.

433. Setting Out in General. Although in most cases the work is very simple in principle, difficulties are commonly encountered in practice, and indirect methods are frequently necessary, especially during the progress of construction. As each case must be dealt with according to circumstances, it is difficult to formulate hard and fast rules, but one or two considerations are of general application.

The need for complete checking of the work is self-evident, as an undetected mistake might have serious consequences. When the work is not such as will enable the closing error to be ascertained, there is a greater sense of certainty if important pegs are located by two different methods based upon independent sets of calculations, than if the same method is repeated.

Instruments should be tested at frequent intervals, and should be so used that errors of non-adjustment are reduced to a minimum. A steel band should be used for linear measurements.

In structural work, a setting out plan should be prepared showing those lines of the work which have to be set out. As it may be impossible to locate these directly owing to obstructions, considerable use may have to be made of parallel lines, and it is therefore desirable at the outset to provide a convenient framework of reserve lines clear of the ground to be occupied by plant. All intersections are marked by pegs, the exact point being indicated by the head of a tack driven into the peg, and full information regarding the linear and angular dimensions to these points should be entered on the plan.

In abstracting the required dimensions from the drawings, the surveyor must be on his guard against possible errors in the figuring of the drawings. Measurements of the drawing should be resorted to as little as possible, and then only for minor features: in the case of skew structures, the skew distances must be calculated. Thus, for setting out a skew bridge the skew span and lengths of the abutments are computed from the given square span and width and the angle of skew. Some of the dimensions necessary to fix the positions of the wing walls may have to be scaled. In the process of setting out such a structure, it is first necessary to locate a centre line and drive a peg thereon to mark the centre point of the bridge. On planting the theodolite over this peg, the angle of skew is set out, and points are established on the faces of the abutments. Pegs are located on the lines of the abutment faces prolonged clear of the work, and their distances from the corners of the abutments are noted. It is unnecessary in the first instance to provide pegs to mark the lines of foundations, back of wall, etc., as such lines can be obtained from that of the face. The end of each wing wall is set out by means of a peg on the abutment face line and an offset distance, but one or more pegs should be driven on the face line of the wing wall clear of the excavation. In the case of battered walls, only the foot of the wall is set out.

CURVES

434. Curve Ranging. This is the setting out, on the ground, of pegs marking a curved line in accordance with a design or plan. The location of the curves will be controlled by the design positions of certain points and straight lines, and by the geometry of the type of curve required. In the case of a highway or railway, the curve will generally be the centre-line, from which widths and other detailed dimensions are to be set out.

435. Circular Curves. Many, though not all, of the curves that have to be set out for engineering work are circular in plan.

It is usual, in this country, to specify a circular curve by giving its radius, which may be in feet or in chains.

A method used in America and some other countries is to give the number of degrees subtended at the centre of the circle by a length of 100 ft. of the curve: if R is the radius in feet, the angle at the centre is $100/R$ radian or $5729 \cdot 6/R$ degrees. See Fig. 224. Thus, $D° = 5729 \cdot 6/R$ or $R = 5729 \cdot 6/D°$. A five-degree curve will have a radius of 1145·9 ft.

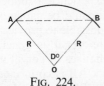

FIG. 224.

Sometimes, the degree of a curve is defined in terms of the length of the chord being 100 ft.: in this case, obviously, $R \cdot \sin\frac{1}{2}D = 50$, and a five-degree curve would have a radius 50 cosec $2\frac{1}{2}° = 1156 \cdot 3$ ft. For curves of over 1000 ft. radius, the difference between the two methods of description is evidently of little consequence.

436. Curvature Considerations. Modern traffic conditions place limitations on the curvatures that can be tolerated, because a vehicle travelling round a curve is subjected to the so-called 'centrifugal acceleration', and this gives rise to problems of mechanical safety and passenger comfort.

Roads and railways are super-elevated, that is, the outer side of the track is at a higher level than the inner side. A particular super-elevation is suitable for a particular speed, and if a vehicle travels at a very different speed, its passengers may feel discomfort or the vehicle may be subjected to undesirable mechanical stresses; or it may, on a road, be in danger of skidding.

The centrifugal acceleration is V^2/R, where V is the speed of the vehicle and R is the radius of the curve. If a road is super-elevated for speed V, and a vehicle travels at twice this speed, the centrifugal acceleration is $4V^2/R$, and the extra $3V^2R$ must be provided by tyre friction.

If W is the weight of a vehicle and g is the 'acceleration of gravity' (32·2 feet per second per second) the force necessary to keep the vehicle on the curve is WV^2/Rg. Taking γW as the maximum force that can be provided by friction, the radius of the curve must be such that $\gamma W > WV^2/Rg$; that is, $R > V^2/g\gamma$. With $\gamma = \frac{1}{4}$ and

V = 88 ft. per sec. (60 miles per hour), we find R > 968 ft.

Some of the required lateral force can be provided by super-elevation; to provide all of it, the cross-section slope would have to be equal to λ, or 1 in 4 in the above example. This is excessive, because such a cross-slope would cause discomfort, and perhaps danger, in a vehicle travelling very slowly.

It is clear, therefore, that high-speed traffic calls for curves of radius not less than 1500–2000 ft.

437. Transition Curves. Where a straight length of track joins a circular track, or two circular tracks of different radii are joined, a travelling vehicle will experience a sudden change of centrifugal force. At high speeds, this could be dangerous and uncomfortable; moreover, it would be impossible to provide correct super-elevation near the junctions. Therefore, it is general practice to provide *transition curves* where changes of curvature have to be made. In Fig. 225, the straight track AB is connected to the circular length CD by the transition curve BC.

FIG. 225.

The essential requirement in a transition curve is that it should introduce a smooth change of centrifugal effect spread over such a distance that the rate of change is tolerable in accordance with some acceptable standards of comfort and safety.

438. Vertical Curves. In high-speed travel, sudden changes of gradient are equally undesirable. The longitudinal sections of roads and railways therefore include vertical curves designed to

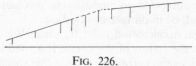

FIG. 226.

effect smooth changes between one gradient and the next, as shown, in an exaggerated way, in Fig. 226. Apart from questions of vertical acceleration, it is necessary to consider also the question of adequate intervisibility; in particular, a change of gradient that is convex upwards must not be too sharply curved. See Fig. 227.

If a driver's eye is h feet above the road, and the radius of the vertical section of the 'hump' is R feet, he cannot see the road surface beyond a distance $(2hR)^{\frac{1}{2}}$ feet. With h = 4, R = 2000, this works out as 127 feet, which is not enough at modern speeds of travel. See section 461.

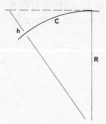

FIG. 227.

439. Elements of a Curve. Considerable portions of most curves will be circular, but there will generally be transition curves between successive circular or straight lengths.

Consider the circular curve AB in Fig. 228. EA and BF are

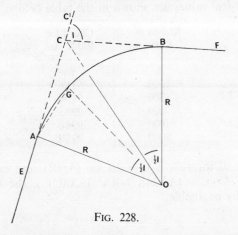

FIG. 228.

tangents, and these meet at C. The external angle I (C'CB) is the *Intersection Angle* or *Angle of Deviation*, and it is the total change of direction along the curve. CA and CB are each the *Tangent Length*, T; obviously, $T = R \cdot \tan \frac{1}{2}I$.   If AG is any chord, the angle CAG is its *deflection angle*, and this is half the angle AOG at the centre. The length of the chord is therefore

$$2R \sin \tfrac{1}{2} \text{ (deflection angle).}$$

Distances along a centre-line are usually reckoned from a defined zero point continuously throughout the work, and for this purpose lengths along a curve are arc lengths, A in Fig. 229. Measuring along a curve is troublesome, and actual setting-out is usually done by means of chord lengths C. Curves are usually pegged out at intervals of either 100 ft. or 66 ft. and intermediate pegs may be placed on sharp curves or non-circular curves.

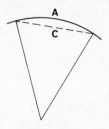

Fig. 229.

If A is 100 ft. on a circular curve, the angle at the centre is 100/R radian and the length of the chord is 2R sin (50/R). Chord lengths can be calculated so as to give 100 ft. arc lengths for various radii of curve; a few values are shown in the table below.

Radius of Curve	Chord for 100 ft. Arc
ft.	ft.
200	98·96
300	99·54
500	99·83
1000	99·96
2000	99·99

The general formula is $C = 2R \sin (A/2R)$, and expanding the sine gives: $A - C = (A^3/24R^2) - (A^5/1920R^4)...$, the second term being usually negligible.

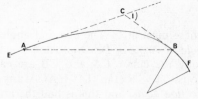

Fig. 230.

The essential geometry of a transition curve is indicated in Fig. 230.

Controlling dimensions will be the curvatures at the ends A and B, and the length of curve along which the change of curvature is to be effected. Given the nature of the curve, the detailed geometry for setting out can be computed. The exact dimensions of the triangle ABC will be required so that the curve can be fitted to the adjoining straight or circular portions.

SETTING OUT CURVES

440. Location of Tangents. In pegging out a line composed of straights and curves, the straights, or tangents, must be set out before the curves connecting them can be located. It is assumed that the surveyor is provided with a copy of the working plan, upon which the line is shown in relation to the controlling traverse of the preliminary survey and to existing features. By scaling distances from the traverse lines or from buildings, fences, etc., several points on each tangent can be obtained on the ground by tape measurements. Such points being temporarily marked, the tangents may then be set out by theodolite, so that, by trial and error, they run as nearly as possible through the marks.

441. Location of Curves. The design should show the locations of circular curves and specify their radii. Types and lengths of transition curves must be specified, with sufficient information to enable the setting-out surveyor to compute the necessary detail measurements.

442. Location of Tangent Points of a Single Circular Curve. For a given pair of straights, there is only one point at which a curve of given radius or degree may leave the first straight tangentially in order to sweep tangentially into the second. The points of commencement and termination of the curve must therefore be determined with greater precision than would be possible by merely scaling their positions from the plan.

(1) Having located the two tangents and defined them by ranging poles, peg out the first tangent EA (Fig. 231) up to about the estimated position of A, the theodolite being placed on EA and aligned on one of the poles. By means of the instrument, produce the straight to align two pegs *a* and *b* a few feet apart, one being placed on each side of C, the position of which is estimated by the chainman from the line of the poles on BF.

(2) Transfer the instrument to some convenient point on the second straight, and produce the latter to meet a string stretched between *a* and *b*. The point of intersection C of the two tangents thus obtained is marked by a peg.

(3) Set up the theodolite over C, and measure the angle ECF.

By subtracting the result from 180°, the value of the intersection angle I is obtained. Calculate the tangent lengths from $T = R \tan \frac{1}{2}I$.

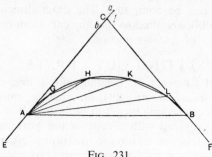

FIG. 231.

(4) From C measure back the lengths CA and CB = T, the tangent points A and B being aligned from the instrument at C. Mark A and B in a distinctive manner, either by painted pegs or by three ordinary pegs, the centre one of which defines the point.

(5) Transfer the instrument to A, and set it over the tangent point peg. Measure the angle CAB, which should equal $\frac{1}{2}I$. This provides a convenient check on the equality of the tangent lengths, which may, however, both be in error by the same amount through a mistake in the measurement of I or in the calculation of T.

(6) The chaining of the first straight may now be completed, the chainage of the point A being noted.

443. Location of Circular Curve by Deflection Angles. In this method, advantage is taken of the geometrical property that arcs of equal lengths subtend equal angles at any point on the circumference of a circle. The angles are half the angles at the centre. Referring to Fig. 231, if each of the arcs AG, GH . . . is 100 ft., and the radius of the circle is R ft., then each of the angles CAG, GAH, . . . is 50/R radian: thus if R is 1200 ft., each deflection angle is $\frac{50}{1200}$ (3437·75) minutes, that is 2° 23′ 14″.

The procedure is to place a theodolite at the tangent point A and with zero reading set on the horizontal circle, line up the telescope in the direction of the tangent AC. Then release the upper plate and set the reading to 2° 23′ 14″. Meanwhile, the zero of the tape is held at A and, with an arrow held at the correct chord length, the forward chainman sweeps round and gets the arrow lined up on the sight line. Then the tape is moved to G, the theodolite reading is set to 2 × (2° 23′ 14″), and the process is repeated. Note that it is desirable to compute the deflection angle to seconds, because a considerable number of multiples of the angle may have to be calculated and it is necessary to avoid undue accumulation of rounding-off error.

As a rule, the first chain peg on the curve will not be 100 ft. from A, if chainage is continuously measured along the centre-line. However, deflection angles are exactly proportional to arc lengths: thus, if the chainage at A is, say, 57827 ft. the arc AG will be 73 ft. and the appropriate deflection angle will be $(0.73) (2° 23' 14'') = 1° 44' 34''$. A convenient way of dealing with this is to set this partial angle backwards on the theodolite, that is, set the circle reading to 358° 15' 26'' for lining up on the tangent. The first chord must of course be calculated for an arc of 73 ft.

If the curve is left-handed, all deflection angles must be subtracted from 360°.

Some checks are available on the setting-out:

(i) the final reading on the other tangent point B should be $\frac{1}{2}I$, after allowing for any back-setting as mentioned above,

(ii) the *Principal Chord* AB can be chained and compared with the correct length $2R \sin \frac{1}{2}I$.

There will generally be a partial chord, or *Sub-chord* at the end B, and the rest of the 100 ft. length is set out along the tangent BF.

SETTING OUT SIMPLE CURVES BY CHAIN AND TAPE ONLY

444. The location of a simple curve may be performed without the use of an angle-measuring instrument in cases where:

(*a*) A high degree of accuracy is not demanded, as in minor roads for instance.

(*b*) The curve is short, as in certain railway and tramway curves, corners of buildings, curved wing walls, etc.

445. Location of Tangent Points. In short curves of small radius, the positions of the tangent points can generally be taken from a large scale plan with sufficient accuracy to permit of the fitting in of the curve between them on the ground. Otherwise, the following method may be used.

(1) Produce the two straights by eye to meet at C (Fig. 232).

(2) Select any pair of intervisible points D and E, one on each tangent and equidistant from C, making CD and CE as long as is convenient.

(3) Measure DE, mark its mid-point F, and measure CF.

(4) Since triangles CDF and COA are similar, $(CA/OA) = (CF/DF)$, so that $T = (CF/DF) . R$.

If CD and CE cannot conveniently be made equal, select any points D and E in suitable positions on the respective tangents.

Measure the three sides of triangle CDE, and solve for angle DCE = $180° - I$. Calculate T from $T = R \tan \frac{1}{2}I$, and measure out the differences DA and EB.

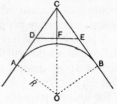

FIG. 232.

If C is inaccessible, angles ADE and DEB may be evaluated by solution of triangles such as A'DF and FEB' from the measured lengths of their sides, A' and B' being convenient points on the tangents. Thereafter the procedure is as in Case 3, section 450.

446. Location of Points on the Curve by Deflection Distances. This is the most useful method for a long curve.

Case (a). When there is no initial sub-chord, or the chainage is not required to be continuous.

(1) With the tangent point A as centre (Fig. 233), swing the chord length AD = C on the chain until the perpendicular offset ED from the tangent = $C^2/2R$. D is the position of the first peg on the curve.

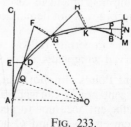

FIG. 233.

(2) By means of poles at A and D produce AD its own length to F. Mark F with a pole or arrow, and, with centre D, swing the chord length to the position DG, so that FG = C^2/R. Peg the point G.

(3) Produce DG to H, and repeat.

(4) To check the work at the second tangent point B, KB being found to be a sub-chord of length c', set out the points L and M as for a whole chord. Bisect LM at N, then KN is the tangent to the curve at K. From KN set off the perpendicular offset PB = $c'^2/2R$, and the error of tangency will be determined.

The above expressions for the offsets from the tangents and from the produced chords are precise. If Q is the mid-point of AD, and A, Q, D, and G are joined to the centre O, the triangles AED and OQA are similar.

$$\therefore \frac{ED}{C} = \frac{\frac{1}{2}C}{R}, \text{ or } ED = \frac{C^2}{2R}.$$

Again triangles DFG and ODG are similar, since both are isosceles and $FDG = 180° - 2GDO = DOG$,

$$\therefore \frac{FG}{C} = \frac{C}{R}, \text{ or } FG = \frac{C^2}{R}.$$

Unless the radius is very small, it is in general sufficiently accurate and more convenient to make $AE = C$, so that the offset $ED = C^2/2R$ is not perpendicular to AE.

In similar circumstances, LN may be set out perpendicular to KL and equal to $C^2/2R$ without the necessity of locating M.

Case (b). When there is an initial sub-chord, c.

A slight modification is necessary in this case on account of the circumstance that the second peg G could not be located in the usual manner from the production of the short chord AD (Fig. 234).

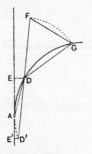

Fig. 234.

(1) Obtain the position of D by swinging the length c about A until $ED = c^2/2R$.

(2) Consider the curve extended as shown dotted, and locate the whole chord DD′ by turning it about D until the perpendicular offset E′D′ has the length given by the approximate formula $(C-c)^2/2R$.

(3) Produce D′D a whole chord length to F, and swing through $FG = C^2/R$ as before.

If the radius is more than, say, three times the chord length, it is usually quite allowable to locate D′ by measuring back AE′ = (C−c) and erecting the offset from E′.

If the sub-chord AD is not too short for reasonably accurate production, an alternative method of locating G is to produce AD to F′, making DF′ a whole chord length C. With centre D, the chord length is swung to DG, the distance F′G being approximately C(C+c)/2R.

447. Location of Points on the Curve by Tangent and Chord Offsets. These methods are useful for short curves. In such cases it is usually unnecessary that the pegs should be equally spaced, and the field work requires no explanation. Formulae from which to calculate the lengths of the offsets are as follows:

(a) Offset o at a distance l on a tangent from the tangent point (Fig. 235).

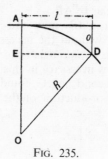

FIG. 235.

From triangle EDO,

$$R^2 = (R-o)^2 + l^2$$
$$\text{whence } 2Ro = o^2 + l^2$$
$$\text{and} \quad o = l^2/2R, \text{ approximately.}$$

(b) Offset o_1 from the mid-point of a chord of length C (Fig. 236).

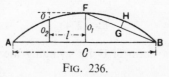

FIG. 236.

In the same manner,

$$o_1 = R - \sqrt{(R^2 - \tfrac{1}{4}C^2)}$$
$$= C^2/8R \text{ approximately.}$$

(c) Offset o_2 from the chord AB at a distance l from its mid-point.

Since a tangent to the curve at its mid-point F is parallel to

AB, o_2 is less than o_1 by the offset o above formulated, so that

$$o_2 = o_1 - R + \sqrt{(R^2 - \tfrac{1}{4}C^2)}$$
$$= o_1 - \frac{l^2}{2R} \text{ approximately.}$$

(d) Having obtained F by the offset o_1, an alternative method of locating other points on the curve consists in bisecting chord FB and erecting an offset GH = FB^2/8R from its mid-point. By dealing similarly with the chords FH and HB, etc., any number of points may be determined.

The approximate expressions above should not be used unless l or C is small in relation to the radius.

PROBLEMS IN RANGING SIMPLE CURVES

448. Case 1. When the Complete Curve cannot be Set Out from the Starting Point. It has hitherto been assumed that the curve is sufficiently short and the ground so flat and free from obstructions that all the required pegs on the curve can be set out from the one position of the instrument at A. It is, however, very commonly impossible to do so.

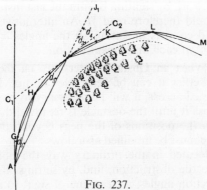

FIG. 237.

In Fig. 237, the pegs G, H and J have been located by the deflection angles d_1, d_2 and d_3 from A, but let it be supposed that, on setting off d_4, the line of sight AK is found to be obstructed.

(1) Transfer the instrument from A, and centre it over J.

(2) Set and clamp the vernier to the angle it read at A when sighting C, *i.e.* either zero or (360° − the deflection angle for the first sub-chord) according to the method used. Sight back on A.

(3) Transit the telescope. Set the vernier to d_4, the tabulated deflection angle for the point K, and the line of sight is now directed

along JK, for, if C_1JC_2 represents the tangent at J,

$$C_1JA = J_1JC_2 = d_3,$$
$$\therefore \ J_1JK = d_3 + \delta = d_4.$$

(4) Continue the setting out from J in the usual manner.

It may not be possible to complete the curve from the station J, in which case the further procedure is as follows:

(1) Set up over L, the last point located from J.

(2) Sight back on the last point occupied by the instrument or on any peg on the curve, with the vernier reset to read the deflection angle for that peg. Thus, if J is sighted, the vernier must first have been brought to read d_3.

(3) Transit the telescope. Set the vernier to d_6, the tabulated angle for the next peg M. The line of sight is now along LM, and the setting out may be continued.

Points which are to serve as instrument stations, or to which backsights are to be taken, should be marked with particular care, as otherwise it will be found troublesome to check in at the end of the curve.

The above method is arranged to possess the advantage that the checked tabular values, d_1, d_2, etc., are employed throughout. On completing the work by sighting B from the last instrument station, the vernier should therefore read $\frac{1}{2}I$. An alternative, but less convenient, method consists in setting off the angles δ, 2δ, etc., anew from the tangent at each instrument station.

449. Case 2. When an Obstacle Intervenes on the Curve. If an obstruction, whether it can be seen over or not, is such that it cannot be chained across, it will be necessary to omit the location of the line across it until the obstacle is removed during construction. To obtain the positions of the pegs beyond the obstacle the usual procedure must be modified as follows:

(1) Having located in the ordinary way the points G and H (Fig. 238) up to the obstruction, find, by setting off the successive tabulated deflection angles, a clear line of sight to a point on the curve. Let AL be this line, the deflection angle being d_5. It is supposed that the point K, although clear of the obstacle, cannot be seen from A.

(2) Calculate the length of the long chord AL from the formula,

$$\frac{\frac{1}{2} \text{ chord}}{\text{radius}} = \sin \tfrac{1}{2} \text{ (central angle subtended by chord).}$$

i.e. AL = $2R \sin d_5$.

(3) Measure out this distance from A, aligning the chaining from the instrument at A, and peg the point L.

(4) Continue the curve from L in the usual manner.

If necessary, pegs such as K can be located by offsets from the long chord by the methods of section 447, or, if the instrument is transferred to L, they may be set out by deflection angles from L. Otherwise they may be left over until the obstruction has been removed.

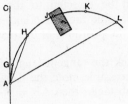

FIG. 238.

It may happen that no clear line AL can be obtained on account of obstacles of the type of Case 1. In such circumstances, having calculated the length of the curve, and so determined the length of the final sub-chord, the curve may be set out from the end B in the reverse direction up to K.

450. Case 3. When the Point of Intersection of the Tangents is Inaccessible. This difficulty is of frequent occurrence. Since the intersection point C is employed both in the measurement of I and as the starting-point from which the lengths T are measured back, the field work must be arranged to supply the twofold deficiency when C is inaccessible.

(1) Select any convenient intervisible points E and F (Fig. 239) on the straights or on the tangents.

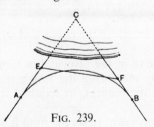

FIG. 239.

(2) Measure angles AEF and EFB; then $I = 360° - \text{AEF} - \text{EFB}$. Calculate the tangent lengths from $T = R \tan \frac{1}{2}I$.

(3) Measure EF, and solve triangle CEF for CE and CF.

(4) To locate A and B, measure from E the distance EA $= (T \sim \text{CE})$, and from F the distance FB $= (T \sim \text{CF})$.

(5) Complete the pegging out of the curve in the usual way.

If no clear line EF can conveniently be obtained, in which case it is necessary to run a traverse between E and F, the required angles AEF and EFB and the distance EF are obtained by calculation.

451. Case 4. When the First Tangent Point is Inaccessible. The field work must first be directed to the determination of the chainage of the inaccessible tangent point, since the length of the initial sub-chord and the positions of the pegs on the curve cannot otherwise be known.

(1) On chaining back the tangent length from C (Fig. 240) and finding that A is inaccessible, note the measurement from C to a point F clear of the obstacle; then AF = T − CF.

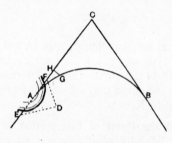

FIG. 240.

(2) By any method of measuring past an obstacle, *e.g.* by solution of a triangle such as FDE, determine the distance from F to some convenient point E on the straight and at the other side of the obstacle.

(3) Obtain the chainage to E; then the chainage of A = the chainage of E + EF − AF.

(4) Compute the length of the curve, and so obtain the chainage of B.

(5) Set off the curve in the reverse direction from B. The result may be checked by measuring the length of the offset from the last peg located to the tangent, its required length being $AH^2/2R$ approximately.

If it is found inconvenient to set out from B, the method of Case 6 may be employed.

452. Case 5. When the Second Tangent Point is Inaccessible. In this case the continuation of the chainage along the second straight forms the difficulty. Let A (Fig. 240) be the second tangent point.

(1) On chaining the second tangent length, establish a point F at a known distance from C.

(2) Obtain, as in the last case, the distances $FA = (T-CF)$, and FE to a convenient point E on the straight.

(3) The chainage of A having been computed from that of the starting-point and the length of the curve, the chainage of E = the chainage of $A+FE-FA$.

(4) From the point E locate the first accessible chainage peg on the second straight.

453. Case 6. When both Tangent Points are Inaccessible. (1) Having obtained the length AF and the chainage of A by the method of Case 4, compute, from $FD = R-\sqrt{(R^2-AF^2)} = R-\sqrt{[(R-AF)(R+AF)]}$ the value of the perpendicular offset from F to the curve (Fig. 241).

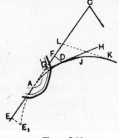

FIG. 241.

(2) With the theodolite at F, set out the point D very carefully. D will be employed as an instrument station from which the curve may be continued.

(3) Calculate the length of arc AD from the circumstance that it subtends at the centre an angle $\sin^{-1}(AF/R)$; then the chainage to D = the chainage of $A+\text{arc } AD$.

(4) Draw up a table of deflection angles referred to the tangent DH.

(5) Set the instrument over D. To lay the line of sight along DJ towards the first peg J, distant c from D, a backsight may be taken on F, and the vernier turned through

$$FDJ = 90° + \sin^{-1}\frac{AF}{R} + \frac{c}{C}\delta.$$

But, unless DF has been made reasonably long, it is much better to secure a long backsight by regarding the curve as extended beyond A to E_1, the point E_1 being established by an offset EE_1. On sighting E_1 from D, and turning through angle

$$E_1DJ = 180° - \tfrac{1}{2}\left(\sin^{-1}\frac{AE}{R} + \sin^{-1}\frac{AF}{R}\right) - \frac{c}{C}\delta,$$

the line of sight will be directed towards J.

356   PLANE AND GEODETIC SURVEYING

(6) Continue the location until the obstacle at B is reached. Check the work by measuring an offset from the curve to the second tangent, and use the method of Case 5 in order to continue the chainage.

The whole of the field work involved in this method requires considerable care in order that the closure may be effected. The instrument stations should be carefully aligned and marked by a tack on the peg. Checks on the progress of the work on a long curve may be made at intermediate instrument stations, such as K, by noting where the tangent KL cuts the first tangent. The distance CL, for instance, should be found to equal

$$\left[T - R \tan \tfrac{1}{2}\left(\frac{180° \times \text{arc AK}}{\pi R}\right)\right].$$

Having determined the chainage of A, the following alternative method of setting out the curve is preferable when the intersection point C is readily accessible. Set up the theodolite at C, and bisect angle ACB. Compute the apex distance, or distance from C to the mid-point of the curve, and, by measuring out this distance along the bisector, locate M, the mid-point of the curve. The chainage of M = the chainage of $A + \tfrac{1}{2}L$. Transfer the instrument to M, backsight on C, and set out the curve in both directions from M.

A third method consists in locating an instrument station on the curve by measuring out from a point, such as E, on the straight. Assuming that D is a point of known chainage to which a measurement can be made from E, the distance ED and the angle CED are obtained by solution of triangle EAD since, the chainage of A having been ascertained, EA, AD, and angle EAD can be evaluated. Alternatively, triangle EGD may be solved. The point D having been pegged, the setting out can be continued from D after back-sighting on E.

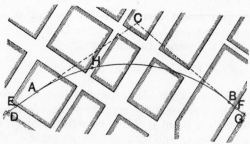

Fig. 242.

The foregoing difficulties may occur simultaneously, particularly in the course of setting out in cities. Thus, in Fig. 242, which represents the centre line of part of an underground railway, the setting out is based upon the points D and E on one tangent and F and G on the other. Very careful traversing must be executed between DE and FG to determine the intersection angle and the positions of A and B. To obtain any point on the curve, such as the centre of a shaft H, the coordinates of the point are computed from, say, E as origin and DE as meridian. A traverse is conducted from E towards H, the length of the last line being computed so that it ends on the required point. The bearing of the tangent at H is also calculated, and is set out from the traverse to form a base for the location of the curve underground.

454. Horse Shoe Curves. If, as in Fig. 243, the angle I subtended at the centre of a curve exceeds 180°, the point of intersection C of the tangents EA and BF lies on the same side of the curve as the centre O. The angles at C are as marked, and evidently from triangle AOC

$$T = R \tan (180° - \tfrac{1}{2}I).$$

The deflection angles may be laid off from AD, the production of the first straight, and they are tabulated in the usual way up to the maximum angle, DAB $= \tfrac{1}{2}I$.

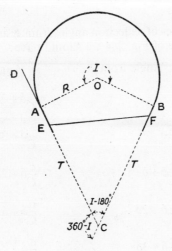

FIG. 243.

In country necessitating the introduction of such curves it will usually be impracticable to obtain the value of I at the intersection point C, and the method of running a series of traverse courses, or, if possible, single transversal between points on the two straights, is generally required. If EF be the traverse closing line or the transversal, then $I = \text{AEF} + \text{EFB}$.

455. Setting out Simple Curves by Two Theodolites. The positions of pegs on a curve may be determined, without the necessity of chaining the chord lengths, by locating them as the points of intersection of the lines of sight from two theodolites placed one at each end of the curve. Since two instrument men are required, the method is not in general use, but might be warranted in cases where the curve lies on ground of such character that there would be difficulty in performing the chaining with the required degree of accuracy.

(1) Having determined the positions of A and B, the chainage of A, and the length c of the initial sub-chord, calculate the length of the curve and from it the chainage of B and the length c' of the terminal sub-chord.

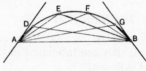

FIG. 244.

(2) Draw up tables of deflection angles from A and B respectively. Thus, for the curve of Fig. 244 we should have:

Point	Angle from A	Angle from B
A	—	$\dfrac{c'}{C}\delta + 3\delta + \dfrac{c}{C}\delta = \tfrac{1}{2}I$
D	$\dfrac{c}{C}\delta$	$\dfrac{c'}{C}\delta + 3\delta$
E	$\dfrac{c}{C}\delta + \delta$	$\dfrac{c'}{C}\delta + 2\delta$
F	$\dfrac{c}{C}\delta + 2\delta$	$\dfrac{c'}{C}\delta + \delta$
G	$\dfrac{c}{C}\delta + 3\delta$	$\dfrac{c'}{C}\delta$
B	$\dfrac{c}{C}\delta + 3\delta + \dfrac{c'}{C}\delta = \tfrac{1}{2}I$	—

The fact that the sum of the deflection angles to each point on the curve equals $\tfrac{1}{2}I$ provides a check in addition to the usual one.

(3) With the instruments at A and B respectively, set off these angles consecutively, the chainman being guided to each point until the pole held by him is simultaneously in the two lines of sight.

Should it be found necessary to shift one or both instruments to sub-stations on the curve, the tabulated angles will be employed throughout if the orientation of the instrument at each setting is performed as in Case 1, section 448.

TRANSITION CURVES

456. A typical situation for a transition curve is as a connection between a straight portion of track and a circular curve. In Fig. 245, the transition curve is AB. Occasionally, a transition curve will connect two circular curves of different radii. The radius of curvature at each end of the transition curve must be equal to that of the curve it joins.

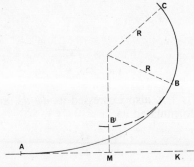

FIG. 245.

If V is the speed of a vehicle travelling from A to B, the centrifugal acceleration increases from zero at A to V^2/R at B. An ideal transition curve will effect this change at a uniform rate.

An additional requirement in practice is that the change of centrifugal acceleration must not occur too rapidly. If S is the total length of transition curve, the time taken in travelling along it is S/V and the rate of change of acceleration is therefore V^3/RS, feet per second per second per second. A recommended reasonable value for this rate is one unit, and this gives the relation: $RS = V^3$, which will determine a lower limit for S when R and V are specified. Thus, with V = 100 ft./sec., R = 2000 ft., we find S = 500 ft.

The geometry of the curve will be largely determined by calculations of this kind, and the setting-out surveyor's task will be to peg out the curve accurately in correct position on the ground in relation to the straights and circular curves indicated on the plan.

In connection with the setting-out, it is often useful to know the *shifts* of the circular curve in relation to the tangent point A and the produced tangent AK. If the circular curve is produced backwards there will be a point B′ where its direction is parallel to AK: the shifts are the quantities AM and MB′.

457. The Clothoid. The curve that exactly fits the requirement mentioned above is the *Clothoid*, or *Euler Spiral*.

The usual notation of differential geometry will be used; *s* is the length of curve measured from a fixed point, ψ is the angle made by the tangent with a fixed direction taken as the *x*-axis, and ρ is the radius of curvature at a point on the curve: further, rectangular coordinates *x* and *y*, or polar coordinates *r* and θ, may be used to denote the position of a point, as indicated in Fig. 246.

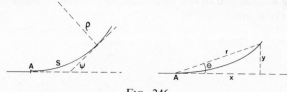

FIG. 246.

The curvature is $1/\rho$, also expressed as $d\psi/ds$, and the desired property can be formulated as:

$$\frac{d\psi}{ds} = \frac{2s}{a^2},$$

where *a* is a linear parameter introduced to preserve correct dimensions. This differential equation is directly integrable and the introduction of the factor 2 makes for the simple solution: $\psi = s^2/a^2$, with the proviso that ψ and *s* are zero at the same point, which of course is A.

In this formula, ψ is to be reckoned in radian measure, and it is now seen that the parameter *a* is the length of curve to the point where ψ is a radian (about 57·3 degrees).

The radius of curvature is $ds/d\psi$, or $a^2/2s$. If S is the total length of the curve to the point where it joins the curve of radius R, then $R = a^2/2S$, or $a^2 = 2RS$. Given R and S, *a* is thus determined: with the numerical values used above, $a^2 = 2,000,000$. Then $S^2/a^2 = 1/8$, and this is the radian value of the angle turned by the whole curve (about 7°).

It seems, therefore, that practical transition curves for high speed traffic will not comprise large deflections, and the values of *s* actually to be considered will be much less than the parameter *a*.

For actual setting-out, the ordinary rectangular coordinates x and y can be used, being distances along and offsets from the tangent AK. It is convenient to calculate these in terms of s, because pegs can then be placed at specific chainages along the curve: however, it is not possible to write finite formulae for x and y of the clothoid, and expressions in series must be derived.

We have: $s = a(\psi)^{\frac{1}{2}}$; and hence: $ds = \frac{1}{2}a(\psi)^{-\frac{1}{2}}d\psi$. Then: $dx = ds \cdot \cos\psi = \frac{1}{2}a \cdot d\psi \cdot (\psi)^{-\frac{1}{2}}(1 - \frac{1}{2}\psi^2 + \frac{1}{24}\psi^4 \ldots)$ and this can be directly integrated, giving,

$$x = a \cdot \psi^{\frac{1}{2}}\left(1 - \frac{1}{10}\psi^2 + \frac{1}{216}\psi^4 \ldots\right)$$

A similar operation starting with $dy = ds \cdot \sin\psi$ leads to:

$$y = \frac{1}{3}a \cdot \psi^{\frac{3}{2}}\left(1 - \frac{1}{14}\psi^2 + \frac{1}{440}\psi^4 \ldots\right)$$

These series in ψ can be immediately expressed in terms of s by the substitution of $\psi = s^2/a^2$. Thus:

$$x = s\left(1 - \frac{1}{10}\frac{s^4}{a^4} \quad \frac{1}{216}\frac{s^8}{a^8}\ldots\right)$$

$$y = \frac{1}{3}\frac{s^3}{a^2}\left(1 - \frac{1}{14}\frac{s^4}{a^4}\ldots\right)$$

In practice, the terms of order s^5 will be small and succeeding terms quite negligible. Taking the numerical values used above, the rectangular coordinates for the total curve are:

$$X = 500\left(1 - \frac{1}{640} + \ldots\right) = 500 - 0.78 = 499.22 \text{ ft.,}$$

$$Y = \frac{500}{24}\left(1 - \frac{1}{896}\ldots\right) = 20.83 - 0.02 = 20.81 \text{ ft.}$$

Other geometrical quantities are easily found from the series for x and y, thus:

$$r = (x^2 + y^2)^{\frac{1}{2}} = a \cdot \psi^{\frac{1}{2}}\left(1 - \frac{2}{45}\psi^2 + \frac{2}{2835}\psi^4 \ldots\right),$$

$$\theta = \frac{1}{3}\psi\left(1 - \frac{8}{945}\psi^2 \ldots\right),$$

and $s - r = \frac{2}{45}a \cdot \psi^{\frac{5}{2}}\left(1 - \frac{1}{63}\psi^2\ldots\right) = \frac{2}{45}\frac{s^5}{a^4}\left(1 - \frac{1}{63}\frac{s^4}{a^4}\ldots\right).$

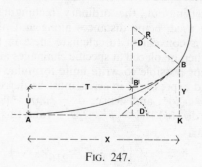

FIG. 247.

These can be used for setting out the curve by deflection and chord method.

The shifts, U and T in Fig. 247, are to be expressed in terms of the total length S and the total deflection D which is equal to S^2/a^2. We have:

$$T = X - R \cdot \sin D = X - \frac{a^2}{2S}\left(D - \frac{1}{6}D^3 \dots\right)$$

$$= \tfrac{1}{2}S\left(1 - \frac{1}{30}\frac{S^4}{a^4} + \dots\right), \text{ and}$$

$$U = Y - R(1 - \cos D) = \frac{1}{12}\frac{S^3}{a^2}\left(1 - \frac{1}{28}\frac{S^4}{a^4} + \dots\right).$$

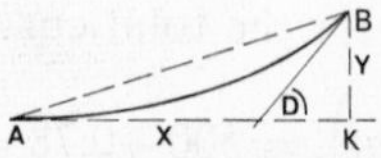

FIG. 248.

The dimensions of the triangle ABK, Fig. 248, and the value of the total angle of turn D, are calculable from formulae derived above, and these quantities are obviously the key dimensions required for placing the transition curve in correct relation to the other curves joined to it.

458. Other Transition Curves. If it is considered unnecessary to maintain *exact* proportionality between distance and curvature, other curves approximating to the clothoid geometry may be used, and various forms of curve have been proposed on grounds of computational simplicity.

For instance, taking only the first terms of the series for x and y of the clothoid, and eliminating ψ, gives the relation: $3a^2y = x^3$.

This is the *Cubic Parabola*.

We have:

$$\frac{dy}{dx} = \frac{x^2}{a^2}, \quad \frac{d^2y}{dx^2} = \frac{2x}{a^2}.$$

The last equation shows that the acceleration in the y direction is proportional to x, so the transition condition will be approximately satisfied so long as the deflection angle ψ remains small.

The radius of curvature can be calculated from the general formula thus:

$$\rho = \left[1 + \left(\frac{dy}{dx}\right)^2\right]^{\frac{3}{2}} \bigg/ \frac{d^2y}{dx^2} = \frac{a^2}{2x}\left(1 + \frac{x^4}{a^4}\right)^{\frac{3}{2}}.$$

This formula can be expanded and inverted to give the series:

$$x = \frac{1}{2}\frac{a^2}{\rho}\left[1 + \frac{3}{32}\cdot\frac{a^4}{\rho^4} + \frac{75}{2048}\frac{a^8}{\rho^8}\cdots\right]$$

A possible way to proceed with this curve is to use the value of a^2 as calculated for the clothoid, put this, and the terminal radius of curvature R, into the above series, and calculate the total co-ordinate X: thus, with $a^2 = 2{,}000{,}000$ and $R = 2000$, we have $a^2/R^2 = \frac{1}{2}$ and

$$X = 500\left(1 + \frac{3}{128} + \frac{75}{32768}\cdots\right)$$

$$= 500 + 11{\cdot}72 + 1{\cdot}14\ldots = 512{\cdot}9 \text{ ft.}$$

The total offset Y is then immediately found from the original formula, and is 22·5 ft.

The total curve length S will be greater than X, so the rate of increase of centrifugal acceleration is not likely to exceed the desired limiting value.

The general formula:

$$\left(\frac{ds}{dx}\right)^2 = 1 + \left(\frac{dy}{dz}\right)^2$$

leads to

$$\frac{ds}{dx} = \left(1 + \frac{x^4}{a^4}\right)^{\frac{1}{2}},$$

which is expanded and integrated to give:

$$s = x\left(1 + \frac{1}{10}\frac{x^4}{a^4} - \frac{1}{72}\frac{x^8}{a^8}\cdots\right).$$

If setting-out data are required at specific values of s, this last series can be reversed to give:

$$x = s\left(1 - \frac{1}{10}\frac{s^4}{a^4} + \frac{23}{360}\frac{s^8}{a^8}\cdots\right)$$

hence $y = \frac{1}{3}\frac{s^3}{a^2}\left(1 - \frac{3}{10}\frac{s^4}{a^4}\cdots\right).$

The direction of the tangent is of course given by

$\tan\psi = \dfrac{dy}{dx} = \dfrac{x^2}{a^2}$, which is $\dfrac{X^2}{a^2}$ at the end of the curve.

Thus, data for fitting the curve into the plan are calculable.

In the cubic parabola, y is proportional to x^3. For small deflections, that is small values of ψ, there is not much difference between x and s, so one would expect that a curve with y proportional to s^3 would also be a serviceable approximation to a strict transition: such a curve is called a *cubic spiral*.

With $y = \dfrac{s^3}{3a^2}$, we have $\dfrac{dy}{ds} = \sin\psi = \dfrac{s^2}{a^2}$.

Thus, $s = a\sqrt{(\sin\psi)}$ and $\dfrac{ds}{d\psi} = \rho = \dfrac{a\cos\psi}{2\sqrt{(\sin\psi)}}.$

The coordinate x can be expressed as a series in s, from the relation:

$$\left(\frac{dx}{ds}\right)^2 = 1 - \left(\frac{dy}{ds}\right)^2 = 1 - \frac{s^4}{a^4};$$

expanding dx/ds and integrating gives:

$$x = s - \frac{s^5}{10a^4} - \frac{s^9}{72a^8}\cdots$$

The formulae for s and ρ combine to give

$$\frac{a^4}{s^4} = 1 + \frac{4\rho^2}{s^2},$$

and using the values of R and S given above, we find:

$$\frac{a^4}{S^4} = 65, \qquad a^2 = 2015563,$$

and the total deflection is:

$$\text{arcsin}\ \frac{S^2}{a^2} = 7°\ 07\tfrac{1}{2}'.$$

459. Lemniscate. Another look at the clothoid formulae will reveal that ψ is nearly equal to 3θ. The curve in which this relation holds exactly is a *lemniscate*. Referring to Fig. 249, it is seen that the angle between the tangent and the polar radius must be 2θ, and the infinitesimal triangle shows that $r \, . \, d\theta/dr = \tan 2\theta$. The solution

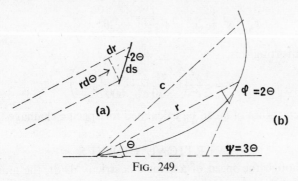

FIG. 249.

of this differential equation is $r^2 = c^2 \sin 2\theta$, where c is a linear constant, and is in fact the maximum value of r, reached when $\theta = 45°$ and $\psi = 135°$. The curve has petal-like loops, and, as before, the similarity to the clothoid only holds for small values of ψ.

From the elementary differential triangle, we have:

$$ds = r \, . \, d\theta \, . \, \operatorname{cosec} 2\theta = r \, . \, \tfrac{1}{3}d\psi \, . \, \frac{c^2}{r^2},$$

hence

$$\frac{ds}{d\psi} = \rho = \frac{c^2}{3r}, \text{ or } 3r\rho = c^2.$$

Again taking the clothoid length S for the limiting value of r, and $\rho = 2000$ ft. at the end of the curve, we have:

$$c^2 = 3 \, . \, 500 \, . \, 2000 = 3,000,000$$

whence $c = 1732$ ft.

Plotting data can be calculated from:

$$x = c \cos \theta \, (\sin 2\theta)^{\frac{1}{2}}, \quad y = c \sin \theta \, (\sin 2\theta)^{\frac{1}{2}},$$

using a series of values of θ.

At any point of the curve the angle turned through is 3θ; for the whole curve, $\sin \tfrac{2}{3}D = (500)^2/3,000,000 = 1/12$, whence D = $7° \; 10'$.

The relation $ds/d\theta = r \operatorname{cosec} 2\theta = c (\operatorname{cosec} 2\theta)^{\frac{1}{2}}$ can be expanded and integrated to give:

$$s = c(2\theta)^{\frac{1}{2}} \left(1 + \frac{1}{15} \theta^2 + \frac{1}{90} \theta^4 ...\right)$$

or

$$s^2 = 2c^2 \left(\theta + \frac{1}{15}\theta^2 + \frac{2}{75}\theta^4 ...\right)$$

By inverting this,

$$\theta = \frac{1}{2}\frac{s^2}{c^2}\left(1 - \frac{1}{30}\frac{s^4}{c^4} - \frac{1}{600}\frac{s^8}{c^8} ...\right)$$

and then values of θ can be calculated for specific chainages along the curve.

VERTICAL CURVES

460. Though the speed of a vehicle is strictly ds/dt, the gradients on railways and motorways are so gentle that it is sufficiently accurate to take the speed as dx/dt, where x is measured in the horizontal direction. The vertical acceleration is d^2y/dt^2 and its rate of change is d^3y/dt^3: if the rate is constant then so is d^3y/dx^3, and the shape of the desired vertical curve can be derived from a relation:

$$\frac{d^3y}{dx^3} = \frac{1}{a^2},$$

in which the linear parameter is introduced to preserve correct dimensions.

Since we can write $dx = V \cdot dt$ (on the above assumption), the differential equation also takes the form:

$$\frac{d^3y}{dt^3} = \frac{V^3}{a^2}.$$

For comfort of travel a limit must be set for this rate of change of acceleration; if, as before, it is taken as one unit, with $V = 100$ ft./sec., the corresponding value of a^2 is 1,000,000, and any value greater than this may be used.

Integration of the relation $d^3y/dx^3 = 1/a^2$ leads to:

(i) $\dfrac{d^2y}{dx^2} = \dfrac{x}{a^2}$, if x is measured from the beginning of the curve;

(ii) $\dfrac{dy}{dx} = \dfrac{x^2}{2a^2} + \lambda,$ giving the gradient at any point, λ being the given initial gradient;

(*iii*) $y = \dfrac{x^3}{6a^2} + \lambda x$, as the equation of the curve in which the first

term describes the shape of the curve and λx is the vertical separation of the curve and the tangent at A. See Fig. 250.

FIG. 250.

If the whole transition curve is symmetrical about its middle point and each half is of horizontal length X, the total change of gradient will be $2X^2/2a^2 = X^2/a^2$ and this will define X if the two gradients to be connected are given. For instance, to connect a horizontal section to a gradient of 1 in 10, we have $X^2/1,000,000 = 1/10$, $X^2 = 100,000$, $X = 316 \cdot 23$ ft. : total length of curve $632 \cdot 46$ ft.

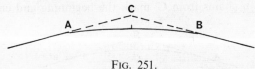

FIG. 251.

Where there is considerable change of gradient between ends of a curve convex upwards, the curvature must not be allowed to become too great for adequate distance of visibility. The curvature is d^2y/dx^2 or x/a^2, so the radius of curvature at the middle point is a^2/X. With the numbers given above, this is 3162 ft. The visibility from a height of 4 ft. is $(8R)^{\frac{1}{2}}$, or 159 ft. in this case. In conditions of high speed traffic, this is much too small for safety, and a larger value of a^2 would have to be used in designing the curve. See next section.

For actual setting out, heights at points along the tangents can first be computed, and then values of $x^3/6a^2$ added or subtracted according to whether the curve is a 'hollow' or a 'hump'.

461. Visibility Considerations. As mentioned in section 438, the distance of visibility on upward convex curves (summits) is an important practical consideration. To estimate required distances, a knowledge of the maximum practical retarding force that can be applied to a vehicle is needed. If the speed is V and the retarding force produces a deceleration of f (ft./sec.²), the distance S in which the vehicle can be stopped is given by $2fS = V^2$. For example, suppose the maximum coefficient of friction that can be used is 2/5, that is, the retarding force is 2/5 of the weight of the vehicle, then we have $V^2 = \frac{4}{5}gS$ or approximately $V^2 = 25 \cdot S$: thus, if V is 100 ft./sec., S is 400 ft.

This example shows that, for modern speeds of traffic, the geometry of vertical curves on summits will be determined by considerations of visibility rather than of vertical accelerations.

In practice, further allowance must be made for time lag in driver's reaction to visual signals; on the other hand, most vehicles are much higher than 4 ft., and are seen before their drivers become visible.

On one-way roads, visibility specifications may be approximately halved, because it is not necessary to allow stopping space for two vehicles.

462. Parabolic Transition Curve. A curve of this shape may be used where traffic is not fast, and it is easy to set out.

In Fig. 252, let two gradients meet at C, and let it be required to join them with a parabolic vertical curve *n* chains long. Pegs A and B, situated $\frac{1}{2}n$ chains from C, mark the beginning and end of the

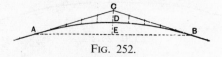

Fig. 252.

curve ADB. By taking the average of the known elevations of A and B, obtain that of E, the mid-point of AB, which is situated on the vertical through C. The elevation of C being given on the profile, the dimensions CE is now known, and since the parabola bisects CE at D, the elevation of D is obtained. Now, from the property of the parabola that offsets from a tangent are proportional to the squares of distances along the tangent, the elevations of points on the curve may be computed. Thus, let $\frac{1}{2}n = 5$ chains, the chainage of C being integral as is usual, and let it be supposed that CD, $= \frac{1}{2}$ CE, has been found to be 0·70 ft., then the offsets in order from A, as shown dotted, are respectively $\frac{1}{25}$, $\frac{4}{25}$, $\frac{9}{25}$, and $\frac{16}{25}$ of 0·70 ft., the corresponding offsets from CB having the same value.

MISCELLANEOUS OPERATIONS IN SETTING OUT

463. Setting Slope Stakes. The operation of slope staking consists in locating and pegging points on the lines in which proposed side slopes intersect the original ground surface.

In the case where the cross sections have been plotted and the new work is shown thereon, all that is necessary is to scale the horizontal distances representing the side widths or 'half-breadths' (section 407) and to set out these measurements on the ground. Slope stakes can, however, be located without the use of plotted cross sections by a trial and error process, which is best followed from an example.

In Fig. 253, let the formation level at a certain point on an embankment, as derived from the profile, be 123·60, the top width being 30 ft., and the side slopes $1\frac{1}{2}$ horizontal to 1 vertical. Let the

FIG. 253.

instrument height of a level commanding the ground be, say, 121·3. The difference of 2·3 ft. between instrument height and formation level represents a quantity, called the grade staff reading, which in this case falls to be added to readings of the staff when held on ground points in order to give the depths below formation of those points. If, for example, the centre height of the earthwork is not known, by reason of a local deviation or otherwise, and the staff is held on the ground at the line peg C, giving a staff reading of, say, 5·6 ft., the centre height of the embankment will be 7·9 ft.

To locate the slope peg A, the staffman must estimate where the foot of the slope will come, and he holds the staff there, measuring also its distance from the centre peg C. Let us suppose the staff reading to be 8·3 ft. This represents a depth from formation level of $8·3 + 2·3 = 10·6$ ft., and, if the point selected is really on the toe of the slope, the corresponding side width ought to be $1\frac{1}{2} \times 10·6 + 15 = 30·9$ ft. If the measured distance happens to have this value, the staffman has estimated the position correctly, but it will usually require a second or third trial before the correct point is found. Thus, if the staff reading of 8·3 ft. was obtained at a distance of 27 ft. from C, the surveyor, on finding that the computed side width of 30·9 ft. does not correspond with the measurement, would direct the staff holder to proceed farther from the centre. The next results might be 9·2 ft. on the staff and 32 ft. on the tape, which would be practically correct. In the same manner, the results of the final trial in setting the right-hand slope stake B might be 3·7 ft. on the staff at 24 ft. from C.

This method is very convenient for taking and recording cross sections, and is invariably adopted by American railway engineers.

The results in the case of the three-level section of our example are conveniently written,

$$\frac{-11\cdot5}{32\cdot0} \qquad -7\cdot9 \qquad \frac{-6\cdot0}{24\cdot0}$$

the upper figures in the case of the side points representing the vertical distances between formation and ground levels, reckoned positive for cuttings and negative for embankments, and the lower figures their distances from the centre. Additional points between the centre peg and the slope stakes may be recorded by observing the staff reading and distance from the centre and applying the grade staff reading positively or negatively to the former. From the side widths and centre and side heights, the area may be computed by the formulae of Section 407.

464. Setting Out Tunnels. The setting out of a tunnel is an operation demanding a high order of precision throughout. A simple type of theodolite may not be sufficient, and an instrument reading to 1 second is often necessary.

The simplest case occurs when the two ends of the tunnel are intervisible or visible from an intermediate point, so that its direction can be laid out on the ground without difficulty and continued beyond the entrances. Permanent ranging marks are constructed clear of the work, and from these the line may be projected into the tunnel at each end. If the tunnel is also to be driven from shafts, these must be carefully aligned on the surface, and, when the shaft is sunk, the alignment at the bottom is commonly obtained by suspending two plumb lines from a frame above the shaft. Each consists of copper, brass, or steel wire, and carries a rather heavy weight immersed in a pail of water for steadiness. The plumb lines should be capable of fine adjustment into line, and the upper end of each should therefore be mounted in a manner permitting lateral movement by means of a screw. The distance between the wires forms a short base which is prolonged underground by setting the theodolite in line with them, numerous face right and face left observations being taken to reduce instrumental and observational errors to a minimum. A lamp, screened by having a sheet of tracing cloth pasted over the glass, is placed behind the far wire when the instrument is being aligned, and a similar lamp is used behind any mark being set out. When the shaft is not on the centre line of the tunnel, the plumb lines are usually placed so that the line through them is roughly normal to the centre line. This line is prolonged on the surface to meet the centre line, and the intersection angle is measured, as well as the chainage of the intersection point and its distance from the plumb lines. By setting out the same dimensions underground, the corresponding point on the centre line and the direction of the tunnel are obtained.

Similar care is required on the levelling. The relative levels of formation at the two ends of the tunnel must be carefully determined by running two or more lines of levels between them, the precautions against error following those adopted in precise levelling (Vol. II, Chapter VI). The formation levels at the various shafts are computed, and are established underground by means of vertical steel tape measurements from level marks at the surface.

The more usual case occurs when the ends of the tunnel are not intervisible, and their relative positions must then be carefully determined. This is generally done by triangulation extended from a carefully measured base line (Vol. II, Chapter III), and, in very rough broken country, this may be the more convenient method, great care being taken to see that the signals are truly vertical and properly centered, as failure to exercise the greatest possible vigilance with regard to this point may be a serious source of error in triangulation when the sides are short. In flat country, or in built-up areas, traversing may be the more convenient method, and there is no reason why it should not be used, even for very long tunnels, provided the taping is very carefully done with a properly standardised steel or invar tape, the legs are sufficiently long, and the angular measurements made with a good micrometer theodolite, reading directly to 10″ or less, and, if necessary, each angle measured on several different zeros. The length of the tunnel is ascertained, and intermediate points are located, by coordinate calculation from the triangulation or traverse, and the centre line, whether straight or curved, may then be set out on the surface. Instrument stations are established, and by observations from them the plumbing wires at the shafts are aligned as before. In the case of curves, the centre line is best located underground by offsets from a series of chords.

The procedure in setting out tunnels naturally varies according to local conditions. Short of actual experience, an adequate knowledge of the methods employed can be derived only by a careful study of descriptions given in publications on engineering and mining.

The gyro-theodolite can be very useful for setting out bearings underground: true, or astronomical, bearings are so obtained.

465. Setting Out Viaduct Piers on a Curve. In setting out a river viaduct, the location of points on the piers and abutments has generally to be performed several times at various stages of the work. The first operation during construction may consist in the location of staging surrounding each pier, and, according to the class of foundation adopted, cofferdams, cylinders, or caissons will subsequently have to be aligned. On completion of the foundations, the outlines of the piers and abutments are set out upon them, and it may at this stage be practicable to employ instrument stations on the piers. The methods will be sufficiently indicated by describing the location of the centre lines of the piers of a curved viaduct by observations from the river banks only.

Two cases may occur: (1) The tangent points may fall on the banks behind the abutments, so that the whole viaduct is on the curve; (2) The tangent points may be situated in the river, so that only a part of the viaduct is on the curve.

466. Case 1. Tangent Points Accessible. The preliminary data will include: (*a*) the radius of the curve; (*b*) the chainage to the face of the first abutment; (*c*) the angle of skew of that abutment relatively to the tangent to the curve at the point at which the latter cuts the face; (*d*) the angles between adjacent piers, or piers and abutment faces, if they are not parallel; (*e*) the lengths of the square or the skew spans; (*f*) the dimensions of the piers and abutments. The situation fixed for the first abutment governs the position of the other lines to be located, and calculation may be made of the further items; (*g*) the chainage to the centre of each pier and to the second abutment face; (*h*) the angles of skew made by the centre lines of the piers and by the face of the second abutment with the curve. These quantities may be derived as follows.

In Fig. 254, NPQ and STU represent the parallel centre lines of two piers or an abutment face and a pier situated on a curve of radius R.

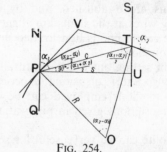

Fig. 254.

Let α_1 and α_2 = the angles of skew made by these lines with the tangents at P and T respectively.

    s = the square span PU,

    c = the chord length PT = the skew span.

Given R, s, α_1, and the chainage of P, it is required to calculate α_2 and the chainage of T.

The angle subtended at the centre O = the change in direction of the curve from P to T = $(\alpha_2 - \alpha_1)$,

$$\therefore c = 2R \sin \tfrac{1}{2}(\alpha_2 - \alpha_1).$$

But PTU = $\tfrac{1}{2}(\alpha_1 + \alpha_2)$, $\therefore c = s \operatorname{cosec} \tfrac{1}{2}(\alpha_1 + \alpha_2)$.

Equating these values of c, we have:

$$2R \sin \tfrac{1}{2}(\alpha_2 - \alpha_1) \sin \tfrac{1}{2}(\alpha_1 + \alpha_2) = s,$$

$$i.e.\ R(\cos \alpha_1 - \cos \alpha_2) = s,\ \text{whence } \alpha_2.$$

Chainage of T = chainage of P + arc PT,

$$= \text{chainage of P} + \frac{\pi(\alpha_2 - \alpha_1)R}{180°}. \quad (\alpha_1, \alpha_2 \text{ in degrees})$$

If NPQ is not parallel to STU, there is no square span, and c is substituted for s in the data. To α_2, as calculated on the assumption of parallelism, it is then necessary to apply the known angle between NPQ and STU.

Setting Out. (1) C being inaccessible, locate the tangent points A and B (Fig. 255) by the method of Case 3, section 450, the length

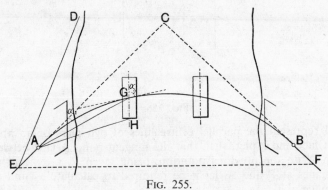

FIG. 255.

of a transversal EF being obtained by measuring out a base line such as ED, observing two or all angles of the triangle EDF, and solving for EF.

(2) A and B, being found accessible, are to be used as instrument stations from which the required points can be located by simultaneous observations. In setting out a point on the centre line,

such as G, the deflection angles CAG and CBG are readily computed from the lengths of the arcs AG and GB obtained as above.

(3) The same principle applies to the setting out of points such as H, the angles CAH and CBH being calculated. Thus, CAH = CAG + GAH, of which CAG is known, while triangle AGH can be solved for GAH, since GH will be known, AG is the chord subtending a known arc, and AGH = $(\alpha_2 - CAG)$. Any point can be located in this manner provided its position is specified relatively to some point of known chainage on the centre line.

If A and B, although on shore, are unsuitable for observing from, it will be necessary to establish more convenient instrument stations of known chainage. If these are on the curve, the above methods apply, each deflection angle being calculated and measured from the tangent through the instrument station. If the stations have to be selected on the straights, the methods of the following case must be used.

467. Case 2. Tangent Points in River. In this case the only difference in the data will be that the angle of skew of the first abutment is measured from the first straight. In Fig. 256, NPQ and

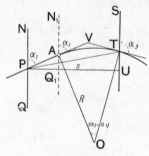

FIG. 256.

STU represent the parallel centre lines of two piers or an abutment face and a pier, such that the tangent point A falls between them. With the previous notation, being given R, s, α_1, and the chainages at P and A, let it be required to calculate α_2 and the chainage of T.

Through A draw N_1AQ_1 parallel to NPQ; then $N_1AV = \alpha_1$, and the methods of the previous case apply to the lines N_1AQ_1 and STU, the perpendicular distance between which is $Q_1U = (PU - PQ_1) = (s - PA \sin \alpha_1)$. Corresponding to the previous results, we therefore have

$$R(\cos \alpha_1 - \cos \alpha_2) = (s - PA \sin \alpha_1),$$

and chainage of T = chainage of $P + PA + \dfrac{\pi(\alpha_2 - \alpha_1)R}{180°}$

$\qquad\qquad\qquad\qquad\qquad\qquad$ (α_1, α_2 in degrees)

Setting Out. (1) Having solved triangle CEF as before, and found that A and B are inaccessible (Fig. 257), select two instrument

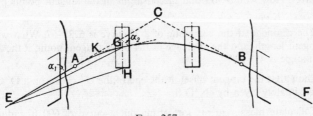

FIG. 257.

stations, such as E and F, from which to set out simultaneous deflection angles. The chainage of E being known, that of F is readily found since EA, arc AB, and BF are known.

(2) Calculate the deflection angles: for a centre line point such as G, these angles are CEG and CFG. KG being the tangent to the curve at G, CEG is obtained by solution of triangle EKG, since EKG = $(180° - i)$, where i is the central angle subtended by arc AG, KG = $\tan \frac{1}{2}i$, and EK = $(EA + R \tan \frac{1}{2}i)$. Similarly for CFG.

(3) The value of angle CEH required in locating H is obtained by solution of triangle EGH, in which EG is obtained from triangle EKG, GH is known, and EGH = $(\alpha_2 - KGE)$. Similarly for CFH.

Modern electromagnetic distance measuring instruments, such as the Mekometer, may be very useful in problems of setting out. For instance, with these instruments a distance between marks on two piers or abutments can be directly measured to an accuracy comparable with that obtainable from precise angle measurements and base line.

EXAMPLES

(1) Find the deflection angle of the chord for setting out a length of 200 ft. of a curve of radius 2300 ft. What will be the offset from the chord to the middle point of the curve?

(2) The chainage at the beginning of a curve of radius 1000 ft. is 3411 + 09: what will be the tangent length and the chainage at the other tangent point, if the intersection angle is 52°?

(3) A circular curve of radius 1400 ft. has a length of 926 ft. Calculate the angle at the centre, the length of the whole chord between the tangent points, and length of each tangent.

(4) A circular curve is to be constructed to pass through three points A, B and C. By survey methods, the chord lengths $AB = a$, $BC = b$ and the angle $ABC = \theta$, are measured. Show that the radius of the required curve is $\frac{1}{2} \operatorname{cosec} \theta \sqrt{(a^2 + b^2 - 2ab \cos \theta)}$.

(5) A new road approaches, on bearing 292°, an existing road having bearing 245°, and is to be connected to it by a left hand curve of radius 600 metres. How would you find the positions of the tangent points on the roads?

(6) The chainage at the beginning of a 2° curve is $628 + 57$. What will be the tangent length, and the chainage at the other tangent point, if the intersection angle is 49° 22'?

(7) Show that the tangent offset at the end of the Nth chord on a D° curve is approximately given by $\frac{7}{8}N^2D$ ft.

(8) Calculate measurements for setting out a curve of 600 ft. radius by offsets from the tangent, to points 25, 50, 75 and 100 ft. from the tangent point.

(9) The tangent at the commencement of a circular curve of radius 2500 ft. has bearing 98° 16', and the chainage at the tangent point is 86 141 ft. The second tangent has bearing 83° 53'. Find the length of the curve and the chainage at the second tangent point. Give a list of bearings, correct to 20", of chords for setting out pegs at each 100 ft. point along the curve.

(10) Two straight portions AB and CD of the centre line of a new minor road have been marked out: the relevant data of a traverse run between them are:

Line	Bearing	Length
AB	65° 46'	
BC	100 37	352·6 ft.
CD	140 54	

Find which straight must be extended, in order to equalise the tangents, and by how much, and find the radius of the necessary circular curve.

(11) Using the same data as in the previous question, find where the tangent points must be if the curve is to have radius 300 ft.

(12) Two parallel tracks 84 links apart, centre to centre, are to be connected by a pair of reverse curves of 4 chains radius, and equal lengths, with a straight of 50 links between them. Show that the length of each curve will be 161·8 links, and find the distance between the connected points, measured in the direction of the two tracks.

(13) To find the radius of an existing curve, two 100 ft. chords AB and BC are marked out on it, and the distance of point B from the chord AC is found to be 5 ft. 8 in. What is the radius of the curve? What angle would you set out from the line AC in order to get a line to the centre of curvature?

(14) A curve of 120 ft. radius and turning through 90° is to be set out by offsets from the principal chord: calculate offsets and distances for 5 intermediate points equidistant along the curve.

(15) Particulars of three lines of a traverse round a bend of a river are:

Line	Bearing	Length (ft.)
AB	46° 09′	563·2
BC	85 28	443·1
CD	110 55	308·5

A circular curve is to be set out so as to be tangent to each of the three lines. Find the radius of the curve, and calculate measurements for setting out the points where it touches the three lines.

(16) A curve of 60 chains radius has been pegged out to connect two railway tangents, the difference between the bearings of which is 18° 00′. It is found necessary to alter the alignment by introducing a straight length of 6 chains midway between the existing tangents and connecting it to them by two arcs of equal radius. If the original tangent points are to remain in the same positions, compute the radius of the required curves and the maximum distance between the original and amended centre lines.

(17) A clothoid transition of length 500 ft. joins a straight to a circular curve of radius 2000 ft. What is the angle turned by the clothoid? Calculate distances and offsets for setting out points at 100 ft. intervals along the transition.

(18) Two straights making an intersection angle of 30° are connected by a circular curve of radius 2000 ft. The bend is to be improved by substituting a pair of clothoid transitions having the same curvature at the apex point. Calculate the positions of the new tangent points and apex in relation to their present positions.

(19) The centre-line of a new road has a portion with rising gradient 1 : 18, and this is to be followed by a circular vertical curve of radius 4000 ft. until the line becomes level. The starting point of the curve is at altitude 177·91 ft. above datum. Find the length of the curve, and the altitude when the level section is reached.

(20) In approaching a summit point, the centre-line of a new road has rising gradient 1 : 14 and at a certain point A the altitude is 321·83 ft. The gradient is 1 : 25 down the other side and a point B on this part of the line has altitude 332·64 ft. AB = 414 ft. Find where the two gradients would meet if produced, and the altitude of the meeting point. A circular vertical curve is to be set out, extending 200 ft. on either side of the gradient intersection. Find the radius of the curve, and the position and altitude of the actual summit point.

(21) A new road is being constructed over a summit. On one side, excavation is along a gradient of 1 : 22 and has reached a point A at altitude 217·3 ft : on the other side the gradient is 1 : 16 and has reached point B at altitude 224·9 ft. The distance AB is 944 ft. Find the position of the gradient intersection point. The vertical curve is to consist of two transitions of length L connected by a circular curve of radius R. In this case, L is 200 ft. and R is 6000 ft. Each transition is a curve $y = x^3/6RL$, where x and y are referred to the produced gradient line as x axis: thus, each transition changes the gradient by the amount $L/2R$. Find the length of circular curve which, with the two

transitions, will give the necessary total change of gradient, and find the positions, in relation to A and B, of the end points of the curved section.

REFERENCES

AITKEN J. AND BOYD J. 'A consolidation of vertical curve design'. *Jour. of the Inst. of Civil Engineers*, Vol. 26, Dec. 1945.

CRAM I. A. 'Standards of design for rural roads'. *Jour. of the Inst. of Highway Engineers*, Vol. VIII, No. 4, Oct. 1961.

FITT R. L. AND INCE B. 'Surveying for dam and reservoir'. *Conference of Commonwealth Survey Officers*, 1967.

GLOVER J. 'Survey for land to be irrigated'. *Conference of Commonwealth Survey Officers*, 1967.

HICKERSON T. F. 'Route location and design'. Fifth Edition, 1967. *McGraw-Hill*.

HYDES W. S. AND JACOMB A. W. 'Highway Design'. *Jour. of the Inst. of Municipal Engineers*, Vol. 88, Apr. and May 1961.

LEEMING J. J. AND BLACK A. N. 'Road curvature and superelevation; a final report on experiments on comfort and driving practice'. *Jour. of the Inst. of Municipal Engineers*, Vol. LXXVI, No. 8, Feb. 1950.

LOACH J. C. AND MAYCOCK. 'Recent developments in railway curve design'. *Proc. of the Inst. of Civil Engineers*, Oct. 1952.

McCAW G. T. 'A curve of Transition'. *Empire Survey Review*, No. 37, July 1940.

McVILLY R. B. 'Parabola as Transition Curve'. *Survey Review*, Vol. XIX, No. 149, July 1968.

ORCHARD D. F. 'A Survey of the present position in road transition-curve theory'. *Jour. of the Inst. of Civil Engineers*, Vol. II, 1938–39.

THORNTON-SMITH G. J. 'Almost exact closed expressions for computing all the elements of the clothoid transition curve'. *Survey Review*, No. 127, Jan. 1963.

— 'A new transition curve'. *Empire Survey Review*, No. 117, July 1960.

— 'A family of transition curves'. *Empire Survey Review*, No. 120, Apr. 1961.

TYSON A. G. 'Transition curve design'. *Jour. of the Inst. of Municipal Engineers*, Vol. LXXVI, No. 6, Dec. 1949.

— 'Highway transition curve design'. *Empire Survey Review*, Nos. 75, 76 and 77, Jan., Apr., and July 1950.

WARREN H. A. 'Highway transition curves; a new basis for design'. *Jour. of the Inst. of Civil Engineers*, Vol. 14, 1939–40.

WASSERMAN W. 'Control surveys for the construction of the Snowy Mountains Scheme'. *Conference of Commonwealth Survey Officers*, 1967.

CHAPTER XI

TACHYMETRIC SURVEYING

---

468. Tachymetry, also called Tacheometry, is generally understood to refer to a process of surveying by use of an instrument which is in effect a transit theodolite, the telescope of which is designed so that the surveyor can take readings on a graduated staff and therefrom derive the distance of the staff from the instrument. The instrument is called a Tachymeter or Tacheometer. It seems that *tachymeter* is the etymologically correct spelling, and this will be used henceforth.

469. Scope. An instrument of the type referred to above, giving measurements of distance as well as horizontal and vertical angles, is clearly capable of providing a complete three-dimensional survey of an area around it. For instance, an area could be surveyed and contoured from observations taken at one instrument position. Moreover, in certain circumstances, there will be obvious advantages as regards speed and convenience, in elimination of the direct measurement of distances by chain or tape, and tachymetry is therefore extensively used by surveyors and engineers.

In recent times, developments in the design of tachymeters have not only improved accuracy but more especially have been aimed at minimising the amount of computation to be done after the readings of the instrument have been recorded.

470. Systems. The measurement of horizontal and vertical angles has already been described in earlier Chapters. The characteristic feature of tachymetry is the determination of distance by other than direct measurement, and in the present Chapter a number of methods of indirect distance measurement will be described, which are usually included under the general title of 'optical distance measurement'.

All these methods are based on the geometry of a very elongated triangle, as illustrated in Fig. 258.

If the triangle is isosceles, and the base side s and the angle α are known, the distance D can be calculated — it is evidently

$D = \frac{1}{2}s \cdot \cot\frac{1}{2}\alpha$: or, the length of either side of the triangle is $\frac{1}{2}s \cdot \operatorname{cosec}\frac{1}{2}\alpha$.

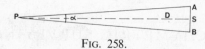

FIG. 258.

The geometry of the above figure is obviously very weak, but the method succeeds in practice because of two features:

(*i*) distances measured by these methods are comparatively short, and they are confined to such values as can in practice be determined to the accuracy desired.

(*ii*) both *s* and α can in practice be measured to an accuracy much higher than that required in the final result; for instance, if *s* is only a few feet it can evidently be measured to an accuracy of 1/1000 ft., and then, even if the triangle is as much as 100 times as long as its base, the value of *D* is still determined with an absolute accuracy of 1/10 ft., which will be good enough for most purposes.

THE STADIA METHOD

This is the most extensively used method, the principle of which seems to have been used by James Watt, nearly 200 years ago.

471. Optical Principle. Consider a simple telescope, with object glass and diaphragm as illustrated in Fig. 259. Assuming the object

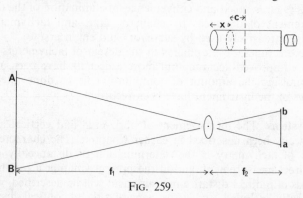

FIG. 259.

glass to be set at a distance from the diaphragm slightly greater than the focal length *f*, the points *a* and *b* on the diaphragm will be the optical images of points *A* and *B* respectively at some distance f_1 in front of the object glass. The lines *aA* and *bB* pass through the centre of the lens.

Suppose a and b represent lines engraved on the diaphragm, and suppose AB represents an ordinary graduated levelling staff; the observer looking through the telescope will see the two stadia lines against the picture of the staff (see Fig. 70), and will be able to read the staff at the points A and B. The difference of the two readings is the length AB; call this S and let the distance ab be s.

From the lens formula we have:

$$\frac{1}{f_1} + \frac{1}{f_2} = \frac{1}{f},$$

and from similar triangles:

$$\frac{S}{s} = \frac{f_1}{f_2}.$$

To focus on the staff, the lens has to be moved forward from the position for 'infinity' focus when it is of course at a distance f from the diaphragm; we suppose that $f_2 = f + x$. Multiplying the lens formula by f_2 gives:

$$\frac{f_2}{f_1} + 1 = \frac{f_2}{f} = \frac{f+x}{f} = 1 + \frac{x}{f}, \text{ hence } x = f \cdot \frac{f_2}{f_1} = f \cdot \frac{s}{S}.$$

Again, multiplying by f_1 gives:

$$1 + \frac{f_1}{f_2} = \frac{f_1}{f} = 1 + \frac{S}{s}, \text{ hence } f_1 = f + f \cdot \frac{S}{s}$$

Now we suppose that the axis of the instrument is at a distance c from the 'infinity' position of the lens, as shown in Fig. 260, then

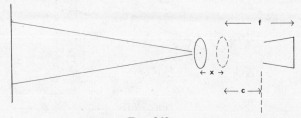

FIG. 260.

the distance of the staff from the instrument axis is $f_1 + x + c$, and reference to the results given above shows that this required distance is

$$f \cdot \frac{S}{s} + f + c + f \cdot \frac{s}{S}.$$

The only variable quantity in the above formula is S, the staff intercept. Now, the manufacturers of the telescope try to engrave the stadia lines on the diaphragm so that the distance ab is a simple fraction, usually one hundredth, of the focal length f, that is, $f/s = 100$. On this assumption, let us consider some dimensions that a practical telescope might have. In foot units, suppose $f = 0.6$ (i.e. 7.2 inches) and $c = 0.4$ (4.8 inches). To obtain the factor 100 mentioned above, s must be 0.006 (about $1/14^{th}$ inch): then $fs = 0.0036$, and the formula for the distance becomes:

$$100 . S + 1.0 + \frac{0.0036}{S} .$$

The distance between staff and instrument is about 100 times the staff intercept: suppose the intercept is 0.15, then the formula gives $(15.0 + 1.0 + 0.024)$ ft. for the distance.

At longer distances, the third number in the bracket will be even smaller and quite negligible in practice.

Thus, the tachymetry formula may be taken as:

Distance = (factor)(staff intercept) + (instrument constant), and the factor is usually 100.

472. Tachymetry Observations. In surveying by tachymeter, the inclination of the line of sight may be considerable, since the sights are generally quite short. Fig. 261 shows the geometry when the

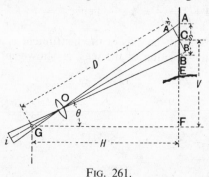

FIG. 261.

telescope is inclined at elevation θ. The tachymetry formula will give the distance D from the axis of the telescope G to the point C on the staff where the central line of the diaphragm appears to cross it, but the staff intercept S is equal to AB, the difference of the two stadia line readings. The quantity to be entered into the formula should obviously be A'B', the distance perpendicular to the line of sight.

The angle ACA' is θ and AA'C is practically $90°$, so $A'B' = S \cdot \cos \theta$, with negligible error in practice.

Thus, we have $\quad D = \dfrac{f}{s} \cdot S \cdot \cos \theta + f + c$.

The horizontal distance from instrument to staff is therefore

$$D \cdot \cos \theta = \frac{f}{s} \cdot S \cos^2\theta + (f+c) \cos \theta,$$

and the vertical difference between G and C is

$$D \sin \theta = \frac{f}{s} \cdot S \cos \theta \cdot \sin \theta + (f+c) \sin \theta.$$

By subtraction of the staff reading EC, the height of the ground at B is obtained. It is often convenient in tachymetry, to set the telescope so that the reading EC is equal to the height of the instrument axis.

It is possible to do tachymetry with the staff held perpendicular to the line of sight: similar formulae are obtained, but this method is now hardly ever used, and it will not be considered further.

473. Anallactic Telescopes. The derivation given above for the tachymetry formula is applicable to the type of telescope which is focused by moving the object glass. It can also be seen that, if the object glass is at a fixed distance from the telescope axis, and focusing is done by moving the diaphragm and eyepiece assembly, then the small quantity denoted by x does not enter into the formula and the third term shown as fs/S does not exist.

However, telescopes of these types are practically obsolete, and almost all telescopes of levels and theodolites nowadays are of the internal focusing design as described in section 23.

The tachymetry formula for the internal focusing telescope is complicated; the derivation is not given here, but the result is, with the notation used in relation to Fig. 17,

$$\text{Distance} = \frac{f(f'+l-x_0)}{f' \cdot s} \cdot S + \left[f + c - \frac{f^2}{l+f-2x_0} \right] + \dots \text{small terms} \dots$$

The point now is, that the various dimensions in the formula can be chosen so that the factor multiplying S can be made equal to a convenient number such as 100, and the expression in square brackets can be made exactly zero.

For example, take the values $f = 10$, $f' = 12$, $l = 11$, $x_0 = 7$, together with $s = \frac{2}{15}$, and $c = 4\frac{2}{7}$. Then, ignoring the small terms which can be shown to be negligible in all practical cases, the tachymetry formula for the internal focusing telescope can be taken as:

$$D = 100 . S$$

A telescope designed to have this property is usually called *anallactic*.

It is interesting to note that a truly anallactic telescope, giving a formula $D = 100 . S$ exactly with no other terms, is theoretically possible. It consists of two convergent lenses at a fixed distance apart, consequently the diaphragm and eyepiece assembly must be on a movable slide for focusing purposes. On this account, the design is not satisfactory, and such telescopes are not used.

474. Special Tachymetric Instruments. Even with the anallactic telescope, some computations have to be done after readings have been taken, in order to find the horizontal and vertical components of the measured distance. It would obviously speed up the work if these quantities could be obtained directly from instrumental readings.

Several designs have been evolved for this purpose, and modern self-reduction tachymeters are of high accuracy.

475. The Beaman Stadia Arc. The reductions would be simplified if the only values of θ used were those for which either $\cos^2 \theta$ or $\sin \theta . \cos \theta = \frac{1}{2} \sin 2\theta$ is a convenient figure. The former varies too slowly for the small angles usually required, but a list of values of θ for which $\frac{1}{2} \sin 2\theta = 0.01$, 0.02, etc., can be prepared as follows:

$\frac{1}{2} \sin 2\theta$	θ to nearest second			$\frac{1}{2} \sin 2\theta$	θ to nearest second		
	°	′	″		°	′	″
·01	0	34	23	·06	3	26	46
·02	1	8	46	·07	4	1	26
·03	1	43	12	·08	4	36	12
·04	2	17	39	·09	5	11	6
·05	2	52	11	·10	5	46	7
				etc.	etc.		

If these particular angles were used, the vertical component V, for a multiplying constant of 100 and no additive constant, would be 1S, 2S, 3S, etc.

The idea is applied in the Beaman Stadia Arc, which has been used to a considerable extent in tachymeters and plane table alidades, and has increased in popularity in recent years. It is fixed to the vertical circle, and consists of a scale engraved with the above angles on either side of zero up to about 26° 33′ 54″, for which

$\frac{1}{2}\sin 2\theta = 0.40$. The scale is figured in terms of $100 \times \frac{1}{2}\sin 2\theta$, and is read against a fixed index mark. To avoid possible confusion between elevations and depressions, the zero is commonly marked 50, so that 50 must be subtracted from every reading. No fitting is required to enable fractional parts of the scale to be read, since for every sight a graduation of the scale is to be brought opposite the index by means of the vertical circle tangent screw.

Let it be supposed that a staff is sighted with an instrument having the tachymetric constant 100 and the stadia arc zero 50, and that the tangent screw is adjusted to bring a graduation of the stadia arc, say 36, exactly opposite the index. Let the observed centre-line reading be 6·10, and the staff intercept 4·95.

The stadia arc reading $= 36 - 50 = -14$.

The vertical component $= -14 \times 4.95 = -69.3$ ft., and the staff position is below the transit axis by $69.3 + 6.1 = 75.4$ ft.

To facilitate the calculation of horizontal distances, the stadia arc also carries a scale of percentage reductions to be applied to the distance readings. In the above example, the distance scale will read 2·0, so that the horizontal distance $= 495 - 2 \times 4.95 = 485$ ft.

476. Self-reduction Tachymeters. Although the introduction of the anallactic telescope considerably simplified the calculation of the horizontal and vertical distances in tachymetric surveying, it was an obvious development to design an instrument by which the required quantities could be obtained directly from staff readings.

One of the earliest such instruments was the Jeffcott Direct Reading Tachymeter. In place of the diaphragm, this instrument had three horizontal pointers, of which the central one was fixed. The others were moved by cams which operated as the telescope was elevated or depressed, in such a way as to introduce the factors $\cos^2\theta$ and $\cos\theta . \sin\theta$ automatically. In fact, the staff intercept between the centre pointer and one of the movable pointers, multiplied by 100, gave the horizontal distance directly, while the intercept between the centre pointer and the third pointer, multiplied by 10, gave the vertical distance, FC in Fig. 261.

A more recent realisation of the same principle takes the form of a glass diaphragm with three non-parallel lines engraved on it. These lines move across the field of view as the telescope is tilted, thus changing the vertical intervals separating the lines.

A typical system is illustrated in Fig. 262. It is usual with these instruments to have a specially graduated staff and to set the telescope so that the apparently lowest line of the three is at a particular staff reading; this facilitates mental subtraction work to derive the interval values. The metric system is very suitable in tachymetry, as can be seen from the Figure; a distance over 100 metres will not often occur in practice, so the lowest line is set at a whole metre, and the decimetre readings, with the proper multipliers, give the required quantities easily. The readings illustrated are 36·6 m. horizontal and 1·68 m. vertical.

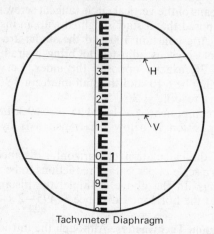

Tachymeter Diaphragm

FIG. 262.

It is evidently necessary, with these moving reference lines, to direct the telescope carefully so that the staff is accurately centred in the field of view, and this can be indicated by vertical lines on a fixed diaphragm.

Instruments of the type described above are extensively used nowadays.

477. Micrometer Setting Tachymeters. All the tachymeters described in previous sections are subject to one inevitable limitation of accuracy – the ability of the observer to estimate the staff readings.

With an ordinary surveying telescope, it is possible to take staff readings to 1/1000 of a foot at distances up to about 150 ft., beyond this, a reading to 0·002 may be possible up to about 300 ft., and at the longest possible distances, perhaps up to 500 ft., the reading can be taken to 0·01. This means that the precision of the distance measurement in tachymetry varies from 0·1 ft. at the shortest distances to 1 ft. at the longest distances practicable.

By means of certain optical devices, and using a graduated staff held horizontally, a considerable improvement of accuracy in tachymetry is possible. An advantage of the horizontal staff is that atmospheric refraction effects are eliminated: with a vertical staff

FIG. 263.—KERN DK-RT SELF REDUCTION TACHYMETER.
(By courtesy of Survey & General Instrument Co. Ltd.)

the refraction effects on the two stadia readings might be appreciably different if the lower line of sight is very close to the ground.

The tachymeter illustrated in Fig. 263 is an example of modern instruments for optical measurement of horizontal and vertical distance.

It is used with a special graduated staff which is supported in the horizontal position and set perpendicular to the line of sight by means of a device attached to the staff. When the telescope is at the correct elevation, the observer sees an image of the staff split horizontally; one section bears a series of equal graduations each equivalent to a distance of two metres, the other image has two short sets of marks whose apparent positions in relation to the main scale depend on the distance of the staff and on the elevation of the telescope.

A coarse reading is taken on the staff, and by means of a parallel plate rotating about a vertical axis, one of the two images is apparently shifted so that a coincidence setting can be made, and the precise reading is taken on the setting control drum.

Fig. 264 shows the appearance of the staff and the fine setting drum for a reading of 85·275 m.

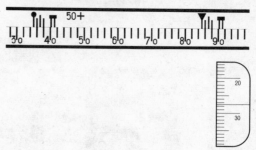

FIG. 264.

Automatic allowance for slope is made by means of two rotating wedges in front of the object glass, geared to the elevation movement of the telescope.

An accuracy of 1/5000 for a horizontal distance of 100 metres is claimed for instruments of this type. A limit to the distance that can be measured is of course set by the length of the staff.

Fig. 265 shows diagrammatically a vertical section of the telescope of a similar tachymeter by another manufacturer. Rays falling on the rhomboid prism 4 are directed, after two reflections, through the upper half of the objective 5 and are brought to focus in the usual way. Rays entering through the lower half of the objective have passed through a slightly wedge-shaped cover 1 and through two achromatic wedges 2 and 3 which can be rotated in planes perpendicular to the optical axis of the telescope.

When the telescope is correctly directed on to the graduated horizontal staff, so as to produce the split image as mentioned above, the picture of the main graduated scale is seen by rays coming through the wedges 2 and 3, and the picture of the short groups of graduations has come via the prism 4.

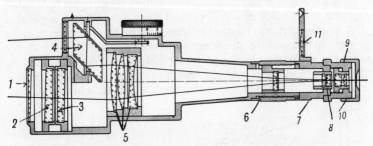

FIG. 265.—DIAGRAMMATIC SECTION OF THE TELESCOPE OF THE RDH DOUBLE-IMAGE SELF-REDUCING TACHYMETER.
(By courtesy of Messrs. Henry Wild & Co. Ltd.)

As drawn in the figure, the two wedges are acting in exact opposition and there is no deflection of the line of sight.

The wedges are geared to the vertical movement of the telescope: when the telescope is elevated or depressed to slope θ, the wedges are simultaneously rotated by equal angles θ in opposite senses. Now, if the deflection angle of each wedge is α, there will be a horizontal component $\alpha . \sin \theta$ of deflection by each wedge but the vertical components of deflection (each equal to $\alpha . \cos \theta$) will cancel out, thus, the line of sight will be deflected horizontally by $2\alpha . \sin \theta$.

As seen in the eyepiece, the main scale of the staff will appear to have been shifted laterally, in relation to the short reference scales, by a distance subtending this angle at the instrument, and the shift can be read against the reference scale.

With $2\alpha = \arctan 0.01$, the full combined deflection of the two wedges would give the direct distance of the staff from the instrument, and the operation of the wedges as described above therefore gives a direct reading of the difference of level between the staff and the transit axis.

For horizontal distance measurement, suppose the telescope is set to horizontal and the two wedges placed so that their effects are combined and produce a total horizontal deflection of 2α; in this situation, the large faces of the wedges are truly vertical. Now, as the telescope is elevated or depressed, the wedges are again caused to rotate through the same angle θ in opposite senses. The horizontal components of the deflections will now total $2\alpha \cdot \cos \theta$ and the two vertical components will cancel out. The staff intercept will now give the reduced horizontal distance.

A mechanism for changing the arrangement of the wedges, either for horizontal or vertical distance reading, is provided.

Further, a micrometer coincidence setting is provided by a small rotation of prism 4 about a vertical axis. This acts as a parallel plate micrometer and causes the image of the short reference scales to shift in relation to the main scale. In this way, a coincidence setting is made, the reading to whole metres only is taken from the main scale, and the rest of the reading is taken on the micrometer setting drum, on which the graduations represent centimetres.

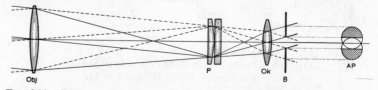

FIG. 266.—DIAGRAM SHOWING THE PASSAGE OF RAYS FROM THE OBJECTIVE TO THE SLOT DIAPHRAGM IN THE WILD RDH DOUBLE-IMAGE SELF-REDUCING TACHYMETER.

Fig. 266 shows diagrammatically the formation of the split images of the staff by use of a double prism and a stop in the telescope. Rays entering the objective from the rhomboid prism fall on the lower half of the prism P and are deviated upwards to the eyepiece Ok. It will, however, be seen that any which fall on the lower half of the objective are cut off after emersion from the eyepiece by the stop B while those which fall on the upper half of the objective are transmitted. Similarly, rays entering the objective from the wedges fall on the upper half of the prism P, and are deviated downwards to the eyepiece Ok. Any rays which enter the upper half of the objective are intercepted by the stop B, while those entering the lower half of the objective pass through the stop. The exit pupils for both sets of rays are shown at the right in Fig. 266. The lower part of the upper circle only contains rays which enter the objective through the rhomboid prism, while the upper part of the lower circle only contains rays which enter the objective after passage through the wedges. All other rays are intercepted by the stop B. Hence, one sees in the telescope two halves of images separated by a horizontal line represented by the apex of the prism P. In observing, this horizontal line is set to view the middle of the staff. The main divisions of the staff, the image of which is formed by rays passing through the rotatable wedges, appear above the line, while the short scale, the image of which is formed by rays passing through the rhomboid prism, appears below the line. (The appearance is similar to that in Fig. 264, but the other way up.) Fine setting of coincidence is done with the drum shown above the telescope in Fig. 265.

THE TANGENTIAL SYSTEM

478. Distances and elevations may be deduced from staff readings taken by a theodolite without any additional fittings whatever.

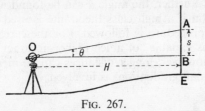

FIG. 267.

Fig. 267 shows the simplest case, that in which the ground is sufficiently level that the staff may be read with a horizontal line of sight. The reading at B is observed with the telescope levelled, and that at A by means of a sight inclined at θ to the horizontal. Denoting the difference of the readings by S, then

$$H = S \cot \theta.$$

The elevation of E is derived from the reading B as in ordinary levelling.

If the situation is such that a horizontal sight cannot be taken, as shown in Fig. 268, the intercept between two elevations (or

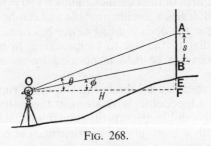

FIG. 268.

depressions) can be observed. Referring to the figure, it is obvious that $S = D \cdot \tan \theta - D \cdot \tan \phi$: thus $D = S/(\tan \theta - \tan \phi)$.

A similar procedure can be carried out with a tilting Level having a graduated drum on the tilting screw. For instance, one turn of the screw may tilt the line of sight through 1/100 radian, so if two staff readings are taken the method is exactly equivalent to tachymetry with a factor of 100.

SUBTENSE METHOD

479. If, in the triangle of Fig. 258, the base s is of fixed length and the angle α is measured, the method is usually called 'subtense'. In practice, the base is invariably in the horizontal position.

The fixed base, or *subtense bar*, is usually 5 to 10 ft. long, and its length should be accurately known; for highest precision, it can be made of invar. Further, the angle α can be found with almost unlimited precision if a high class theodolite is used and a suitable observational procedure is followed. For these reasons, the subtense method is capable of much higher accuracy than ordinary tachymetry; moreover, longer distances can be measured because no reading of staff graduations is involved.

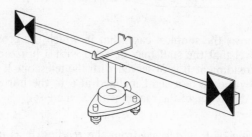

FIG. 269.

A typical subtense bar is illustrated in Fig. 269. It is 6 ft. long between the centres of the two signal marks, and it is supported on a simple tripod. It has a levelling bubble and can be clamped to its vertical axle after the open sighting device has been used to set the bar perpendicular to the line to be measured. In windy weather, the bar may have to be held in position by an assistant, or by light guy ropes.

The angle subtended by the bar at the other end of the line is measured with a theodolite. With a vernier theodolite or an optical scale-reading theodolite, the repetition method should be used (see section 86): with a more precise instrument, it is best to measure the angle a large number of times (say 20) and change the zero after each measurement; it is not necessary to change face, as the telescope remains at a fixed elevation.

The required distance is given, as indicated in section 470, by the formula $D = \frac{1}{2}s \cdot \cot \frac{1}{2}\alpha$. It is important to note that this is the *horizontal* distance, because the angle is measured in the horizontal plane; there is no correction for slope.

480. Accuracy of Subtense Method. Error in the calculated distance can arise from error in S or error in α, or both. An error in s is obviously reproduced proportionately in D: thus, if s is 10 ft. and its length is known to 1/1000 ft., the calculated value of D will be accurate to the same fraction, 1/10,000, of D.

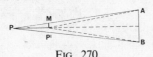

FIG. 270.

The effect of an error in the measured angle is illustrated in Fig. 270. Let APB be the correct angle α, and AP′B the erroneous angle $(\alpha + \varepsilon)$; the resulting error in the distance is PP′. The angle PAP′ is $\frac{1}{2}\varepsilon$ and the length of AP′ is practically D; thus, the perpendicular P′M is $D \cdot \frac{1}{2}\varepsilon$ if ε is in radian measure. Angle MPP′ is $\frac{1}{2}\alpha$, so PP′ is $\frac{1}{2}\varepsilon \cdot D \cdot \operatorname{cosec} \frac{1}{2}\alpha$.

The same formula can obviously be derived by differentiating the distance formula with respect to α.

It is however simplest to consider the effect of an angle error as being a proportional quantity, just as with an error in s. Thus, if the angle is N seconds and it has an error of ε seconds, the accuracy of D is the fraction ε/N of D. For example, if the angle is $\frac{1}{2}°$, or 1800 seconds, and D is required to an accuracy of 1/10,000, the angle must be measured to 0·18 second.

It is seen therefore that in normal circumstances the subtense method could be capable of an accuracy of 1/5000 provided that the angle is at least a substantial fraction of one degree. An accuracy of 1/2000 should present no difficulties at distances up to 1000 ft. with a 10 ft. subtense bar.

A long line can be measured in two pieces by setting up the theodolite near the mid-point and the subtense bar at each end (or using two subtense bars).

481. Use of Subtense Method. In rough country, or across water, rivers or swamps, the subtense method can be useful in running a traverse. Some instrument makers provide a subtense bar which can be fitted to the tripods of special traversing outfits using three tripods. The subtense method is, however, rather slow, and traversing nowadays is much easier and faster with electromagnetic distance measuring apparatus, if that is available. A theodolite is of course still required for measuring the traverse angles.

482. Rangefinding Methods. A rangefinder has a fixed base from the ends of which the distant object is observed, and the degree of convergence of the two directions, AP and BP in Fig. 258, gives the distance required. As illustrated in Fig. 271, two telescopes A and B are fixed at the ends of a tube which in portable instruments is usually one yard, or one metre, long. In an eyepiece at the centre, the observer sees superimposed images from the two objectives; the picture of a vertical object such as a ranging pole appears to be split and coincidence can be restored by horizontal movement of the prism.

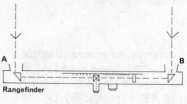

FIG. 271.

The prism position is indicated by a scale seen in another eyepiece, and the scale is graduated directly to read the required distance. A one-metre rangefinder may be graduated from 250 up to 10,000 metres, but the accuracy beyond 1000 metres is rather poor. Rangefinders are not much used in ordinary surveying.

483. Telemeter. Rangefinding can also be done with variable base and fixed angle. Referring to Fig. 272, the telescope A is fixed and

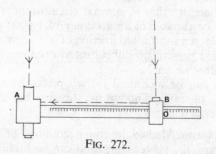

FIG. 272.

the telescope B slides along the graduated arm. The lines of sight of the telescopes are at a fixed small angle with each other, and the picture from B is directed into the eyepiece of A by means of reflecting prisms.

Coincidence of split images from the two telescopes is obtained by finely controlled movement of telescope B along the graduated arm, and the tachymetric triangle is thus completed. If, for instance, the lines of sight are set at the fixed angle of *arctan* 0·01, each centimetre on the scale corresponds with a metre of distance from A.

The range of this type of instrument is strictly limited by the length of graduated arm provided for the moving telescope. The accuracy is the same at all distances within the range, and is said to be about 2 cms.

USES OF TACHYMETRIC SURVEYING

484. In the simplest case, an ordinary Level fitted with stadia lines can be used in levelling a section, and the distances as well as the heights can be obtained if the instrument is located at a point on the line. An observer and a staffman constitute the field party.

If the Level is fitted also with a horizontal divided circle, a detail survey with contours can be made over a limited area, either by locating and levelling a network of spot-heights or by following out individual contours and obtaining positions on them by directions and distances. The area of survey can be increased by shifting the instrument to an adjoining position and surveying at least two of the previous points from the new station. For obvious reasons, this procedure should not be carried very far unless the instrument stations are controlled by a traverse or other independent survey.

The observations on tachymetric surveys with a Level can be recorded in an ordinary levelling book provided with extra columns for stadia readings, calculated distances, and horizontal circle readings. The field work can be carried out by a surveyor and a staffman, but if the work is extensive, it can be speeded up and perhaps done more economically, by employing two staffmen and by having a recording assistant so that the surveyor is not constantly transferring his sight between the book and the eyepiece.

Tachymetric surveying with a theodolite requires additionally the booking of vertical angles. As there is a fair amount of computing to be done, it is better to use a self-reducing tachymeter for any extensive work involving sloping sights.

The tachymeter is an excellent instrument for survey of detail by a traversing procedure, especially for large scale surveys of built-up areas. In such circumstances, the legs of the traverses are necessarily short, and at any traverse station it is easy to take distances and directions to prominent nearby points of detail, with little or no interruption by traffic.

To survey a point like the corner of a building, the direction is of course taken by setting the telescope on the required object, and the distance is obtained by taking appropriate readings on the horizontal staff held perpendicular to the line of sight and with one end on the point of detail.

The field party may consist of the surveyor and two staffmen, one of whom sets up at traverse stations and the other attends to the surrounding detail points. The recording should be done on a form designed to suit the type of tachymeter in use.

Tachymetric surveying of detail is used by the Ordnance Survey Department for the preparation of the new large-scale surveys of towns.

485. Computation and Plotting. A tachymetric traverse run between control points should be computed and adjusted by methods applicable to traversing generally. After the instrument stations have been plotted, it is best to add the detail or spot-height points by protractor and scale.

EXAMPLES

(*N.B. Unless otherwise indicated, it is to be assumed that the telescope of the Level or Theodolite has stadia lines with factor* 100, *and no additive constant.*)

(1) A Level was set up at one side of a road and the readings given below were taken on a staff held on the kerbs at each side:

	Centre line	Stadia Lines	
Near side	4·172	4·274	4·070
Far side	6·205	6·510	5·922

What was the width of the road and what was its cross gradient?

(2) In a tachymetric survey of a cross-section of a stream, the following readings were taken on a staff held at the water's edges and various points on the bottom; the telescope of the instrument was set level.

Station	Readings			Notes
A	4·05	3·97	3·88	edge of water
B	6·97	6·81	6·65	
C	9·98	9·73	9·48	
D	11·53	11·22	10·88	
E	12·85	12·41	11·98	
F	12·13	11·63	11·08	
G	10·43	9·80	9·17	
H	7·44	6·72	6·01	
J	4·76	3·96	3·16	edge of water

Find the area of the cross section of the water.

(3) A Level was set up on the line between two points A and B, and the following staff readings were booked:

	Centre line	Stadia lines	
Staff at A	5·174	5·708	4·642
Staff at B	9·854	10·246	9·464

Find the length and slope of the line AB.

(4) The undermentioned observations were taken with a theodolite from one station of a tachymetric survey, to a staff held vertically on two points A and B:

	Bearing	Ver. Angle	Centre line	Stadia lines	
Point A	274° 15′	−7°	4·16	4·84	3·48
Point B	320° 38′	+8°	5·07	6·18	3·95

Find the horizontal distance and the gradient between the points A and B.

(5) A subtense bar has a calibrated length of 5·998 ft. It subtends a horizontal angle of 28′ 56″ measured at a theodolite station. Find the distance from the bar to the theodolite. What will be the error in this result if the angle is in error by 2″?

(6) A subtense bar 2 metres long is set up near the middle of a traverse line AB. At A, the angle subtended is 50′ 38″, and at B the angle is 44′ 13″. Find the length AB.

(7) A theodolite is set up on level ground, near the bottom of an embankment and 8·8 ft. from the bottom of the slope face: the telescope axis is 5·2 ft. above ground. A 14 ft. levelling staff is laid on the ground along the top edge of the embankment and perpendicular to the line of sight to the theodolite. The whole length of the staff subtends a horizontal angle of 11° 48′, and the vertical angle to it is +23° 06′. Find the height and slope of the embankment.

(8) Observations are taken on to a vertical staff with a theodolite, of which the transit axis is at elevation 154·3 ft. When the line of sight is directed to staff reading 3·00 ft., the elevation angle is 4° 59′, and when the line of sight is directed to staff reading 11·00 ft. the elevation angle is 5° 44′. Compute the reduced level of the staff station and its distance from the theodolite.

(9) To find the distance across a river, between marked points A and B, a surveyor sets up a signal on B, and another at C on the same side of the river, so that the angle ABC is 90°. The distance BC is carefully measured and found to be 91·27 m. The horizontal angle BAC is found by theodolite to be 2° 17′ 55″. Find the distance AB. What would be the error in this result,
 (i) if the angle is wrong by 5″,
(ii) if the distance BC is wrong by 0·05 m.?

(10) The telescope of a theodolite is 4·82 ft. above the ground station. A staff is held vertically on a benchmark of altitude 107·83 ft., and the readings of the staff against the cross-line of the telescope are 8·07 ft. when the telescope is at depression angle 9°, and 6·84 ft. when the telescope is at depression angle 10°. Find the altitude of the theodolite station.

(11) A traverse is being measured by theodolite and subtense bar. At one station, the theodolite transit axis is 5·1 ft. above the ground mark and the

horizontal angle subtended by a 6-foot subtense bar placed at the next station is 45′ 24″. The subtense bar is 3·8 ft. above the ground mark, and the vertical angle measured to it on the theodolite is 13° 38′ elevation. Find the horizontal and vertical distances distances between the two station marks.

(12) A Level with stadia lines and a horizontal divided circle is set up near the middle of a small plot of land ABCD, of nearly rectangular shape. The following readings were taken on a staff held at the corners:

Point	Direction	Staff readings		
A	38°·2	4·95	4·64	4·32
B	137 ·0	8·47	8·07	7·68
C	212 ·0	8·76	8·41	8·06
D	310 ·6	5·21	4·92	4·63

Draw a plan of the plot at a suitable scale, find its area, and estimate the gradient in the direction of greatest slope.

(13) A theodolite with stadia lines is set up near the foot of a slope and placed with the telescope at elevation 18°. A staff is taken to points on the slope, and held vertically, and the following readings are taken:

Staff station	Readings (ft.)			Notes
A	5·31	5·17	5·02	bottom of slope
B	5·02	4·72	4·42	
C	4·38	3·91	3·44	
D	4·76	4·12	3·47	
E	7·69	6·88	6·06	
F	12·28	11·31	10·35	top of slope.

Draw a section of the slope at suitable scales.

REFERENCES

ANGUS-LEPPAN P. V. 'Indirect methods of distance measurement, and a new rangefinding device'. *Survey Review*, No. 135, Jan. 1965.

BABBAGE G. 'The subtense bar–its uses and errors'. *Canadian Surveyor*, June 1965.

BIRD R. G. 'The accuracy of subtense bar measurements in relation to the number of observations of the subtended angle'. *Survey Review*, Vol. XIX, No. 148, Apr. 1968.

EVANS S. E. 'Hints on tacheometry with an ordinary theodolite'. *Empire Survey Review*, No. 30, Oct. 1938.

GEISLER M. 'Precision tacheometry with vertical subtense bar applied to topographical surveying'. *Survey Review*, No. 134, Oct. 1964.

HURST G. 'Sub-tense methods of measuring distances in underground traverses'. *R.I.C.S. Journal*, June 1955.

MATHER R. S. 'Errors in height traversing with reduction tacheometers'. *Survey Review*, No. 135, Jan. 1965.

MICHALSKI Z. M. 'Notes on traversing with subtense bar'. *Empire Survey Review*, No. 93, July 1954.

REDMOND F. A. 'The use of "even angles" in stadia surveying'. *Empire Survey Review*, No. 17, July 1935.

SAASTAMOINEN J. 'Tacheometers and their use in surveying'. *Canadian Surveyor*, Oct. 1959.

SWEETING E. K. G. 'Use of tacheometry in the Ordnance Survey'. *Commonwealth Survey Officers Conference*, 1955.

TAYLOR E. W. AND HOGG F. B. R. 'The application of the surveyor's telescope applied to precise tacheometry'. *Conference of British Commonwealth Survey Officers*, 1947.

WATSON L. H. AND SHADBOLT C. H. 'The subtense bar applied to mine surveying'. *Chartered Surveyor*, May 1958.

WATTS B. 'Application of direct-reading tacheometers to large-scale surveys'. *R.I.C.S. Journal*, July 1954.

CHAPTER XII

HYDROGRAPHIC SURVEYING

486. Hydrographic surveying is that branch of surveying which deals with bodies of water. It embraces all surveys made for the determination of water areas, volumes, and levels, rate of flow, and the form and characteristics of underwater surfaces. The usual methods of applying the fundamental principles of surveying and levelling have to be adapted to the conditions, and some of the operations and apparatus are of a specialised character.

487. Scope. Extensive hydrographic surveys, directed to the preparation of charts for the use of navigators, are undertaken by various countries. Such surveys are conducted from specially equipped surveying vessels, and comprise the determination of depths available for shipping, the survey of currents, the location of shoals and other dangers, buoys, anchorages, and lights, as well as the mapping of conspicuous land features which will guide the navigator. An examination of the operations forming the routine on board a surveying vessel is outside the scope of this volume, but the civil engineer uses similar methods on a smaller scale in connection with the design and maintenance of certain classes of works. The relationship of hydrographic surveying to those branches of engineering which deal with harbours, docks, navigable waterways, coast protection, etc., is evident; but the application of the subject to civil engineering is wider than might at first sight appear, since works connected with water supply, water power, irrigation, sewage disposal, flood control, land reclamation, viaducts, and river works generally, also involve the practice of hydrographic surveying.

That part of the subject which relates to the measurement of the discharge of rivers and streams, as required in water supply and similar projects, is dealt with in the latter part of the chapter. The more general branches of the subject may be classed as marine surveying, but they are just as applicable to inland waters as to the sea.

TIDES AND SEA LEVEL

488. Marine surveyors, and engineers concerned with work on coastlines, in harbours, or in tidal estuaries, must take account of the tidal changes of sea level, and should have some knowledge of the mechanism of the tides.

The periodical oscillations of sea level seen at a place on the coast are caused by the variation of attractive force of the Moon and the Sun, as the positions of these bodies, in relation to the Earth, change by virtue of the rotation of the Earth and the monthly revolution of the Moon.

In theory, given the relative positions of the Earth, Sun and Moon and supposing the Earth to be completely covered by a deep ocean, it would be possible to calculate the exact form of the ocean surface as a solution of a problem in statics. The effects on the water level, due to the presence of the Sun and Moon, would be only 2 or 3 feet.

On the real Earth, the situation is so different that this *equilibrium tide* solution is of no practical use in tide calculations. The Earth's irregular land-masses, the variations of depth of the water, and the dynamical nature of the problem, cause the tidal phenomena to be widely different at different places; at some points on the coast there is almost no tidal change, and at others the tidal oscillation exceeds 50 feet. Moreover, the relation in time of the tidal phenomena to the actual positions of the Sun and Moon is very variable.

The influence of the Moon is predominant, so the cycle of tidal phenomena at a place has, as a rule, a principal period equal to the apparent period of rotation of the Moon around the Earth — about 24 hours 50 minutes on average.

In practice, tidal phenomena are studied empirically by prolonged observation, and the predictions of future tides are based on analysis of past records. To put it more precisely, the periods of variation of the forces involved are known with great accuracy from astronomy, but the amplitudes and phases of the oscillations at any place can only be found by harmonic analysis of numerous, preferably continuous, records of sea level actually observed over periods of years.

Any predictions of tide levels and times are therefore products of extrapolation, and the predicted values may be expected to become increasingly unreliable as the interval of extrapolation is more and more extended. Predictions of times and heights of tides are published for a very large number of ports and harbours all over the world in the form of Tide Tables. At places where records of sea level are currently kept, the observations can be used to correct earlier predictions.

There are certain organisations, one of which is the Institute of Coastal Oceanography and Tides, Bidston, Liverpool, which are specially equipped to undertake tidal analysis and prediction work.

The most noticeable feature of the tidal oscillation, after the

$12\frac{1}{2}$-hour periodicity, is the half-monthly alternation of spring-tide and neap-tide phenomena. At full and new Moon times, the Sun and Moon are acting in conjunction to produce the tidal 'bulge' and the spring tides at such times are generally much higher than the neap tides that occur at the times of the Moon's first and third quarters.

Since the Moon is the principal tide-producing influence, it is to be expected that high tides at a place will occur at about the times when the Moon is at lower or upper transit: in fact, there is usually a lag between the Moon's passage and the time of high water; this lag is called the *lunitidal interval*. The lunitidal interval at the time of full or new Moon is called the *establishment* of the place, and the average of the lunitidal intervals is called the *mean establishment*.

From a study of the behaviour of the lunitidal interval at a place, it is possible to make up a set of rules by which the times of high and low tides can be calculated with quite reasonable accuracy in relation to the times of the Moon's meridian passages. This procedure does not require harmonic analysis and prediction.

At a place where no tidal information is available, it may be possible to estimate times and heights of tides fairly accurately by consulting the predictions published for the nearest points along the coast.

If knowledge of the actual level of the sea surface at a particular place is of importance in connection with work in hand, such as hydrographic surveying or harbour engineering operations, it will generally be necessary to set up some sort of tide gauge in the vicinity and have the levels recorded during the work, if no such gauge is already in action.

Sea level is sometimes influenced by forces other than gravitational attraction. Currents, abnormalities of barometric pressure, and persistent winds can cause appreciable deviations of sea level from predicted values.

MEAN SEA LEVEL

489. This is simply the mean level of the water surface derived from observations taken frequently, or continuously, over a sufficiently long period—at least a year for a reasonable determination. Mean Sea Level is adopted by most, but not all, countries as the zero surface for heights of benchmarks and contour values.

490. Selection of Site. To determine mean sea level, a Tide Gauge must be erected in a suitable situation. If it is required to record tides at a particular place the gauge must be set up there, but if a datum surface for an extensive area is to be established the selection of the site is a matter of some importance. Abnormal conditions must be avoided; obviously unsuitable would be sites inside enclosed bays, in narrow straits or estuaries, or adjacent to extensive areas of shallow water, or in a place exposed to very rough seas.

The site of the principal tide gauge established by the Ordnance Survey is in a sheltered spot close to deep water and open ocean at Newlyn in Cornwall.

491. Tide Gauges. These are of two classes. Self-registering tide gauges produce continuous records automatically; the usual mode of operation is by means of a float controlling the position of a pencil on a rotating drum covered by a paper chart. These gauges require daily or weekly attention for chart changing and general maintenance. Other tide gauges require regular attendance of an observer who notes water levels and times at frequent intervals.

FIG. 273.—STAFF GAUGE.

The Staff Gauge (Fig. 273) is simply a graduated board, 6 in. to 9 in. broad, firmly fixed in a vertical position. Its length should be more than sufficient to embrace the highest and lowest tides known in the locality. The graduations and figures should be very bold, as it may be necessary to read them from a distance. The division is sometimes carried to tenths of a foot, but, for ordinary engineering purposes, there is no advantage in closer graduation than quarter feet.

Piers, etc., form suitable supports for the staff, since the zero must always be under water. If such a support is not available, the gauge must be erected below low water mark, and securely strutted or guyed. In such a case, if the tidal range necessitates a long staff, it may be preferable to arrange the gauge as a series of posts from above high water mark outwards, these being set by levelling so as virtually to constitute one staff. In setting any gauge, the zero may be fixed at a predetermined level, or, more usually, having erected the gauge, the elevation of its zero is observed by levelling in relation to a permanent mark on land.

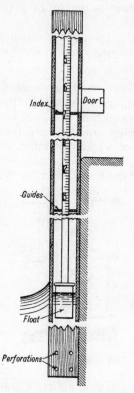

FIG. 274.—FLOAT GAUGE.

The *Float* or *Box Gauge* (Fig. 274) is designed to overcome the objection that accurate reading of a staff gauge is at times difficult on account of the wash of the sea. It is enclosed in a long wooden box about 12 in. square, in the bottom of which are bored a few holes. The surface of the water thus admitted is comparatively smooth, particularly when the holes are small and well below the surface. The float carries a vertical rod, which may itself be graduated or which may carry a pointer over a fixed scale. In the former case the graduations increase downwards, and the reading is obtained against a fixed index. With the latter arrangement the rod may be made quite short, and the attached pointer is brought through a narrow continuous slit in the face of the box so that readings are obtained on a staff gauge attached to the outside of the box.

The *Weight Gauge* is suitable for situations where the above gauges would be liable to disturbance or would not be conveniently accessible for reading. In the simplest form, a weight is attached to a graduated wire, chain, or tape, and observations are made by lowering the weight to touch the surface of the water and reading against a fixed mark. Alternatively, the gauge may take the form shown in Fig. 275, in which the chain carrying the weight passes over a pulley and along the surface of a graduated board, to which it is hooked when not in use. The chain is furnished with an index, and readings are obtained by unhooking the chain, lowering the weight to touch the water surface, and noting the graduation opposite the index. The length of chain must be such that readings can always be obtained at the lowest state of the tide, but the graduated board need not be unduly long as a second and third index can be attached to the chain at intervals of, say, 10 ft. In referring to these, the reading is increased by 10 or 20 ft., as the case may be.

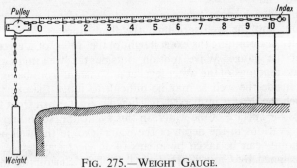

FIG. 275.—WEIGHT GAUGE.

The reduced level of the water surface corresponding to zero reading must be determined. This can be obtained by reading a levelling staff with its foot opposite the bottom of the suspended weight and noting the corresponding gauge reading. Otherwise it may be deduced from the level of the pulley, the length along the chain from the bottom of the weight to the index, and the horizontal distance between the vertical part of the chain and the zero graduation. The value of the zero may change through stretching of the chain, which should be tested occasionally by steel tape, and adjusted if necessary.

492. Self-registering Gauges. Many types of automatic gauges have been constructed, but, while they vary in detail, the essential parts are similar in all forms. A float, protected from the action of wind and waves, is hung from a wire or cord which is coiled round a wheel and is maintained at a constant tension by means of a counterweight. The vertical movement of the float, transferred through the float wheel, is reduced in scale by means of gearing, and is finally communicated to a pencil or pen which traces a curve on a moving sheet of paper. In the commonest form the paper is mounted upon a drum which is rotated once in 24 hours by clockwork. A week's record can be received on the sheet without confusion between the different curves. In more elaborate gauges the recording mechanism is arranged to accommodate a band of paper sufficiently long to contain a month's record. The paper is paid out from one cylinder, passes over a second, against which the pencil or pen is pressed, and is wound upon a third drum.

In the housing of automatic gauges it is frequently necessary to lay piping from below low water level to a well, constructed under the building and in which the float operates. The effect of wave action outside is communicated but slightly to the water in the well because of friction in the pipe and the provision of a grid at its seaward end.

Sometimes, when a tide gauge has to be erected in an exposed position, and the well in which the float rides is deep, there is difficulty in obtaining the exact height of the water for a given reading on the gauge. Wave motion outside the well may prevent accurate readings of the water level at any instant from being taken, while, inside the well, it may be difficult to judge the exact depth to which the float is immersed or when a rod let down from the top touches the surface of the water. In this event, the 'soundings'—that is, observations of the depth of the water level in the well below a fixed point—can be taken by means of an electrical device, which was designed by Capt. C. H. Ley, R.E., for determining the zero correction of the Ordnance Survey tide gauge at Newlyn, and which is described in *The Second Geodetic Levelling of England and Wales*, 1912–21. This device consists of an electrical buzzer connected to a pair of ear-phones and a dry battery. One terminal of the apparatus is earthed and the other is connected, by means of a long flexible lead, to a weighted pointer or index. When the pointer is lowered and makes contact with the water in the well a loud noise is heard in the buzzer. This apparatus, which is very simple, is extremely sensitive and enables the exact level of the water in the well to be determined with great accuracy.

Variations of water level can also be registered through the variations of pressure on a diaphragm suitably fixed in a device placed on the bottom; gauges operating on this principle have not been found to be entirely successful, however.

MEAN SEA LEVEL AS A DATUM SURFACE FOR HEIGHTS

493. For small local surveys it is often quite sufficient to assume an arbitrary horizontal plane as the reference or datum plane from which all heights are reckoned, and to define this plane as being a certain distance below some fixed point or benchmark. But if a benchmark of known height is within or close to the area of the survey, any levelling work should be connected to it and thus placed on the general level datum of the country. Extensive levelling works should always be connected to the national system.

Most countries have adopted mean sea level as datum for heights of points on land: in a few countries, some other datum such as mean low water, is in use. The advantage of mean sea level is that it is a natural level surface (the Geoid) extending over the whole Earth and perpendicular everywhere to the direction of gravity.

The adoption of a datum means in practice that the level of the sea has been recorded at some place over a long period, usually several years, and on the basis of these observations a definite height is ascribed to a permanent benchmark near the gauge station. It is important that, once a datum has been adopted it should remain unaltered as long as possible.

494. Variations of Mean Sea Level. The definition of a datum by reference to a fixed benchmark is necessary because the position of mean sea level at a place is not quite constant. If mean sea level at a place is found separately for each of a series of consecutive years, for instance, the annual values will probably differ by several tenths of a foot. The variations may be quite random: more likely, they will show also a tendency indicating a slow secular relative movement between land and sea. For example, during the past few decades, sea-level has apparently been rising at Southend at a rate of about 3·3 millimetres a year, and at a place on the east coast of Sweden it has been falling at 8·4 millimetres a year.

495. Chart Datum. Depths below sea level, shown on charts to be used for navigation purposes, are normally referred to Mean Low Water Ordinary Spring Tides (L.W.O.S.T.). This is a safety measure, because then the depth of water at any point at any time can be expected to be greater than the value indicated on the chart. On new editions of British charts, the datum for soundings is to be Lowest Astronomical Tide, that is the lowest level predictable from terms in the harmonic analysis having astronomically determined periods.

Therefore, when soundings are being made offshore for the purpose of preparing such charts, it will usually be necessary to record the time of each sounding as well as to set up a temporary tide gauge on the adjacent coast and arrange for readings on it to be taken during all periods when soundings are being made. The tide gauge must be kept in action during at least one period of spring and neap tide.

SHORE LINE SURVEYS

496. Surveys of shore-lines, whether along the sea, lakes or rivers, may be done by any convenient land surveying methods. If only the shore line is to be surveyed, the method of traversing will probably be best. Along narrow rivers, both banks may be surveyed from a traverse along one side, and distances across the water can be found by tachymetric methods. Along wide rivers, a network of triangulation may have to be set up, to connect traverses run along both banks. The survey along the shore may be used to fix the positions of prominent landmarks or of specially erected signals to be used for finding the positions of soundings and other observations made from boats.

A survey along the sea coast will probably include the location of the high-water and low-water lines. The former is usually easily recognised from deposits and marks; survey of low-water mark presents difficulties because of the short time available, and other reasons.

Air photographs can be extremely helpful in coastal survey work particularly if they are taken at the time of low-water, when complicated inter-tidal channels, rocks, etc., are easily seen, as well as the low water mark.

POSITION FIXING

497. When soundings or other observations are being taken from a boat, the principal survey problem is to fix the position of the boat at each observation. If an area is to be systematically covered, there will also be the problem of navigating the boat so that the observation stations are in fact located in the desired positions.

Surveying ships, specially equipped to carry out marine survey work, may fix their positions by use of radio navigational aids such as Decca: in the open ocean, astronomical fixation methods may have to be used. Marine surveying of this type is not discussed in this book: for further information, the reader may refer to the Admiralty Manuals of Hydrographic Surveying and of Navigation.

The fixation of the positions of boats in relation to adjacent land can be done by various methods some of which are peculiar to hydrographic work. The observations may be taken (*a*) entirely from the boat, or (*b*) entirely from land, or (*c*) both ways.

498. Fixation by Bearings Observed in the Boat. If bearings are taken in the boat to two visible features or signals on the land, as indicated in Fig. 276, it is obviously possible to plot the boat's position on the chart by reversing the bearings and drawing lines through the two shore points: due allowance must be made for magnetic declination, and this is best done by observing a few bearings on land and drawing lines of magnetic meridian on the chart.

Bearings may be observed with a hand-held compass of a type used by yachtsmen.

This method is evidently not very precise, but it has the advantage that the boat can be kept on a definite line by steering it so that the bearing to some prominent mark or feature on land remains constant.

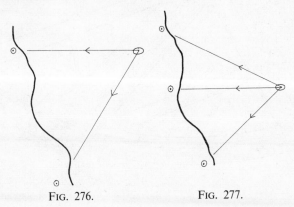

FIG. 276.   FIG. 277.

499. Fixation by Resection, using the Sextant. Using three features or signals on the shore, two angles can be measured as indicated in Fig. 277, and the boat's position is thus determined by resection. The angles will usually be observed by sextant: if the boat is large enough, two sextants can be used with advantage, because each angle will change by small amounts between successive boat stations, and if there are two observers each can concentrate on one of the angles and will not have to make large alterations to his sextant setting.

The plotting of positions resected by sextant angles will usually be done mechanically, see section 511.

500. The Sextant. A typical sextant is illustrated in Fig. 278, and its geometrical principle is indicated in Fig. 279. A sextant is an instrument for measuring a subtended angle in any plane. The observer holds the instrument so that he sees the object A directly through the telescope T, which has a small magnification perhaps $4 \times$ or $6 \times$. The mirror O fixed to the arm OM rotates about the geometrical centre of the graduated circular scale. When the sextant is moved into the plane containing both the observed points, the pivoted arm can be turned so that object B is seen in the telescope after reflection in mirror O and in the half-silvered mirror P. The arm is adjusted until object B appears to coincide with A, and the reading of the scale is taken.

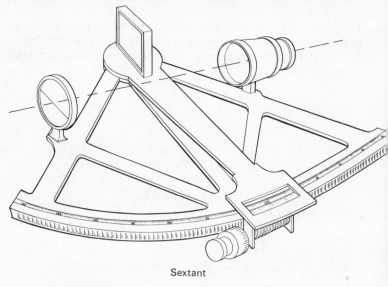

Sextant

FIG. 278.

It is evident that the position of the zero mark on the scale must be such that when the reading is zero the mirror O is parallel to the mirror P: then OB will be parallel to PA. Also, as the arm OM is rotated, the line of sight OB will rotate through twice the angle; therefore the graduation numbered $N°$ must actually be $\frac{1}{2}N°$ round the graduated arc about the centre O. The graduations usually go up to $120°$.

A sextant should read zero when a very distant object is made apparently to coincide with itself. This test should frequently be done: if the reading is not zero, the simplest thing to do is to remember or note the correction. Thus, if the reading is $2\frac{1}{2}'$ when the test is made, all readings must be reduced by $2\frac{1}{2}'$. Alternatively, it is possible to remove the error by adjusting the screws that hold the mirror O in its frame. The test must be done on a distant object, otherwise there will be a slight parallax effect due to the distance of O from the line PA, an inch or two, as a rule.

Fine adjustment of the arm OM may be done by the usual clamp and tangent-screw mechanism. However, some sextants have accurately machined teeth round the arc, of such size that the spacing of successive teeth is exactly $\frac{1}{2}°$ round centre O. A worm-wheel is used to make the fine setting, and a drum on the worm-wheel axle is divided into $0' - 60'$, perhaps with half minute divisions also. There is provision for disengaging the worm when a large movement of the arm is required.

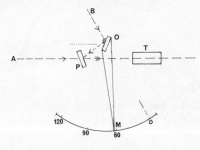

FIG. 279.

A zero error in a sextant of this type can be removed by loosening the drum on its axle and setting it to read zero when the coincidence test is made.

The accuracy of a good sextant is about $\frac{1}{4}'$. As there is no means of eliminating eccentricity effects, errors from this source may be appreciable in a much-used instrument.

501. Fixation on Ranged Lines. Soundings are usually taken on lines more or less perpendicular to the shore. One way to keep a boat on a desired line is to set up two signals on the land as shown in Fig. 280. The helmsman's job is to keep the boat on line by watching the signals: an observer takes bearings, or measures one angle with a sextant, using another shore mark as reference.

502. Location of Positions by Time Intervals. If a motorboat is in use, and an echo-sounding apparatus, the boat may be steered along the marked line at a slow speed and soundings recorded at regular intervals of, say, one minute. This will be a reasonably accurate method provided the beginning and end of each run are fixed by bearings or angles.

503. Fixation by Intersections from the Shore. If two theodolites are set up as shown in Fig. 281, simultaneous observations to the boat will fix its position. It will be convenient either to set the lower plates of the theodolites so that the readings are zero when the telescopes are directed on each other, or to place the plates so that bearings referred to the coordinate system of the map are read on the instruments.

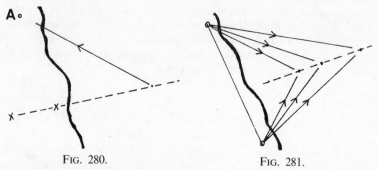

Fig. 280.   Fig. 281.

The two observers must make sure that the observations are numbered or otherwise identified in correct pairs.

Instead of theodolites, it is possible to use plane-tables with copies of the chart on them, oriented correctly: lines are drawn through boat positions, and either chart can be completed by means of a tracing from the other.

A signalling system must be arranged to inform the observers on land when an intersection should be taken.

These intersection methods of course need two observers on shore as well as the boat party. One observer can be dispensed with if the boat is run along a marked line as indicated in Fig. 280, and an observer at A takes bearings or angles. If the sounding is being done in tidal waters, an observer is in any case needed on shore to take tide-pole readings.

504. Fixation by Cross-rope. In harbours and on narrow rivers, if there is not much traffic on the water, it is possible to locate sounding positions by stretching a rope across from bank to bank and hauling the boat along it and taking observations at intervals marked on the rope by knots or tags. Each end of the rope is fixed by reference to detail on the banks. If the width is considerable, it is best to float the rope by attaching pieces of cork or light wood to it after it is first stretched across.

This method is of course very accurate, and is suitable for surveying detailed cross-sections of rivers and channels.

505. Fixation by Depression Angles. Where there is a cliff or steep slope from the water's edge, a convenient way to locate boat positions is by observing depression angles θ with a theodolite as shown in Fig. 282. The height H of the instrument above the water must be obtained by levelling or other suitable method. The horizontal distance of the boat from the theodolite station is then H . cot θ. Readings of the horizontal circle will also give the direction of the boat, so a complete survey from one theodolite station is possible.

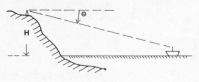

FIG. 282.

In surveying on tidal waters, the correct value of H must be used after reference to the tide-pole observations.

An error $\delta\theta$ in the angle will cause an error H . $\delta\theta$. cosec$^2\theta$ in the calculated distance: for example, if $\delta\theta$ is one minute (about 1/3400 radian) and H is 200 ft., and cot θ is 10, then cosec$^2\theta$ is 101 and the positional error is 101/17 or about 6 ft. This is the error at 2000 (i.e. H cot θ) feet distance.

SOUNDING

506. If only a few soundings are to be made, a simple lead-line or sounding-rope can be used. This is a length of thin rope with a cylindrical lead weight, about 10 lbs., at one end. The rope is marked off by coloured tags of cloth or leather at 1 ft. intervals measured from the bottom of the weight. Larger distinctive tags are placed at 5 ft. or at 6 ft. (1 fathom) spacing.

To take a sounding, the weight is thrown overboard and the other end of the rope held firmly: if the boat is moving, the weight should be thrown ahead and the depth reading must be taken when the rope is judged to be vertical.

Sounding by rope is a slow and inconvenient method.

507. Echo Sounding. For any extensive sounding, the work can be greatly speeded up by use of echo-sounding equipment. By this apparatus, super-sonic pulses are transmitted to the water and reflected from the bottom: the time taken for the double journey is used to give the required indication of depth.

Sound travels in sea-water at nearly 5,000 feet per second. The pulses are regularly repeated at intervals of a fraction of a second, so the sounding is almost continuous. Many ships and craft of all sizes have sounding gear permanently fitted, and in most cases the transmitter and detector are attached to suitable places on the hull. Echo-sounding apparatus for temporary use in small boats is now available, which is easily portable and can be operated by dry batteries.

FIG. 283.

Echo-sounders of the more powerful types usually operate so as to draw a profile of the sea-bed on a moving strip of paper. The principle of operation is shown in Fig. 283. The rotating arm moves at a controlled rate across the strip of specially treated paper which is wound along slowly between two drums. When the arm is in position A, a pulse is transmitted into the water, and when the pulse returns after reflection the arm has moved to B. Both the transmission and the return of the pulse are caused to send electrical signals to the end of the rotating arm and produce a mark on the paper. Thus, the distance AB is, on some known scale, proportional to the depth of the water. Depth can be read off the chart by use of a curved scale, which may be printed on the chart.

A small portable sounding apparatus is illustrated in Fig. 284; these small machines must be read by means of a meter or a scale, because they do not have the power to move and mark a paper strip. In the apparatus illustrated, a rotating arm has a small neon flash lamp at the end. When the arm is at zero reading, a pulse is sent out and the lamp flashes; similarly when the pulse returns. The observer must watch for the return flashes and read their position on the scale.

The transducer, which serves both to transmit and receive the pulses, must be fitted on the boat so that it is not at any time above the water surface, because it can be damaged if it is operated without water round it. The instrument illustrated will indicate depths down to 100 feet.

Since the speed of sound in water depends on the temperature, and on other factors, it is advisable to compare some readings with accurate lead-line observations at two or three different depths, to find if there is any appreciable zero error or scale calibration error to be allowed for.

Records of soundings must be written down along with the clock

FIG. 284.—ECHO SOUNDER.
(By courtesy of Marine Electronics Ltd.)

time and any angles or bearings taken to fix the position of the boat.
There should also be a column for corrections that will have to be
made, based on tidal observations, to reduce the depths to standard
datum.

PLOTTING SOUNDINGS

508. Reduced soundings may be plotted in section form in the
ordinary manner, provided the soundings have been taken in lines.
This method of representing the variation in level of the bottom is
required in river work, and is sometimes used in marine surveying
for the design of engineering works. For navigational and general
engineering purposes the reduced soundings are shown in plan.
They are exhibited as spot depths, the values of the reduced soundings being written at the points representing their positions. The
interpolation of contours increases the value of the chart to the
navigator, and is a necessity for engineering purposes, as the plan
then exhibits the topography of the bottom, and is available for
the calculation of volumes of dredging. For engineering purposes,
depths should be referred to the same datum level as heights on
land.

509. Locating Soundings in Plan. Having laid down the shore survey and from it the positions of landmarks and range signals, the section lines are drawn (in the case where the soundings have been taken in lines), and the positions of the soundings are spaced out. In plotting cross rope soundings to a large scale, vertical and lateral sag of the rope may occasion difficulty in spacing the soundings when the total length of section is known. In such a case, the distance should be divided up into a number of equal parts corresponding with the number of soundings. If the recorder has noted the rope distance at which the line of section is cut by the range of two landmarks represented on the plan, the spacing out can be performed in two parts with increased accuracy.

The plotting of sounding stations on charts or plans is almost invariably done by mechanical or graphical methods, because soundings are generally so numerous that the quickest method for getting them located on the paper must be used.

When positions have been fixed by bearings or by intersection observations from land stations, the location of positions on the plan needs no further explanation: it is only necessary to draw lines at the correct bearings through the appropriate points, and the boat stations are then fixed at intersection points. Points spaced out on cross-ropes or on ranged lines also present no problem.

Further explanation is necessary when the station is fixed by two sextant angles from the boat. This is the well-known *three point problem* or *resection*.

510. The Three-point Problem. The problem may be stated: Given three points A, B, and C, representing the shore signals, and the values *a* and *b* of the angles APB and BPC, subtended by them at the boat P; required to plot P. The problem may be solved mechanically, graphically, and analytically.

511. Mechanical Solutions. (1) *By Station Pointer.* The station pointer or three-arm protractor (Fig. 285) consists of a graduated circle with one fixed and two movable arms, the fiducial edges of which pivot about the centre of graduation. The edge of the fixed arm passes through the zero division, and the movable arms can be set so that their fiducial edges subtend the observed angles with the fixed arm. For plotting on large-scale charts the arms can be extended by clamping on lengthening pieces.

To use the station pointer, set off the angles *a* and *b*, and clamp the movable arms. Move the instrument over the plan until the bevelled edges of the three arms simultaneously touch the points

A, B, and C. The centre of the protractor then marks the position
of P, which is recorded by a prick mark.

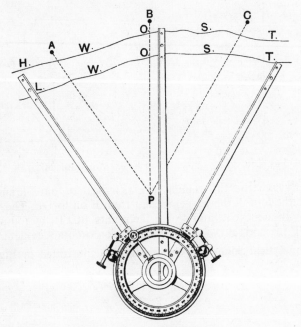

FIG. 285.—STATION POINTER.

(2) *By Tracing Paper Protractor.* On a piece of tracing paper
protract *a* and *b* between three radiating lines from any point.
Apply the tracing paper to the plan, and move it about until A, B,
and C simultaneously lie under the lines; then prick through the
point P at the apex of the angles.

512. Graphical Solutions. The object is to use the angles *a* and *b* in
a geometrical construction which will directly locate the position of
P. There are several methods, of which two are described below.

(1) Perhaps the simplest is illustrated in Fig. 286. Join AC and
construct the triangle ACD by setting out the angles *b* and *a* at
A and C as shown. Then the required point P is on the line DB:
it is in fact at the point where the circumcircle of ACD cuts DB
produced, but probably the easiest way to locate P is to set out
angles *a* and *b* at some point on DP, as shown by the broken lines,
and then use a set-square and ruler to find P so that PA and PC
are parallel to the broken lines.

(2) Another method is illustrated in Fig. 287. Construct the triangle BAD having a right-angle at A and the angle $90° - a$ at B:

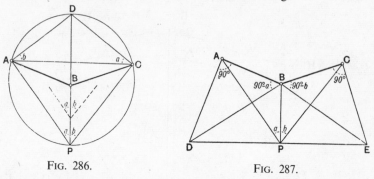

FIG. 286.   FIG. 287.

similarly construct the triangle BCE as indicated. Join DE and then P is at the foot of the perpendicular from B on to DE. The points BADP are concyclic, and so are the points BCEP; by joining AP and CP the correctness of the construction becomes evident.

513. Analytical Solutions. One method is illustrated in Fig. 288.

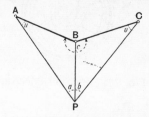

Fig. 288.

The object is to calculate angles u and v, which can then be pro-tracted on the plan to fix P by intersection. The angle c is assumed to be known, and will probably have been measured in the survey of the shore points. The internal angles of the quadrilateral ABCP total $360°$, so we must have :

$$u + v + a + b + c = 360°,$$

or

$$u + v = 360° - a - b - c = s, \text{ say.}$$

From the sine formulas for the two triangles :

$$BP = \frac{AB \cdot \sin u}{\sin a} = \frac{BC \cdot \sin v}{\sin b}$$

hence

$$\frac{\sin u}{\sin v} = \frac{BC \cdot \sin a}{AB \cdot \sin b}$$

Assuming that the lengths of AB and BC are known from the survey observations, this last expression can be calculated; call it $\cot w$. Then

$$\frac{\sin u}{\sin v} = \cot w$$

therefore

$$\frac{\sin u - \sin v}{\sin u + \sin v} = \frac{\cot w - 1}{\cot w + 1} = \frac{1 - \tan w}{1 + \tan w} = \tan(45° - w)$$

$$= \frac{\cos \frac{1}{2}(u+v) . \sin \frac{1}{2}(u-v)}{\sin \frac{1}{2}(u+v) . \cos \frac{1}{2}(u-v)} = \frac{\tan \frac{1}{2}(u-v)}{\tan \frac{1}{2}s}$$

Hence,

$$\tan \tfrac{1}{2}(u-v) = \tan \tfrac{1}{2}s . \tan(45° - w)$$

Thus, with $u-v$ calculated and $u+v \ (= s)$ known, the values of u and v are found.

There are very many methods for calculating coordinates of P when the positions of A, B and C are given in a plane coordinate system. (See Vol. II, Chapter V).

514. Indeterminate and Weak Fixations.
If the point P happens to be on the circumcircle of ABC, the resection problem is insoluble because any point on the arc APC will subtend the same angles a and b at AB and BC respectively. All the methods for constructing P will fail by becoming indeterminate; for instance, point D in Fig. 286 will coincide with B.

If P is near the circumcircle, the fixation will be very weak, and the surveyor must always be on the watch to select reference points that will not lead to this situation. Obviously, there will be danger if the angle ABC is concave towards the sounding stations, and in such situation it is advisable to select a more distant reference point in place of either A or C.

515. Finishing the Soundings Plan.
The values of the reduced soundings are written neatly at the correct positions. If soundings are very close together, it is advisable to write the numbers diagonally in relation to the line of sounding so as to avoid confusion of one number with its neighbours. If soundings are expressed to tenths of a foot, the decimal points may be put at the sounded positions: if soundings are in fathoms and feet, it is customary to use a suffix notation; thus 7_4 is a sounding of 7 fathoms 4 feet.

If the scale does not permit showing all the soundings, a selection must be made but it is important to show any sounding indicating a shallow spot (shoal).

Contours can be interpolated in the usual way. Admiralty Charts show contours at certain depths, such as 3, 5, 10 and 20 fathoms, but on engineering plans the interval is usually 1, 2, or 3 feet.

Areas between contours may be tinted blue, in successively deeper shades for deeper water.

THE SURVEY OF TIDAL CURRENTS

516. Observations of the direction and velocity of tidal currents have to be undertaken by the engineer in connection with certain harbour and coast protection projects. They are also required to aid in the selection of suitable points at which to discharge crude sewage into tidal waters, to ensure its being carried seawards. Such observations must be made at all states of the tide, and a complete set should therefore extend over a series of spring and neap tides.

517. Methods. The most satisfactory method of ascertaining the direction and velocity of the tidal stream over an area is by immersing floats, which drift with the current and are located from time to time. If, however, it is required to determine the characteristics of a current at a particular point, the measurements may be made by current meter.

518. Floats. The requirements of a float for tidal current observations are that (1) it should be carried along by the subsurface current, and therefore should present as little surface as possible to the action of wind and waves, (2) it should be easily identified from a distance.

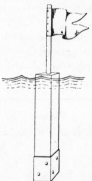

Fig. 289.—Rod Float.

Fig. 289 shows a wooden rod float, 3 or 4 ft. in length, and weighted at the bottom with sheet lead. The double float of Fig. 290 consists of a surface float from which is suspended a perforated cylinder. The float and cylinder are of sheet iron, and the chain or wire connecting them is made adjustable in length so that the direction and velocity of current at various depths may be indicated. A pail is a useful substitute for the cylinder shown. It may be loaded with stones until the float is nearly submerged. Flags attached to floats

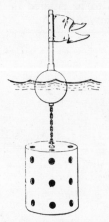

FIG. 290.—DOUBLE FLOAT.

should not be larger than is strictly necessary: different colours should be used as a means of distinguishing the various floats of a series.

519. Surveying Course of Float. The position of a float at various points of its course can be measured: (1) by two angles from a boat alongside the float; (2) by an angle from the boat and one from the shore; (3) by two angles from the shore. The first method is the most satisfactory, and is that usually adopted. The others are not recommended, as it is difficult to secure simultaneous observations unless the sea is very smooth.

If the object of the survey is to ascertain the characteristics of the current at a particular point, successive floats are placed in the water on the upstream side of the point at intervals of about half an hour. If the survey is to extend over a considerable area, the floats must be placed at different distances from the shore. In either case, the boat or launch takes up to a position alongside each in turn, and sextant angles to three shore objects are observed at each position, the time of each observation being noted.

Currents over an extended area may be surveyed by floating large pieces of white paper on the water and taking a series of aerial photographs at timed intervals. The photographs must of course include portions of coast line or other fixed objects for reference purposes.

422   PLANE AND GEODETIC SURVEYING

520. Note Keeping. The records of angles taken at float positions should also include the time of each observation and remarks on the weather and the water conditions. A column may be provided for tide gauge readings to be entered up from the gauge records or the tide reader's observation book.

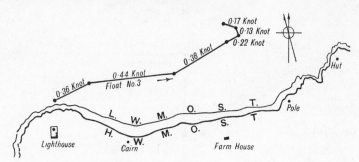

FIG. 291.

521. Plotting and Reducing. The sextant fixes are plotted mechanically or geometrically. From the scaled distances between the points so obtained and the watch times of the fixes, the speed of the current from point to point is computed, the results usually being expressed in knots (of 6,080 ft. per hour). The speeds are recorded in the current book, and are also written up in plan as shown in Fig. 291. It is useful to have the tide curves for the period of the observations plotted for reference.

STREAM MEASUREMENT

522. The measurement of stream discharge is an operation which falls to be performed by the engineer in connection with the design of water supply, irrigation, and power schemes. For certain special purposes a single measurement may be all that is required, but, since stream flow fluctuates from day to day, costly mistakes may occur by basing designs on insufficient observations. A complete investigation should extend over a considerable period, and should include measurements of the maximum and minimum flow and the duration of each at a time.

523. Methods. The methods of measuring the discharge of a stream at a place may be classed as follows:

(1) By measuring the area of cross section of the stream at the place and determining the mean velocity of flow.

(2) By introducing a weir or dam across the stream and observing the head.

(3) By chemical means.

The results are most commonly expressed in cubic feet per second, or gallons per minute or per day.

524. Gauges. For the comparison of measurements made at different times, the result of an observation should be accompanied by a statement of the stage of the stream in terms of the elevation of its surface at the place. This is obtained from the reading of a

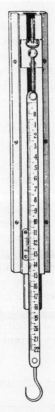

FIG. 292.—HOOK GAUGE.

gauge erected at the discharge station and connected by careful levelling to a permanent bench mark.

In its simplest form the gauge may consist of a stake firmly driven into the bed and carrying a nail of known level from which measurements are made to the water surface. Otherwise, the gauge takes one of the forms described in sections 491 and 492, but with the graduation sufficiently close to enable readings to be estimated to 0·01 ft. if required. The float gauge is particularly useful. Oscillations of the water are most effectively damped out by admitting the water to the float chamber through a cock, and refined readings may then be made against a vernier fixed on top of the chamber. The self-registering form is sometimes used to obtain a continuous record of stage. The most precise reading is possible by the use of the hook gauge (Fig. 292) which is employed in refined observations for the measurement of small heads over weirs. The sharp-pointed hook is carried by a scale which can be moved relatively to a fixed index and vernier by means of a screw. To take a reading, the hook is lowered into the water, and is then slowly raised by the screw until the point just touches the surface. The reading can be estimated to the nearest 0·001 ft., if the water surface at the hook is thoroughly protected from wave action.

AREA-VELOCITY METHOD OF DISCHARGE MEASUREMENT

525. To avoid abnormal results, the site at which the measurement is to be made—called the discharge station—must be selected so that the stream lines are as regular as possible. The stream should be nearly straight for some distance on either side of the station, and the channel should be free from obstructions, and have its bed fairly uniform in shape and character.

The cross section is measured by ordinary levelling or by sounding. Soundings are located by cross rope or by means of a steel tape stretched across the stream, and are spaced at intervals of from 2 ft. upwards. If the width is great, location by the methods used in offshore sounding will be required. It is advisable to check the results by making the measurements twice.

The average velocity of the current is obtained:

(1) By means of floats.

(2) By current meter.

(3) By calculation from the measured inclination of the surface.

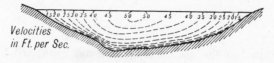

Velocities in Ft. per Sec.

FIG. 293.

526. Distribution of Velocity. The velocity of the stream lines varies throughout the cross section in a manner depending upon the shape of the section, the roughness of the bed, and the depth of water. A typical distribution of velocity over the cross section is illustrated in Fig. 293 by means of curves of equal velocity. The minimum velocity occurs at the bed, due to the retardation produced by friction: the maximum velocity occurs a little below the surface and away from the banks.

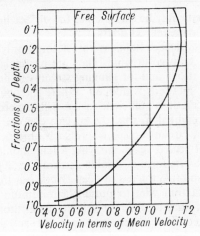

FIG. 294.

Fig. 294 shows the typical manner in which velocity is distributed along a vertical. Observation of numerous streams shows that the curve approximates to a parabola the axis of which is horizontal and coincides with the stream line of maximum velocity. This line is usually situated between the surface and 0·3 of the depth and approaches the surface as the depth decreases. In the case of moderately smooth channels, the mean velocity along the vertical has a value of from 0·7 to 0·95 of that at the surface, the coefficient increasing with increase of velocity and depth. The ratio is less than 0·7 for channels having very rough beds or much obstructed by weeds. The stream line having a velocity equal to the mean velocity in the vertical occurs at about 0·6 of the depth. The mean velocity is also closely given by the mean of the velocities at 0·2 and 0·8 of the depth. These figures are applicable to normal cases only. The distribution of velocity is changed if the stream is is covered with ice or if the water is being drawn through a sluice.

The distribution of surface velocity or of mean velocity in the verticals from one side of the stream to the other depends upon the shape of the cross section. If the bed were very regular, the variation would approach that of a parabolic law with the maximum velocity in mid-stream. Practically, the distribution of velocity across a stream cannot be satisfactorily predicted. In order to measure the mean velocity throughout the cross section by floats or meter, it is therefore necessary to divide the area into a number of sections bounded by verticals and to measure the mean velocity in each.

527. Floats. Floats may be classified as (a) surface floats, (b) double or sub-surface floats, (c) rod floats.

Surface Floats. Surface floats are most commonly employed, but they should not be used in wind. They are particularly suitable for rough determinations and for gauging streams in high flood, when the use of other methods is difficult. Where the course of the float will be easily seen, it is sufficient to use a flat piece of wood or a corked bottle weighted so that the surface exposed to the wind is not too great. In circumstances where such objects are difficult to observe, a larger sealed vessel carrying a flag should be used.

Double Floats. The form of double float of Fig. 290, as employed in sea work, is suitable for use in deep rivers, but the chain should be replaced by a thin wire to present a minimum of area to the action of stream lines of different velocity from those at the depth of the submerged float. Even so, when the sub-surface float is adjusted to the required depth, such as that of mean velocity, the indications are likely to be affected by the speed of the surface float and by the exposed area of the wire when the depth is great. Such floats are unsuitable for use on small streams.

Rod Floats. These consist of wooden rods or metal tubes weighted at the bottom so that they float vertically with only a short length exposed above the surface of the water. The total length is such that the clearance between the lower end and the bed of the stream is as small as can be secured without danger of fouling the bottom. A number of rods of different lengths is required to suit the various depths across the stream. Rod floats are intended to indicate directly the mean velocity in the vertical, but are likely to give too large results because, on account of the necessary clearance at the bottom, they are not acted upon by the stream lines of least velocity. This error is greatest, and rod floats are unsuitable, when the bed is irregular, since the clearance is then excessive over part of the run. Francis investigated the effect of clearance in the case of a rectangular channel, and deduced the formula,

$$v_m = v_r \left(1 \cdot 012 - \cdot 0116 \sqrt{\frac{c}{d}}\right),$$

where v_m = the mean velocity in the vertical of the rod,

v_r = the velocity of the rod,

c = the clearance,

d = the depth of the stream.

Rods are likely to give better results than other types of float. They are very satisfactory if the bed is smooth, as in artificial channels.

528. Measurement by Floats. Field Work. To measure velocity by means of a float, observation is made of the time it takes to travel a measured distance down the stream. A base line is set out on one bank as nearly as possible parallel to the axis of the channel and of length from 50 to 300 ft. At each end a line is set out across the stream at right angles to the base, and is marked by ranging poles or, in narrow streams, by a rope. The cross section of the stream is measured at each of these ranges, and, if the base length exceeds 100 ft., intermediate cross sections are taken at equal intervals of 50 or 100 ft. The area to be used in computing the discharge is the average area over the length of the run.

To determine the mean velocity, a number of float observations is required at different points across the width of the stream. When the ends of the run are marked by ropes, distances from one bank are conveniently marked by means of equidistant tags affixed to the ropes. When rod floats are used it will be necessary to select lengths to suit the depth of water in which each run will be made. In the case of double floats, the wire must be adjusted so that the centre of the immersed float is situated at 0·6 of the depth. The float is placed in the water at some distance above the upper range, and, when it crosses that range, the time is taken, and note is made of the position of the float relatively to the tags. The same observations are made when it reaches the lower range. The time of the run is sometimes taken by stop watch. The operations are repeated with the float at various positions from one bank to the other.

If the stream is too wide to be spanned by a rope, the passage of the float across a range is timed by sighting along the marking poles. A theodolite may be set up with its telescope clamped horizontally in the direction of the range.

In the case of very wide rivers, the location of floats is best performed by sextant observations from a boat as in tide work.

529. Measurement by Floats. Calculation of Discharge. The mean cross section throughout the run is first determined from the results of the section measurements at the upper and lower ranges and at the intermediate positions, if any. Since the width of the stream

428 PLANE AND GEODETIC SURVEYING

should be nearly constant over the base length, the same number of observations usually defines the form of the bed at each cross section. Each cross-sectional area is therefore divisible into the same number of trapezia, the parallel sides of which are the observed depths, and their width the constant distance between the soundings. Triangles may take the place of trapezia at the sides, and the width of these may vary from one section to another. The dimensions of the partial areas for an average cross section are readily obtained by averaging the dimensions of the corresponding areas of the individual sections.

The next step is to obtain the mean velocity throughout each of those partial areas. The velocity of a float is the distance between the ranges divided by the time of run, notwithstanding that the path of the float may not be parallel to the base line. In the case of a surface float, the mean velocity in the vertical throughout its path is obtained by multiplying the velocity of the float by a coefficient of from 0·7 to 0·95 according to the conditions (section 526), but surface floats should not be used unless the value of this factor can be carefully determined, preferably by means of a current meter. The velocity thus derived represents the mean velocity in the average cross section along a vertical whose position in the width of the stream is the corresponding mean position of the float throughout its run. This is obtained by averaging the observed positions of the float at which it passed the two ranges.

On a base representing the width of the stream the average positions of the different runs are set off, and from them ordinates are erected to represent the mean velocities given by the floats. The resulting curve exhibits the variation of mean velocity across the stream, and is utilised to interpolate the value of the mean velocity for each partial area. These are represented by the ordinates corresponding to the positions of the centroids of the areas, which in the case of trapezia may usually be taken with sufficient accuracy at the middle of their width. The product of each partial area by its mean velocity gives the discharge through each section, the sum of which is the discharge of the stream.

530. The Current Meter. As a means of measuring the velocity of flowing water, the current meter (Fig. 295) proves more convenient than floats, and with careful usage gives better results. Various forms differ considerably in detail, but the instrument consists essentially of a spindle mounted on a fork and carrying a wheel with cup-shaped or helical vanes, which is rotated by the action of the flowing water. The spindle is vertical in some forms and horizontal in others, and is constructed to rotate with a minimum of friction. The number of revolutions made by the wheel during the time the meter is in operation is recorded by gearing or other means. The instrument is suspended by a rope, wire, or jointed tube, and the weight of the mechanism is balanced by a two- or four-bladed tail, which keeps the instrument facing the current. A lead weight may be fixed at the bottom to assist in maintaining the meter in position. The weight is sometimes made torpedo shaped, and may be provided with a rear blade. Provision is made for the attachment of additional weights when the instrument is used in swift currents.

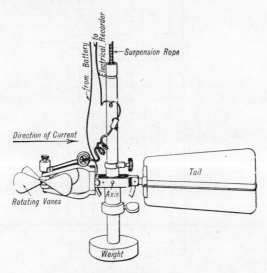

FIG. 295.—CURRENT METER.

Several methods are employed for counting the revolutions of the wheel. In a simple form of meter the number of revolutions is shown on a dial on the instrument itself. When the meter is immersed at the required depth, the recording mechanism is thrown into gear by the operator pulling a cord. After a noted interval it is released by a similar pull, and the meter is then drawn to the surface and read. It is usually more convenient to be able to note the revolutions without having to pull up the meter at every observation, and this is effected in various ways. In the acoustic type the gearing is so arranged that a tap is made at every fifth or tenth revolution, and the sound is communicated to the observer through a tube. In the electrical type the revolutions are indicated by a sounder consisting either of a telephone receiver or a buzzer.

For prolonged observations and high velocities electrical registration is most convenient, the revolutions being recorded upon a dial or dials above the surface.

531. Rating Current Meters. Since the results given by meter represent the number of revolutions in a given time, it is necessary to know the relationship between the velocity of the current and the number of revolutions per second made by the meter. This ratio is obtained by rating the meter.

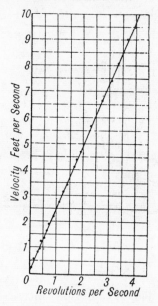

FIG. 296.—METER RATING CURVE.

Rating is performed by running the meter at a uniform speed through still water over a measured distance and noting the number of revolutions and the time taken. Runs are made at speeds varying from the least which will cause the wheel to rotate to the greatest likely to be encountered in gauging. At specially equipped laboratories or rating stations the meter is suspended from a trolley, which is run at a sensibly uniform speed. In ordinary practice, rating is usually performed by hanging the meter from the prow of a dinghy, which is towed repeatedly along the course at different speeds.

On plotting the mean velocities of the runs against the revolutions per second, a curve similar to that of Fig. 296 is obtained. For all meters the curve is practically a straight line, and is taken as such. The slope depends upon the type of meter. The ordinate corresponding to zero revolutions represents the velocity of current required to overcome initial friction and start the wheel. From the rating curve the velocity corresponding to any observed number of revolutions per second can readily be obtained.

In place of using the curve directly in the reduction of observations, it is more convenient to prepare a rating table giving velocities corresponding to revolutions per second. This is compiled from the equation,

$$V = c_1 R + c_2 ,$$

where V = the velocity of the current in feet per sec.,
R = the number of revolutions per sec.,
c_1, c_2 = constants, the former representing the slope of the curve, and the latter the ordinate for zero revolutions.

The constants are obtained from the curve, or their most probable values may be computed directly from the rating observations.

532. Use of Current Meter. In measuring discharge from velocities taken by current meter, one cross section only is required. It is divided into partial areas as before, and the meter is used at different points across the width to give the mean velocity over each area. The velocity determinations are made either by using the meter in the verticals of the centroids of the partial areas or by taking observations at any points across the width, fixing their positions by angles or on a cross rope, and interpolating the values for the partial areas.

If the water is sufficiently shallow, observations can be secured by wading, the meter being fixed to a graduated rod. If there is a bridge conveniently near, and situated on a straight reach of the stream suitable for a discharge station, the meter should be suspended from successive points along it. Otherwise, the observations are made from a boat, or, if repeated measurements are required, a cableway may be thrown across the stream, and observations taken from a suspended car or platform. When a boat is used, the meter is attached to a rod or tube which is held vertically and clear of the prow. When the depth is considerable, the meter is suspended by a wire or rope, preferably from an outrigger provided with a pulley. The boat is headed upstream, and attention must be paid to keeping in the range at each observation. Unless the width is too great, the boat is brought into position and held against the current by ropes from each bank.

In taking an observation with the meter, the sounding is first taken, and the meter is then lowered to the required depth. If the meter is one in which the recording mechanism is thrown into and out of gear by a cord, the interval between the two pulls is measured by means of an ordinary or stop watch and entered against the revolutions shown by the dial. With acoustic or electric meters, the count may be taken over two periods of one or two minutes each.

The object of the meter observations is to ascertain the mean velocity in the verticals at various distances across the width of the stream. Different systems of measurement are available, according as one or several observations are made in each vertical or the integration method is adopted.

Single Observations. Single observations are taken either at the surface or at the depth of mean velocity. Surface measurements are not usually made in normal cases, but the method is useful in swift currents and in times of flood, when it is difficult to maintain the meter in a desired position much below the surface. In taking the observation, the meter is held six inches or more below the surface in order that it may be wholly submerged and below the wave line. The mean velocity in the vertical is deduced by multiplying the observed result by a coefficient having an average value of 0·9.

Mean velocity is obtained directly by immersing the meter to a depth of 0·6 of the depth of the vertical. This method is commonly used in ordinary circumstances, and, although the ratio, 0·6, is only an average value, the accuracy of the results is sufficient for most practical purposes.

Multiple Observations. The best results are obtained by making measurements at a sufficient number of points in the vertical to enable the velocity curve to be plotted similarly to Fig. 294. The area contained between the complete curve and the depth axis divided by the total depth equals the mean velocity. This method is too laborious for ordinary measurements, but is that required in the evaluation of coefficients for the reduction of single point observations.

Good results are obtained by two observations in each vertical. These should be made at 0·2 and 0·8 of the depth of the vertical, the mean of the results giving a close approximation to the mean velocity.

Integration Method. In this method the meter is moved slowly and at a uniform rate along the vertical from the surface and back, the number of revolutions and time being observed. The meter is thus exposed to all the velocities in the vertical, and these are integrated and averaged mechanically in the result. Certain types of meter are not quite suited to the method since the vertical motion, unless very slow, causes the wheel to revolve. The correction to be applied to the indications of such meters may, however, be ascertained by moving the meter vertically at the same uniform speed through still water.

The integration method is sometimes extended to give the mean velocity over an entire cross section. The meter is moved up and down at an angle of about 45° to the horizontal from one bank to the other and back (Fig. 297). This system is suitable for the rapid gauging of streams of moderate depth.

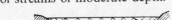

FIG. 297.

533. Measurement by Current Meter. Calculation of Discharge. Except when the method of Fig. 297 is employed, the total discharge is obtained as the sum of the discharges through the partial areas into which the soundings divide the cross section. The mean velocities over those areas, if not directly observed, are derived by interpolation in the same manner as described for float measurements (section 529). In the case where the integration method is applied to give the mean velocity for the entire cross section, the discharge is simply the product of the observed velocity by the cross-sectional area.

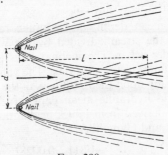

FIG. 298.

534. Thrupp's Ripple Method. If a high degree of accuracy is not required in discharge measurements, rapid determinations may be made from surface velocities observed by the method described by Mr. E. C. Thrupp.* It is based upon the circumstance that, if a small obstruction is placed in the surface of a stream, ripples are formed if the velocity exceeds about 9 in. per sec., and, as the velocity increases, the angle between the diverging lines of ripples becomes

*Min. Proc. Inst. C.E., Vol. CLXVII, p. 217.

more acute. To afford a simple means of measuring the rate of divergence, Mr. Thrupp used two 3-in. wire nails—about $\frac{1}{8}$ in. in diameter—at a fixed distance d apart (Fig. 298), and found that the velocity could be derived from the distance l from the base line so formed to the point of intersection of the last ripples. He obtained for the velocity in ft. per sec.

$$V = 0.40 + 0.206 \; l, \text{ for } d = 6 \text{ in.,}$$
$$V = 0.40 + 0.28 \; l, \text{ for } d = 4 \text{ in.,}$$

where l is measured in inches.

The method would appear to be quite as accurate as that of surface floats and much more convenient.

535. Velocity by Formula. In this method of estimating discharge —sometimes known as the slope method—measurement is made of the average cross-sectional area of the stream on a straight reach as well as of the longitudinal inclination of the water surface. To obtain the mean velocity of the stream, the formula in general use is that deduced by Chezy, *viz.*,

$$V = c\sqrt{ri},$$

where V = the velocity in ft. per sec.,
  r = the hydraulic mean depth,
  i = the inclination of the water surface,
  c = a variable coefficient.

The hydraulic mean depth, or hydraulic radius, is the cross-sectional area of the stream divided by the wetted perimeter or length of bed under water. The inclination is the ratio of the fall in a measured distance to that distance. The value of the coefficient c depends principally upon the roughness of the bed, but also upon the inclination and hydraulic mean depth. In practice, the value of c is commonly derived from tables or diagrams based upon the formula of Kutter and Ganguillet, *viz.*,

$$c = \frac{41.65 + \dfrac{1.811}{n} + \dfrac{0.00281}{i}}{1 + \left(41.65 + \dfrac{0.00281}{i}\right)\dfrac{n}{\sqrt{r}}},$$

in which the coefficient n depends upon the character of the bed. The value of n varies from 0.020 for irrigation canals with well-trimmed bed in perfect condition to over 0.035 for canals in very bad order with much weed and stones. For rivers very uniform in alignment, slope, and cross section, and with a smooth sandy or gravel bed, n is taken as 0·025. As the irregularities of the bed, etc., increase, n increases to 0·035, and may reach 0·050 if there is an excessive amount of weed or in the case of torrents bringing down much detritus.

In estimating discharge by formula, a straight reach of river should be selected with as nearly uniform a cross section as possible. The fall and the distance between the points at which it is measured should be sufficiently great that the inclination can be determined without serious error. Cross sections are taken at intervals along this distance, and an average section is deduced. The slope of the water surface is measured by simultaneous readings of gauges placed one at each end of the reach and similarly situated with respect to the current. Gauge readings are taken to 0·01 ft., and the zeros of the gauges are connected by careful levelling.

The results obtained by the slope method are inferior in precision to those in which the velocity is actually observed, principally on account of uncertainty in assigning a suitable coefficient in the formula and in measuring the slope. The method, however, is useful in making an isolated rough estimate of flood discharge from the flood marks left on the banks.

WEIR METHOD
OF DISCHARGE MEASUREMENT

536. A weir is an artificial barrier built across a stream, which flows over it in a cascade of definite form, from the dimensions of which the discharge is computed. The weir method is specially applicable to the gauging of small streams when accurate results are required. The cost of construction usually prohibits its use on large streams, but in such cases existing dams are sometimes utilised in a similar manner.

537. Varieties of Weirs. The various forms of weirs differ in certain particulars affecting the rate of discharge over them.

Crest. Weirs constructed specially for the measurement of stream flow are of the sharp-crested type. In these, the crest, or edge over which the flow takes place, is virtually a line. In the case of dams or broad-crested weirs, the water passes over a surface, and a different condition of flow obtains.

Shape of Opening. The opening or notch through which the water flows may be of rectangular, triangular, trapezoidal, or stepped form (Fig. 299). The rectangular weir is that most commonly used in stream gauging, the length of the notch being at least three times the head of water over it. Other forms provide for an increase in the breadth of the flow as the discharge increases. The triangular or V-shaped notch has the merit of giving a constant shape to the issuing stream at all heads. It is well adapted for the accurate measurement of small discharges, but is not so suitable as the rectangular form for use in very shallow streams. The trapezoidal form with side slopes of 1 in 4 was proposed by Cippoletti with a view to balancing, by the increase of breadth upwards, the loss due to increased contraction with increased head. The stepped form is designed to yield good measurements in dry weather and in flood.

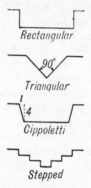

Rectangular

Triangular

Cippoletti

Stepped

FIG. 299.—FORMS OF WEIRS.

The centre notch is sufficiently small that the minimum flow can be measured with greater accuracy than is possible with a very small head over a long crest.

End Contractions. The form of the stream issuing from a rectangular notch depends greatly upon the position and size of the opening relatively to the cross section of the channel on the upstream side. In Fig. 300, which represents the most common arrangement, the breadth of opening or length of crest is less than that of

FIG. 300.—
WEIR WITH END CONTRACTIONS.

FIG. 301.—
WEIR WITHOUT END CONTRACTIONS.

the stream. The stream lines at the sides are sharply curved, with the result that the stream is contracted in width just after passing over the notch. Such a weir is said to have end contractions. When, as in Fig. 301, the breadth of the weir is the same as that of the approaching stream, the weir is without end contractions. Intermediate conditions arise when insufficient room between the sides of the notch and those of the stream prevent the proper development of the curved stream lines, and the end contractions are then incomplete, a condition to be avoided in gauging.

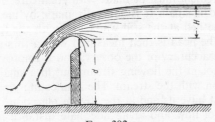

Fig. 302 illustrates diagrammatically the normal form of the issuing stream or nappe in longitudinal section. The stream lines near the crest are curved upwards causing a vertical contraction, which is completely developed when the distance d is greater than $2H$. When the height of the crest is so small relatively to the head of water over it that the tail water is higher than the crest, the weir is known as a drowned or submerged weir.

538. Construction of Weirs. The site for a weir should be selected so that on the upstream side the channel is straight and free from great irregularities in order that the water may approach the weir with as uniform velocity as possible. The weir will be constructed of stout uprights and planking or tongued and grooved boarding, and the wall must be vertical and normal to the flow. Sheeting may have to be used to prevent leakage below or round the sides. The timbers forming the crest and sides of the notch are chamfered on the downstream face, the edge not exceeding $\frac{1}{8}$ in. in breadth. Otherwise, an $\frac{1}{8}$ in. thick metal strip is fixed, as in Fig. 302. The crest, in other than V notches, should be set accurately horizontal. If the weir is one with end contractions, these should be complete, so that the formulae based on that condition may be strictly applicable. This is ensured by making the distance between the sides of the upstream channel and the sides of the opening not less than twice the head. The use of the standard formulae also assumes that the nappe is free from contact with the weir below the crest, to secure which there must be free admission of air under the nappe.

539. Measurement of Head. The dimensions of the weir being known, it is only necessary to measure the head of water over it to enable the rate of discharge to be computed. By the head is meant the difference of level, H (Fig. 302), between the crest, or the bottom of the V in the triangular form, and the surface of the sensibly still water several feet above the weir and beyond the influence of the depression of surface caused by it. This upstream distance should always exceed $3H$.

The measurement must be made with great care, more especially when the head is small. It is generally made by means of a stake driven into the bed at the still water and with the top at the level of the crest. The head is then measured with a thin-edged scale or steel rule. The accuracy of the observation is improved if the head is measured not in the flowing stream but in a small gauge pit in communication with the stream. The connecting pipe, of 1 or 2 in. diameter, must have its end at the stream placed at right angles to the direction of flow and at an adequate distance from the weir. For the best results the hook gauge is used, the instrument being accurately set so that at zero reading the hook is at the same level as the crest. Various forms of float gauge are also used, and recording mechanism may be added so that the varying level of the float can be plotted automatically.

540. Discharge Formulae. There are several formulae available for computing the rate of discharge over weirs, and the reader is referred to text-books on Hydraulics for a discussion of the subject. The simplest and most commonly used formulae are those of Francis.

Rectangular Weirs. For a sharp-crested rectangular weir without end contractions, and neglecting the effect of the velocity of the approaching water, Francis obtained

$$Q = 3 \cdot 33 \, LH^{\frac{3}{2}},$$

where $Q =$ the discharge in cubic feet per sec.,
$\quad L =$ the length of the crest in feet,
$\quad H =$ the observed head in feet.

The effect of the velocity of approach of the stream is equivalent to an increase of head. This velocity, v, need not be measured directly, but is obtained approximately by dividing the approximate discharge obtained as above by the cross-sectional area of the approach channel where the head is observed. The velocity head, $h = v^2/2g$, was allowed for by Francis by putting

$$Q = 3 \cdot 33 L[(H+h)^{\frac{3}{2}} - h^{\frac{3}{2}}].$$

The effect of velocity of approach is more commonly dealt with by increasing the observed head by ah, where a is an experimental coefficient having a value between $1 \cdot 2$ and $1 \cdot 68$, and which may be taken as $1 \cdot 5$, so that

$$Q = 3 \cdot 33 L(H + 1 \cdot 5h)^{\frac{3}{2}}, \text{ approximately.}$$

From the new value of Q a closer value for v can be obtained, and a second approximation may be made for Q, but this is seldom necessary since the velocity effect is small in most cases.

For a weir with end contractions, experiment shows that each complete end contraction shortens the effective length of the weir by about $0.1H$, so that for two end contractions,

$$Q = 3.33(L - 0.2H)H^{\frac{3}{2}},$$

or, allowing for velocity of approach,

$$Q = 3.33(L - 0.2H_1)H_1^{\frac{3}{2}},$$

where

$$H_1 = [(H+h)^{\frac{3}{2}} - h^{\frac{3}{2}}]^{\frac{2}{3}}, \text{ or } = H + 1.5h,$$

according to the method of treating velocity head. The velocity correction is usually insignificant in the case of weirs with end contractions, since the approach channel is of much greater sectional area than the nappe.

Triangular Weirs. In the case of a sharp-edged notch, with apex angle θ, the discharge formula has the form,

$$Q = c \tan \tfrac{1}{2}\theta \cdot H^p,$$

the velocity head being negligibly small. The values of p given by different investigators range from 2.47 to 2.5, and c varies from 2.48 to 2.56. For ordinary measurements it is sufficient to adopt

$$Q = 2.5 \tan \tfrac{1}{2}\theta \cdot H^{2.5},$$

which, in the common case of a right-angled notch, gives the easily remembered expression,

$$Q = 2.5H^{2.5}.$$

Correction for velocity of approach is seldom necessary, but can be made as before.

Trapezoidal Weirs. The discharge from a trapezoidal weir is the discharge over a rectangular weir of the same length of crest, and having end contractions, plus that over the triangular notch which would be formed by the sloping sides. If, therefore, $\tfrac{1}{2}\theta$ is the inclination of the sides to the vertical,

$$Q = 3.33(L - 0.2H)H^{\frac{3}{2}} + 2.5 \tan \tfrac{1}{2}\theta H^{\frac{5}{2}}.$$

By selecting $\tfrac{1}{2}\theta$ to make $2.5 \tan \tfrac{1}{2}\theta = 0.666$, we have

$$Q = 3.33LH^{\frac{3}{2}}.$$

Cippoletti suggested $\tan \tfrac{1}{2}\theta = \tfrac{1}{4}$, and adopted

$$Q = 3.367LH^{\frac{3}{2}}.$$

541. Dams. When an existing dam or broad-crested weir is to be utilised for the measurement of stream flow, the effective length and levels of the crest are measured. The dimensions of the cross section must be ascertained since the coefficient to be used in computing the discharge is derived from published results of experiments on dams of similar form. As with sharp-crested weirs, the head is measured to the surface of the still water behind the dam. If, at the time of gauging, part of the flow is discharged through or round the dam, the quantity so discharged must be separately measured.

The accuracy of the results is much inferior to that attained with sharp-crested weirs. It depends greatly upon the selection of a suitable value for the discharge coefficient and also upon the condition of the dam, particularly as regards uniformity of crest and apron.

CHEMICAL METHOD OF DISCHARGE MEASUREMENT

542. This method consists in introducing into the stream, at a uniform and accurately known rate, a fairly concentrated solution of some chemical, and comparing the analyses of the water before and after its introduction. The percentage of the chemical found in the water (or the increased percentage, if the chemical is already present in the untreated water) bears the same ratio to its percentage in the solution as the volume of solution introduced per second bears to the discharge per second of the stream.

Or, let P = percentage of chemical in the solution used,

p = percentage of chemical in the water samples,

Q = flow of stream in cub. ft. per sec.,

q = flow of solution in cub. ft. per sec.

Every second there is added $\dfrac{Pq}{100}$ cub. ft. of chemical.

Its dilution in the water samples $= \dfrac{Pq}{100Q} = \dfrac{p}{100}$, or $Q = \dfrac{Pq}{p}$.

The success of the method is dependent upon several factors. The chemical should be one for which a very sensitive reagent is available, and a sufficient quantity must be used that the dilution may not be greater than will permit of the analyses of the water samples being made with the required accuracy. Thorough mixing of the solution with the stream is essential. On other than very narrow streams it should be introduced at several points across the width, the flow of solution being continuous during the test. Samples of the untreated water are taken at the station where the solution is introduced, and, at some distance downstream, samples of the much diluted solution are collected from all parts of the cross section after a sufficient interval has elapsed for the flow to arrive there. The distance between the stations must be great enough to ensure complete mixing. Conditions favourable to mixing are irregularity of alignment and bed, and the distance between the stations may then be shorter than in the case of a stream with very uniform flow. The method is particularly suitable for turbulent streams with rocky beds.

CONTINUOUS MEASUREMENTS OF STREAM DISCHARGE

543. Investigations of stream discharge made in connection with the design of water schemes should be continued over several years to ensure a reasonably accurate estimate of the extremes of flow. The necessary observations are made either by means of weirs or by the area-velocity method.

544. By Weir. Continuous records are most easily and accurately secured by the weir method, and simply require the employment of an intelligent and reliable person to read the head daily. A single daily observation is sufficient during periods when the head is changing slowly, but, when it is subject to rapid fluctuations, several readings should be made at intervals throughout the day. Continuity in the record of head is secured by the use of a self-registering gauge.

A continuous measurement of discharge may be obtained mechanically by means of a weir recorder. In this instrument a float rotates a cylindrical drum on which is cut a spiral groove, the curve of which is deduced from the weir formula. A lever engages with the groove and actuates a pen, which traces out a continuous graph of rate of discharge. The apparatus may also be fitted with an integrating mechanism whereby the total flow up to any time may be exhibited.

545. By Area-Velocity Method. In the application of this method, the discharge is measured on several occasions by current meter or floats at times of low, average, and flood stages. With gauge heights as ordinates, and discharges as abscissae, the observed discharges are plotted, and through the points so obtained a curve, known as the discharge or station rating curve, is drawn. This curve, or a table prepared from it, then serves to give the discharges corresponding to gauge readings taken daily by an observer or recorded by an automatic gauge. The method is dependent upon permanence of the stream bed and stability of the gauge. In streams with shifting beds the discharge curve may not remain applicable throughout a long investigation, but may require modification from time to time in the light of additional gaugings.

REFERENCES

FARQUHARSON W. I. 'Datums of hydrographic surveys'. *R.I.C.S. Journal*, Nov. 1951.

GORDON D. L. 'Tidal levels and the land surveyor'. *Chartered Surveyor*, Nov. 1963.

— 'Rationalisation of chart datum'. *Conference of Commonwealth Survey Officers*, 1967.

JACKSON J. E. 'Tidal observations in Ceylon'. *Empire Survey Review*, No. 20, Apr. 1936.

MARGRETT A. D. 'Hydrographic surveying for civil engineering development'. *Chartered Surveyor*, Apr. and May 1956.

MAXWELL P. S. E. 'Echo sounding'. *Empire Survey Review*, Nos. 17, 18 and 19, July and Oct. 1935, Jan. 1936.

— 'Admiralty Manual of Hydrographic Surveying, Vol. I'. *Hydrographer of the Navy*, 1965.

CHAPTER XIII

FIELD WORK FOR AERIAL SURVEYS

546. In the previous chapters of this volume discussion of survey methods available to the engineer-surveyor for the conduct of land surveys of relatively limited areal extent has been confined exclusively to ground methods, where all necessary measurements can be carried out either by, or under the direct control of the field surveyor. With the development of aerial photogrammetry from a technique suitable only for extensive small scale mapping, to a stage when it has become competitive, in both accuracy and economy, with traditional methods of conducting medium and large scale engineering surveys, it is evident that the implications of this alternative survey technique must be seriously considered.

As compared with ground survey, aerial photogrammetry is essentially a group activity with three related but quite distinct specialised branches. These may be classified as:

(a) Aerial photography.
(b) Measurements on the ground (field survey).
(c) Measurements from photographs (photogrammetry).

Descriptions of the specialised techniques required in activities (a) and (c) are outside the scope of this volume. (References to literature in this field will be found in the bibliography.) However, a detailed understanding of them is not required in considering the application of aerial photogrammetry to a particular survey project.

547. A decision to make use of the technique will depend upon an assessment of the three variables, accuracy, speed and economy as they apply to the particular circumstances. If aerial survey methods are contemplated on the grounds of possible speed and economy, the engineer must decide what he wants the survey to give him, particularly as regards accuracy. Tenders can then be called for from photogrammetric contractors to supply the necessary specialist services (a) and (c). Although most contractors retain their own field surveyors and will contract to undertake the entire survey, the engineer may wish to keep all field survey activities under his own direct control: in any case, to do so will reduce the tender prices. After considering the cost and time required, to carry out the whole survey by conventional methods, the air survey alternative may be accepted.

Once a tender which meets the survey specification has been accepted, the field surveyor becomes directly involved. The object of this chapter is to describe the field activities peculiar to photogrammetric mapping which he will be called upon to perform. These may be summarised as:

(*i*) The provision of suitable ground control.

(*ii*) The field completion of the photogrammetric plots.

(*iii*) Field checking of the completed work to ascertain whether the specified accuracy has, in fact, been achieved.

In studying the methods of conducting these field activities two distinct situations are distinguished.

(*i*) Where the surveyor is called in when the photography has already been obtained, or where existing photography is to be used for mapping, and

(*ii*) where the surveyor is called in at the flight-planning stage, and well in advance of the photographic sorties.

The two cases are examined separately.

PROVISION OF CONTROL WHERE THE PHOTOGRAPHS ARE ALREADY TAKEN

548. Before he can proceed, the surveyor will require to know the approximate locations where ground control is needed, together with the accuracy in position and height appropriate to the project. The latter requirement will, in most cases, be governed by the scale of the photography rather than by the scale of the final map. Details for the location of suitable control points are provided by the photogrammetrists, after stereoscopic examination of the photography, in the form of annotated positive prints of the relevant photographs. One commonly used convention is to mark the photographs with coloured circles, say red for points to be fixed planimetrically (i.e. surveyed for position) and blue for points where the height only is needed. Thus, a photograph marked with a double circle will indicate that a control point is required to be fixed in three dimensions within the area bounded by the circles. Normally each location will be marked on only one of the photographs relating to the area in question, although the site may be visible on 3, 5 or 6 photographs of the series to be used in the survey. It is the surveyor's job to identify points (preferably in groups of 3 close together, to minimize the possibility of misidentification) within the circles, and in so doing, he must be quite certain that the points chosen are, in fact, clearly visible and well defined on all of the relevant photographs in the series. Since the positive identification of a point is of equal importance to the accuracy of its fixation, precision in identification must never be sacrificed for the simplification of the fixing survey.

549. Guiding Principles for Photo-identification. The successful selection and photo-identification of points suitable for ground control depends largely upon the surveyor's ability to relate his terrestrial view of the terrain to the much smaller scale aerial view which the photogrammetrist will see through the optical system of his plotting machine. This is not a simple matter and considerable practice is required before a surveyor can, with confidence, choose some object or feature which will be neither too large nor too small at the instrumental viewing scale. A pocket stereoscope is an essential piece of equipment for this stage of the work, but, as the best of these instruments magnify only about 6 times compared with the 10 times magnification of many plotting machines, there may be a tendency initially to select objects which are much too large.

The general principles governing the selection of photo-control points may be set out as follows:

(1) All points of detail chosen must be identifiable on all the relevant photographs, sharply defined and in the same position as when the film was exposed. This is particularly important when man-made features are used and when considerable time has elapsed since the photographs were taken.

(2) For height control it is essential that points should be chosen in areas where the ground is relatively level and reasonably free from vegetation, even if such choice means some sacrifice in the sharpness of definition. It follows from this requirement that, where it is not possible, owing to steep slopes or other topographic factors, to choose a point suitable for both planimetric and height control, a separate height point should be chosen nearby in a more favourable situation.

(3) Since shadows, glare and definition may vary greatly between overlapping photographs in adjacent strips care must be taken to ensure that objects chosen in the common overlap are truly visible on both. Frequently it will be found that on one or other of the photographs only the shadows of certain objects appear.

(4) The design of measuring marks in many plotting machines is such that symmetrical coincidence with points of detail in the stereo-model is much more accurately effected than the 'pointing' at corners more familiar to the field surveyor. It is preferable therefore, to select the centre rather than the corner of a small object.

Examples of good and bad points are illustrated in Fig. 303.

(5) In the search for suitable objects it is well to remember the need for positive identification, and it often helps to identify the smaller points of detail as close as possible to some major feature of the terrain, which is easily and unambiguously distinguished on the photograph. In featureless country this may not be quite such a simple matter, but, providing the surveyor has an approximate idea of his position the plane-table resection procedure described in Chapter VII will frequently prove effective. The table is set up near a suitable point of detail and rays are drawn on a sheet of tracing paper to all surrounding bushes or identifiable features. The tracing paper is then transferred to the photograph and moved around until the rays intersect the terrain details corresponding with those on the ground. The plane-table position may then be pricked through.

550. Recording the Photo-identification. As each point is chosen and identified it is necessary to record adequate details for its subsequent re-location by the photogrammetrist. For small scale mapping it may be sufficient to prick through the point with a fine needle, using a pocket magnifier, and to annotate the reverse side of the photograph. The details to be recorded will include an identifying serial number, a description of the feature and a statement as to whether it is fixed for position or height. For medium and large scale work this system is not recommended. Instead, for each point or group of nearby points which together make up a unit of ground control, a diagram should be prepared to illustrate the relationship of the point, or points, to the surrounding detail. The sketch should illustrate shadows similar to those on the photograph, and must contain a north point to enable it to be oriented more easily. The points should be marked on

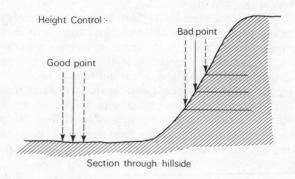

Section through hillside

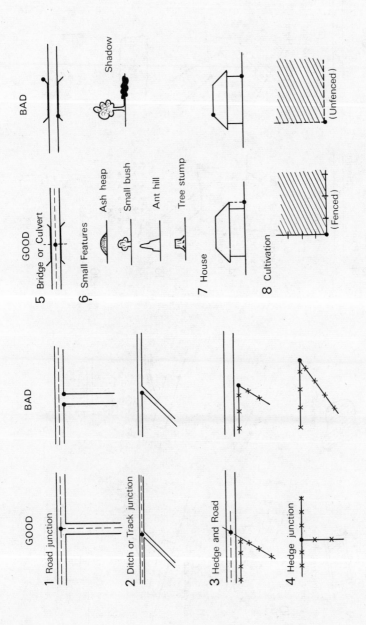

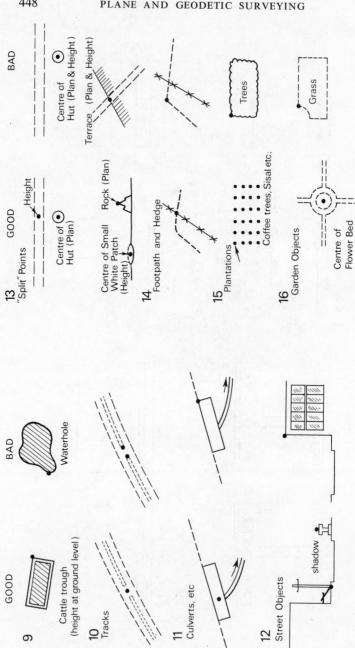

Ground Control Points

both the diagram and the photograph, but on the latter only in such a way that detail in the immediate vicinity is not obscured. As with all survey activities where field records must be interpreted unambiguously by persons other than the surveyor, there is a need for standardisation in the system of recording, and to this end, the majority of survey organisations and contractors require that their own printed forms be used for photo-identification sketches.

An example of a typical form is illustrated by Fig. 304.

AIR SURVEY AREA....KAPTARAKWA...

Station....KAP. CI. and KAP. C2....

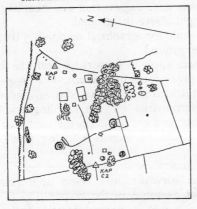

Point identified on Photograph No.....007....

Description of point identified....KAP. CI. Tree Stump.........KAP. C2. Fence junction...

		CI +167 779 · 6	CI. 8570·7
Co-ordinates—	D·O·S	y= C2 +167 592 · 9	Height C2. 8567·1
		CI +198 147 · 3	
(referred to Equator and....35°...E)	x= C2 +197 982 · 4	Identified by....a. n. other....

REMARKS....Heights to G. L....

FIG. 304.

551. Fixing the Ground Control Points. Having selected and identified suitable control points we may now turn to the problem of fixing their positions and heights to the required order of accuracy.

The plotting accuracy of even the most sophisticated plotting machines is a function of the flying height of the photographic aircraft above the ground and therefore the tolerance in the fixation of control will, in general, be related to this factor rather than to the scale of the final map. Although, in most cases the map scale will have been chosen to be compatible with the achievable accuracy of the photogrammetric equipment, this is by no means always the case.

The exact relationship between flying height and plotting accuracy will naturally vary with different combinations of aerial cameras and plotting equipment, but in the absence of more specific data relating to a particular survey project, the following criteria for the allowable tolerances in fixing ground control points may be safely adopted:

(a) For large scale engineering and township mapping $\pm 0.02\%$ of the flying height above the ground.

(b) For medium scale engineering and planning maps $\pm 0.03\%$ of the flying height above the ground.

(c) For small scale topographical mapping $\pm 0.05\%$ of the flying height above the ground.

These tolerances apply to surveys for both the fixing of position and height of ground control points. For maps and plans on which no height data is to be shown, vertical control is still needed by the photogrammetrist for the levelling and scaling of the stereo-models, but in these instances the tolerance in the fixing of the height points can be increased to five times the above percentages.

Once the accuracy requirements have been determined it is evident that the surveyor must select appropriate methods of survey. Thus, for example, in the preparation of a 1/2,500 scale plan for which it would be normal practice to use 1/12,500 scale photography:

The flying height with a standard 6″ focal length survey camera would be 6,250 feet, and $\pm 0.02\%$ of 6,250 feet $= \pm 1.25$ feet.

Therefore, if the points to be used to control the scale of a particular pair of photographs are 8,000 feet apart, the permissible linear misclosure of the fixing survey will be of the order of 1/3,000–1/3,500. This order of accuracy in the determination of position is well within the range of the simpler methods of theodolite traversing discussed in Chapter IV and of the tachymetric methods described in Chapter XI, provided that a reasonable density of established survey control points already exists in the project area. Where these conditions do not obtain, the initial stages of the fixing survey will entail the establishment, observation, and adjustment of a suitable control framework by triangulation or theodolite traversing. In this process it would be false economy for the surveyor to be unduly influenced by the accuracy requirements of the mapping technique. The establishment of the control framework should be regarded as part of the normal process of increasing survey control density, and the ruling standards applicable to the area, and to the order of the control, will apply. If the photo-control points can be incorporated in the framework survey so much the better, but it will quickly be realised that the siting and nature of many of them precludes such direct fixing. The points may fall in any part of the terrain, in valleys, woods or built-up areas. The fixing of the photo-points from the nearest rigorously established survey control points to the lesser standard of accuracy may still require considerable ingenuity in overcoming the problems set by the nature of the terrain and by the nature of the points themselves. It may be found advantageous to survey to a peg placed near the object in the first instance and subsequently establish the final position by bearing and distance.

In all surveys for photo-control it is essential not only that the equipment and method should be appropriate but that the fact of attainment of the required standard can be demonstrated. It follows that the field work should be designed to give an independent check on each fix whenever possible.

Height control in excess of the required precision for the project quoted in this example, and indeed for all large scale plans, is easily achieved by the methods of ordinary spirit levelling, tachymetric levelling, or the observation of vertical angles. These methods tend to be used because of their familiarity and inherent simplicity, and because the intermediate heights produced as a by-product of levelling may prove valuable at a later stage in checking the final map. Where heights are not required on the plan, and for mapping at smaller scales, the time and expense of levelling to a degree of precision which cannot be utilised in the production of the map is no longer justified, and more rapid and cheaper methods must be sought.

552. Height Control by Altimeter. Perhaps the most simple and convenient method of providing height control for surveys in the above category is through the use of aneroid barometers as described in Volume II Chapter VII. The type known as Surveying Altimeters, manufactured by Messrs. Wallace and Tiernan, which are scaled to read directly in feet or metres are amongst the most suitable instruments for the purpose. As the accuracy requirements differ from those of the mapping projects described in Chapter VII of Volume II, some further explanation of the field procedure is necessary.

Four methods of altimeter heighting are in general use, any one of which, when used in the proper circumstances, will yield results of the required accuracy. The methods in increasing order of accuracy are:

(a) Single altimeter method.
(b) Single base method.
(c) Leap frog method.
(d) Double base method.

553. Single Altimeter Method. The utility of this method is confined to the heighting of points within $\frac{1}{4}$ hour's travelling time from a point of known height. The altimeter is first read at a point whose height has been rigorously fixed, and the temperature is also recorded. The observer then moves rapidly to the point to be heighted and again reads the altimeter scale and the temperature. He now returns to the starting point and repeats the observations. Altimeter readings are first corrected for temperature from the nomograms provided with the instrument. The height of the new point can then be determined by applying to the corrected altimeter reading an amount equal to half the difference between the initial and final readings at the point of known elevation.

554. Single Base Method. This method requires the use of two altimeters and two observers with synchronised watches, and is applicable to situations where the difference in elevation between the base (i.e. the point of known height) and the points to be fixed does not exceed 500 feet, and where the horizontal separation of the base from the most distant point is less than ten miles. The procedure is as follows:

One altimeter is sited at the base station of known height and is read at ten or fifteen minute intervals throughout the day. The time, and temperature (wet and dry bulb) are recorded for each reading, together with any relevant observations on the weather, such as change of wind strength, or variation in the amount of cloud cover. At the beginning of the day's work both altimeters are read together at the base station and any index error of the field instrument, relative to the base instrument, is noted. Watches are synchronised, then the field instrument is carried successively to the points to be heighted. At each point, time, temperature and indicated height are read and booked. On completion of these readings the field altimeter must return to the base where a second set of observations of altitude, time and temperature are taken, for comparison with the base altimeter.

If, during the altimeter traverse, the field altimeter can be read at other points of known height a useful check on the consistency of the results can be achieved. Such checks are extremely useful when the method is being used towards the limits of its normal range of application. It is, of course, possible to use more than one field instrument from the same base.

The accuracy of the height of a point derived by the Single Base Method varies with both the horizontal and vertical distance of the point from the base. Provided that both base and field altimeters are read as far as possible under similar conditions, out of doors, and protected from the sun and strong winds, and that no readings are taken when local atmospheric conditions are obviously unstable, the expected *average* accuracy of heights determined with Wallace and Tiernam altimeters should be in accord with the following table:

Max. vertical range	Max. horizontal distance	Accuracy of final height
± 300 ft.	2 miles	± 3 ft.
± 500 ft.	4 miles	± 10 ft.

To maintain similar standards of accuracy outside this range the Leap Frog or the Double Base Method should be used.

555. The Leap Frog Method. Two altimeters are again required and the method is applicable to the determination of heights of points spaced irregularly between two points of known altitude. As in the previous method the altimeters are first compared at the initial base station, and the observers' watches are synchronised to enable all subsequent readings to be taken simultaneously at some specified interval. One observer now moves his altimeter to station 1 in the programme whilst the other remains at the base. At the agreed time both altimeters are read and the base instrument is then moved to station 2 of the programme. Simultaneous readings are once more recorded then the altimeter on station 1 is moved to join the instrument on station 2. After reading both instruments at station 2 the leap-frog procedure is repeated through the remaining field stations until the traverse is completed, when final readings, with both altimeters together, are made at the second base of known height. Observations of wet and dry bulb temperatures must be made each time the altimeters are read, to enable corrections to the indicated heights for differences in temperature and humidity to be determined.

556. The Double Base Method. To extend the vertical range of the single base method to approximately 1000 feet, two bases should be selected, the heights of which straddle the expected height range of the photo-control points. The bases should, if possible, fall within the survey area and should not be more than 10 miles apart. Altimeters located at the two bases are read at 10 or 15 minute intervals throughout the day in exactly the same manner as in the single base method. The field procedure is also identical with the single base method and, although temperature and humidity corrections are not necessary in double base altimetry, it is sound policy to record the wet and dry bulb temperatures at each station to allow for reduction of heights by the single base method should either set of base observations be suspect on account of local variations in atmospheric conditions, or for any other reason.

ALTIMETER SINGLE BASE COMPUTATION FORM

BASE *NYOKA*

HEIGHT OF BASE *5356 ft*

Page *1*
Date *6·10·54*
Field Book *7667*
Pages *87·88*

Point Number	Time	Temperature °F			Mean Humidity %	Field Reading	Base Reading	Indicated Difference	T. & H. Correction Factor	Corrected Difference	Final Height	References
		Field	Base	Mean								
NYOKA											*5356*	*See Comps Page 27*
134/2/1	*11·30*	*76*	*78*	*77*	*70*	*5512*	*5542*	*– 30*	*1·061*	*– 32*	*5324*	
134/2/2	*12.00*	*74*	*74*	*74*	*74*	*5600*	*5552*	*+ 48*	*1·054*	*+ 51*	*5407*	
134/2/3	*12.15*	*76*	*74*	*75*	*65*	*5659*	*5556*	*+103*	*1·056*	*+109*	*5465*	
134/2/4	*12.45*	*80*	*80*	*80*	*61*	*5979*	*5566*	*+413*	*1·067*	*+441*	*5797*	
134/2/5	*13·00*	*76*	*76*	*76*	*66*	*5994*	*5570*	*+424*	*1·058*	*+449*	*5805*	
134/2/6	*13·15*	*76*	*76*	*76*	*59*	*5698*	*5578*	*+120*	*1·057*	*+127*	*5483*	

FIG. 305.

Notes.

(1) *The Mean Temperature (Column 5) is the mean* dry bulb *temperature of the base and field stations.*

(2) *The Field Reading (Column 7) is the field altimeter reading* corrected to the base, *in the field book.*

(3) *The T. and H. Correction Factor (Column 10) is obtained by entering the nomogram with the mean dry bulb temperature (Column 5) and the mean humidity (Column 6) of the base and field stations.*

(4) *Final heights are obtained by applying the corrected differences* in turn *to the height of the base. Care must be taken to use the correct sign.*

Fig.(306) ALTIMETRY LEAP-FROG TRAVERSE COMPUTATION (12)

Point Number	1 Mean Humid	2 Mean Dry Temp	3 T.&H. Corr. Factor	4 Base Reading	5 Field Reading	6 Index Corr.	7 Corrected Reading (5+6)	8 Diff. (7-4)	9 Corr. Diff. (3×8)	10 Computed Height	11 Adj.	12 FINAL HEIGHT
B. Mangu				5577								see comps page 25.
BF. Mangu				5578		(-1)	5577			5818		5818
B. Mangu	81	69	1.044	5582								
F. 106/3/1					5620	-3	5617	+35	+37	5855	0	5855
B. 106/3/1	81	70	1.046	5625								
F. 106/3/2					5559	(-5) (+5)	5564	-61	-64	5791	0	5791
B. 106/3/2				5561								
BF. 106/3/2				5569		(-8)	5561					
B. 106/3/2	81	69	1.044	5568								
F. 106/3/3					5662	-5	5657	+89	+93	5884	+1	5885
B. 106/3/3	85	67	1.040	5672								
F. 106/3/4					5851	(-1) (+1)	5852	+180	+187	6071	+1	6072
B. 106/3/4				5861								
BF. 106/3/4				5859		(+2)	5861					
B. 106/3/4	90	70	1.047	5874								
F. Mangu					5632	+1	5631	-243	-254	5817	(+1)	5818
B. Mangu				5641								
BF. Mangu				5646		(-5)	5641					

FIG. 306.

Notes.

(a) *In the field, the function of each barometer at each station should be noted as follows:*

'F' *denotes the field barometer.*

'B' *denotes the base barometer.*

—*and when the two barometers are together at the same station for indexing purposes:*

'B' *denotes the barometer which will remain behind at the base.*

'BF' *denotes the barometer which will move forward to the next field station.*

(b) *The same particular barometer always moves forward first from the indexing station. Each time the leap takes place, the functions of the two barometers are reversed, e.g. before the leap, 1 is the base barometer (B) and 2 is the field barometer (F); after the leap, 2 becomes the base barometer (B) and 1 becomes the field barometer (F). The Index Correction (Column 6 above) is always applied to the 'F' or 'BF' barometer readings by incremental portions of the change in index difference between the indexing stations and with the opposite sign where the functions are reversed due to the leap.*

557. Reduction of the Observations. (See also Vol. II, Chapter VII.) Surveying altimeters are calibrated in feet or metres using

some standard pressure-altitude relationship. In the case of the
Wallace and Tiernan altimeter this is the standard atmosphere of
Meteorological Table 51 of the Smithsonian Institution. The
mechanism is designed so that the pointer will deflect the same
amount for equal changes in altitude. If the assumed standard
conditions do not exist while the survey is being made, corrections
to the indicated heights must be applied. Temperature has the
greatest effect on the density of the air and is, therefore, the most
significant. A correction for relative humidity is required only
when high humidity occurs with high temperature. Both temperature and humidity corrections are obtained directly by entering the
charts or nomograms, supplied with the instruments, with the
observed wet and dry bulb temperatures. Examples of suitable
standard forms of computation for the different observing methods
are illustrated by Figs. 305, 306 and 307. These examples are taken
from the Survey of Kenya's Survey Manual of Field Practice.

ALTIMETER DOUBLE BASE COMPUTATION FORM

UPPER BASE — HT. OF ALTIMETER A. 6078
DIFFERENCE IN HEIGHT C. 688
LOWER BASE — HT. OF ALTIMETER B. 5390

Page 6
Date 10·2·1959
Field Book 9071
Pages 16·17

Point Number	Time	1. Upper Base	2. Field	3. Lower Base	4. 2–1	5. 2–3	6. 1–3	7. C÷6	Elevation A+(7×4)	Elevation B+(7×5)	References
											FINAL HEIG
122/4/1	11·30	6042	5542	5362	-500	+180	+680	1·012	5572	5572	5572
122/4/2	12·00	6052	5610	5370	-442	+240	+682	1·009	5632	5632	5632
122/4/3	12·15	6056	5659	5376	-397	+283	+680	1·012	5676	5676	5676
122/4/4	12·45	6066	5972	5382	-94	+590	+684	1·006	5983	5984	5983
122/4/5	13·00	6070	5990	5388	-80	+602	+682	1·009	5997	5997	5997
122/4/6	13·15	6078	5692	5396	-386	+296	+682	1·009	5689	5689	5689

FIG. 307.

PROVISION OF CONTROL IN ADVANCE OF THE PHOTOGRAPHY

558. We may now turn to the situation where the field surveyor has
advance knowledge of the flight-lines along which it is proposed to
photograph the overlapping strips which will cover the area under
survey. In engineering surveys of large and medium scales this is of

great advantage in simplifying the task of providing ground control. It is evident that, if suitable sites for control points could be selected and artificially emphasised, in such a manner that their images will appear clearly and unambiguously on the photographs, most of the problems associated with photo-identification would be eliminated. In practice it is not always possible to eliminate the need for photo-identification entirely but aircrews experienced in survey photography are usually able to adhere to the planned flight-lines so closely that the effort of pre-marking control points is well worth while.

Flight-line information, and details of the areas in which control will be required, will normally be supplied to the field surveyor by the photogrammetrists, annotated on the best available map of the survey area and identical in scale with the one supplied to the aircrew. Conventions similar to those already described are used to indicate the location and type of control required. In this case however, the circles are often replaced by rectangles with their longer sides perpendicular to the line of flight. This device is used to draw the surveyor's attention to the fact that more than one point should be marked to ensure that at least one of the group will be correctly sited in the lateral overlap, even if the flight-lines depart slightly from their planned positions.

559. Pre-marking of Control Points. Once the approximate locations are known, pre-marking can begin. The aim is to provide symmetrical marks dissimilar in shape from any natural or artificial features of the terrain, and to exaggerate the contrast sufficiently so that the image of the mark will be clearly distinguishable from its background on an aerial photograph. Clearly the size of marks is of importance and will vary with the scale of the proposed photography. For 1/12,500 scale photography a method which has been proved in practice to be very satisfactory is that of laying out crosses of whitewashed stones according to the dimensions given in Fig. 308.

For photography at other scales the dimensions should be adjusted proportionally (i.e. for 1/25,000 scale photography the dimensions will be doubled). Other materials and other designs of marks can, of course, be used but in general it will be found that granular materials which scatter light are the most likely to increase photographic contrast under normal conditions.

It will be realised from the foregoing remarks that a proportion of the pre-marked points will not be used, either because of variations in the flight-lines or exposure interval, or simply because they do not show up clearly on the photographs. For this reason it is pointless to survey any of them at this stage. Instead, all efforts should be concentrated upon providing an excess of points through- out the annotated areas, leaving the final selection of those to be fixed until after the photographs are available for examination. If time allows, the pre-marking can be continued to good effect by emphasising other important features not directly connected

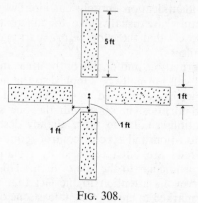

FIG. 308.

with the immediate control problem but, nevertheless, vital to the overall project. Such features may include established survey con- trol points and points for which coordinates could be obtained during the photogrammetric process to assist in later stages of the work. The marks used for such features should be different in design from the photo-control marks. For 1/12,500 scale photo- graphy discs of any suitable whitewashed material roughly 2 feet in diameter show up quite clearly.

560. Recording the Pre-marked Points. As an aid to identification and subsequent survey of the selected points, all pre-marked points should be plotted on the map as accurately as possible and each one given a unique identifying serial number. A fair copy of the map must then be prepared for use by the photogrammetrists.

561. Field Completion of Photogrammetric Surveys. With photo- grammetric mapping at the scales discussed in this chapter there will always be a proportion of the detail which cannot be plotted by the photogrammetrists, either because it is entirely screened from the air, or because it is obscured by glare or dense shadows on the photographs. The proportion of detail which cannot be plotted will vary with the nature of the terrain and the photographic scale. It will indeed be one of the factors to be considered in assessing the economy of the method *vis à vis* ground survey. The amount of plottable detail vital to the survey can be increased to some extent by judicious use of the methods of pre-marking already described, but it is almost inevitable that the services of the field surveyor will be required to fill in the missing detail and to resolve queries of photogrammetric interpretation. No unfamiliar problems are involved. The map scale is now the controlling factor in the surveyor's choice of suitable survey methods.

562. Field Checking of the Completed Survey. In the final analysis the engineer is only interested in whether the results achieved by the photogrammetric methods meet the accuracy requirements originally specified. Here again the role of the field surveyor is vital and once again a different set of accuracy standards will govern his work and methods. Accuracy can only be acceptably assessed, to the satisfaction of all interested parties, if the methods used in conducting the check surveys are not themselves subject to errors equivalent in magnitude to the tolerances permitted by the specification. Furthermore, the checks must be designed to be self consistent. In sample checks on positional accuracy, therefore, the work must be carried out and proved to the same standard as that of the original control supplied to the photogrammetrists. For maximum significance, the check surveys should be carried out in areas most vital to the project and, in general, well away from the photo-control points. In this task the additional pre-marked survey points referred to in Section 559 will prove useful as their established coordinates can be directly compared with the values obtained for them in the photogrammetric process. Wherever possible, height checks, when these are necessary, should be conducted by levelling in closed circuits or between bench-marks. The object in each case is not to prove how inaccurate photogrammetric mapping is, compared with ground methods, but to provide a reliable yardstick by which the suitability of the mapping for a particular purpose can be assessed.

BIBLIOGRAPHY

Aerial photography and photogrammetry
American Society of Photogrammetry, Manual of Photogrammetry, 2nd. Ed. 1952. Photo-Interpretation Manual, 1959.
Bertil HALLERT, Photogrammetry: Basic principles and general survey. McGraw Hill Book Co. Inc. 1960.

Alimetry
C. A. BIDDLE, Heights by Aneroid Barometer. Tellurometer (UK) Ltd., London.

Altimetry Manual and Operating Instructions for W. and T. Altimeters. Wallace & Tiernan, Co. Inc., Belleville 9, New Jersey, U.S.A.

Field Practice
Survey Manual, 1964 Ed. Survey of Kenya, P.O. Box 30046, Nairobi.

REFERENCES

BERE C. G. T. AND WILLIAMS V. A. 'Notes on the pre-marking of control by the Directorate of Overseas Surveys'. *Conference of Commonwealth Survey Officers*, 1967.

BROCKLEBANK R. A. 'Photogrammetry in municipal surveying and mapping'. *Canadian Surveyor*, Sept. 1963.

COOPER J. 'The large-scale mapping of Hong Kong by air-survey'. *Conference of Commonwealth Survey Officers*, 1967.

MOTT P. G. 'Aerial methods of surveying for civil engineering'. *Proc. of the Inst. of Civil Engineers*, Vol. 26, 1963.

MOTT P. G. AND DAWE H. 'Aerial photography on scales 1/1000 and larger'. *R.I.C.S. Journal*, Feb. 1953.

PRYOR W. T. 'Experience of the Bureau of Public Roads in highway surveys'. *Proc. of the American Soc. of Civil Engineers*, Dec. 1956.

SMITH G. J. 'The use of photogrammetric surveys for highway location'. *Canadian Surveyor*, Jan. 1959.

SMITH W. P. 'Ground Control for aerial survey'. *R.I.C.S. Journal*, Dec. 1951.

WIGGINS W. D. C. 'Mapping contract documents'. *Conference of Commonwealth Survey Officers*, 1955.

ANSWERS TO EXAMPLES

Chapter II (page 127)

(1) 1885·183
(2) 199·987
(3) 765·739
(4) 688·64
(5) 1379·289
(6) 0·065
(7) 65°F
(8) 461·536
(9) 1 lb. 13 oz. : 0·024 ft.
(10) 0·442
(11) 0·029 m.
(12) 170·795
(13) 405·74 : 0·07
(14) 1·568 :0·007

Chapter V (page 232)

(1) +411·1, −628·6
(2) 1105·8 :18° 25′
(3) −113·5, −51·9
(4) B 3247·1 9035·4
    C 3255·9 8922·8
(5) AC = 612·1, BC = 690·3
    perpendicular = 541·7
(6) BC = 1636·2   angle A = 99° 05′
    CA = 752·9   B = 27° 02′
    AB = 1338·6   C = 53° 53′
    area = 497,600 sq. ft. = 11·42 acres
(7) 3872   11633
(8) Coordinates before adjustment:
    A   1000·0   1000·0
    B   902·3   1329·8
    C   924·8   1758·2
    D   1110·0   1677·7
    E   1219·0   1468·4
    F   1428·9   1226·9
    G   1186·0   860·8
    A   988·6   1004·2
(9) 3756·9
(10) Bearing adjustment +40″
    Coordinate closure adjustments +1·9, −0·6
(11) 392 along AB
    293 along BC
    245 along CD
     67 along DE
    368 along DE

(12) 79·2, 1070·7
(13) 540·5 ft.
(14) 0° 14′ 09″, 5457·41
(15) 32365·90, 10792·17
(16) 1811 m.
(17) About $2\frac{1}{2}$°W
(18) About 6118, 4810, to nearest foot
(19) 249° 11′, 255·0
(20) 20847·2, 16741·1

Chapter VI (page 266)

(1) 9·14
(2) 265·90, 264·49
(3) 101·14
(4) 5·72, 4·47
(5) 1 : 24, 2° 23′
(6) 4·0, 4·6, 4·6, 4·2
(7) Set to read 6·460 on A,
    then centralise bubble
(8) 85″ up
(9) 48 ft. from A, 51 ft. from B
(10) 5·6 mm.

Chapter IX (page 334)

(1) 2·377 acres
(2) 1 hectare
(3) 4743 sq. ft.
(4) 177,488 sq. ft., 4·075 acres
(5) 509 sq. ft.
(6) 577 sq. ft.
(7) 160/9, 0·2491
(8) 74a 1r 2p : nearly 3 roods
(9) $95\frac{1}{2}$ sq. ft.
(12) 19900 cu. ft. : −278 cu. ft.
(13) +563 cu. ft.
(14) 500 million cu. ft.

Chapter X (page 375)

(1) 2° $29\frac{1}{2}$′ : 2·17 ft.
(2) 487·7 ft. : 3420 + 17
(3) 37° 54′ : 909·2 : 480·7
(6) 13·17 ch : 653 + 25
(8) 0·5 at 25·0, 2·1 at 49·9,
    4·7 at 74·8, 8·3 at 99·5

(9) 627·6 ft., 86768·6

bearings: 97° 35′ 20″
           96° 26′ 40″
           95° 18′ 00″
           94° 09′ 20″
           93° 00′ 40″
           91° 51′ 40″

(10) Extend AB by 27·4 ft.: R = 268 ft.

(11) 2·4 on from B, 25·0 back from C

(12) 361 links

(13) 880 ft. : 83° 29′ 20″

(14) 19·1 at 24·9, 31·1 at 53·8,
     35·1 at 84·9, 31·1 at 115·9,
     19·1 at 144

(15) R = 760 ft. : 271·5 back from B
                   171·6 on   from C

(16) 4082 : 23·6

(17) Deviation 7° 09′ 43″

     Distances 100   200   299·9 399·7 499·2
     Offsets   0·17   1·33 4·50   10·67 20·81

(18) Tangent points 528·5 back; apex 23·5

(19) 222·3 ft.: 184·09 ft.

(20) 245·6 from A : 339·38 : R = 3590 ft.
     302 from A : 334·25

(21) 617 from A : 448 : 193 beyond A : 97 back from B

Chapter XI (page 396)

(1) 38·4 ft.; 1 : 19

(2) 744 sq. ft.

(3) 184·8 ft.; 1 : 39·5

(4) 159·2 ft.: 16·2°

(5) 712·7 ft. : 0·8 ft.

(6) 291·2 m.

(7) 34 ft. : 30°

(8) 204·1 : 606

(9) 2273·6 m.; 1·4 m.; 1·2 m.

(10) 121·94

(11) 454·3 : 111·6

(12) 9030 sq. ft.; 1 : 30

FIG. 1

FIG. 2

METRIC GRADUATION
OF TAPES

The following information, and the two diagrams referred to, have been supplied by Messrs. Rabone Chesterman Ltd.:

'The millimetre has been taken as the basis for graduating and numbering boxwood folding rules and pocket tape rules. Thus these instruments will be figured in millimetres and will be graduated with fine graduations of one millimetre on one edge of the instrument, and coarse graduations of five millimetres on the other edge (see Figure 1 opposite).

Measuring tapes will be divided into metres with the decimal part of the metre figured at every 100 mm. graduation mark. Steel tapes will be further graduated at 10 mm. and 5 mm. intervals and the first and last metres will be finely graduated with 1 mm. graduation marks (see Figure 2 opposite). 5 mm. and 1 mm. graduation marks will not appear on the synthetic (fibreglass) tapes.

Studded band chains will be in lengths of 20 metres, will be divided by brass studs at every 200 mm. point, and will be marked at every metre. The first and last metres will be further divided into 10 mm. intervals. The markings will appear on both sides of the band.

Land chains will also be in lengths of 20 metres, and will be composed of links 200 mm. in length, and will be tallied at every metre.'

METRIC GRADUATION OF FOLDING RULES AND POCKET TAPES.

METRIC GRADUATION OF MEASURING TAPES.

INDEX

The numbers refer to Sections

Aberrations in Lenses, 14
Abney Clinometer, 121, 337
Accuracy of Compass Traversing, 264
Accuracy of Plane Tabling, 369
Accuracy of Traversing, 195, 209
Adding Machines, 266
Adjustment of Bearings in Traverse Computation, 282
Adjustment of Diaphragm, 97
Adjustment of a Level, 117, 118
Adjustment of a Theodolite, 89, 90
Adjustment of Transit Axis, 96
Adjustment of Traverse Misclosure, 286, 293
Adjustment of Traverse Network, 287
Adjustment of Vertical Circle Zero, 98
Air Photographs, 431, 496, 546–
Air Survey, 546–
Alidade, 354, 355
Altimeter Heights, 552–
Amsler's Planimeter, 401
Anallactic Telescope, 473
Angle Measurement in Traversing, 216
Angle of View of a Telescope, 31
Arbitrary Meridian. 202, 223
Area calculated from Coordinates, 404
Area of Closed Traverse, 404
Area Measurement, 394–
Arrows, in Chaining, 49
Astigmatism of a Lens, 18
Astronomical Checks in Traversing, 244
Astronomical Meridian, 198
Automatic Level, 108, 115
Auxiliary Bearing (Computation Check), 279, 285
Axes of Coordinates, 274
Axis of Bubble Tube, 36
Axis of a Lens, 10

Back Bearing, 205
Back Bearing Method in Traversing, 227
Backsight in Levelling, 312
Balancing Line in Earthwork, 427
Barrel Distortion of a Lens, 20
Base Line in Plane Table Surveying, 365
Beaman Arc, 475
Beam Compass, 187
Bearings, 197–206, 218, 219
Benchmarks, 315
Boning Rods, 338
Bowditch Rule for Traverse Adjustment, 286

Box Sextant, 58–62
Bubble Adjustment on a Level, 118
Bubble Adjustment on a Theodolite, 92
Bubble Calibration, 37
Bubble, circular, 38
Bubble: principle of use, 36, 83
Bubble Sensitivity, 37, 341
Bubble Tubes, 34, 36, 71

Calculating Machines, 266
Calibration of Bubble, 37
Capacity of a Reservoir, 421
Capstan Screw, 4
Catchment Area, 388
Catenary method of measurement, 144, 154–156, 158
Centering, optical, 74
Centering Head of Tripod, 73, 82
Centering a Plane Table, 360
Centering a Theodolite, 74, 82
Centering Traverse Signals, 212, 213, 235
Centre of a Lens, 10
Centrifugal Acceleration on Curves, 436
Centroid of a Section, 415
Chains, 43, 48
Chain Surveying, 133–136, 172–182, 193
Chain Surveying: Problems, 183–186
Chaining Arrows, 49
Chaining Pins, 49
Change Point in Levelling, 311, 328
Change of Volume in Earthwork, 428
Channel Cross Section, 504
Chart Datum, 495
Check Lines in Chain Surveying, 174
Check Measurements of Lengths, 152
Checking of Air Surveys, 562
Checking Levelling Field Work, 328
Checking of Setting Out, 433
Checks in Computations, 265, 277, 279, 285
Checks on Levelling Computations, 319
Checks in Traversing, 238, 241, 242
Chemical Methods for Discharge Measurement, 542
Chezy Formula for Stream Discharge, 535
Choice of Formula for Computation, 270
Chords of Angles, 300
Chord of a Curve, 439
Chromatic Aberration of Lens, 15
Circular Bubble, 38
Circular Curves, 435
Circular Measure of Angles, 271

Circumferentor, 103

Clamp; impersonal, 67

Clamps on Theodolite, 67

Clinometer, 119, 121, 337, 368

Closed Traverse, 194, 239

Closing Error in Traversing, 241, 245, 246, 282, 286

Clothoid Curve, 457

Collimation Line of a Telescope, 22

Colouring on Plans, 192

Coma of a Lens, 17

Compass, liquid, 41, 104

Compass, prismatic, 41, 104

Compass; Surveying, 103

Compass; trough, 72, 356

Compass Theodolite, 258

Compass Traverse, 195, 256–264, 291–293

Compensating Errors, 170

Computation of Bearings and Distances, 278

Computation of Compass Traverse, 291–293

Computation of Coordinates in Traverse, 284

Computation Form for Traverse, 284

Computation of Intersection and Resection, 296, 513

Computation of Latitudes and Departures, 283

Computation of Traverse, 280–

Computing Scale for Areas, 398

Conformal Projection, 298

Conjugate Foci of a Lens, 12

Contour Gradient, 384, 387

Contour Interval, 372

Contour Line, 371

Contouring, 370–

Contouring by Levelling, 374–376

Contouring with Plane Table, 368

Controlled Traverse, 194

Convergence between true and grid bearing, 201

Convergence of Meridians, 198

Convergent Lens, 11

Coordinates; Rectangular, 274

Cowley Level, 115

Cross Bearings in Traversing, 240

Cross-Lines of a Telescope, 22

Cross Section; Area of, 407

Cross Sections, 324, 331–

Cross-staff, 53

Cubic Parabola, 458

Cubic Spiral, 458

Cumulative Errors, 169

Current Meter, 530–

Currents; Tidal, 516–

Curvature of Bubble, 35, 37

Curvature Correction of Earthwork, 415

Curvature of Field of a Lens, 19

Curvature Effect in Levelling, 346

Curves, 434–462

Curve Ranging, 434

Datum Surface for Levelling, 308, 314, 327, 493

Declination, magnetic, 41

Deflection Angle of a Curve, 439, 443

Deflection Angles in Traversing, 216–218

Deflection Distances of a Curve, 446

Degree of a Curve, 435

Departures, 275, 276, 283

Depression Angle for Fixing Position, 505

Detail Surveying, 152, 232

Deviation Angle of a Curve, 439

Deviations in Traversing, 210

Diagonal Eyepiece, 27, 79

Diaphragm of a Telescope, 22–25, 94, 97

Diopter, 11

Dip, magnetic, 41

Direct Bearing Method in Traversing, 220, 224–228

Directions, 197

Discharge Formulae for Weirs, 540

Discharge of Stream, 522–

Dislevelment of Theodolite, 102, 235

Dispersion of Light, 15

Distortion by Lenses, 14, 20

Diurnal Variation of Magnetic Meridian, 200

Divergent Lens, 11

Double Base Method in Altimetry, 556

Drainage Area, 388

Drop at middle of a Tape, 166

Drums for Steel Tapes, 146

Dumpy Level, 106

Earthwork Calculations, 392, 408–

Earthwork Contours, 389

Eccentricity of Theodolite, 91

Echo Sounding, 507

EDM, 129, 195, 208

Electro-magnetic Distance Measurement, 129

Engineer's Chain, 43

Equal Sights in Levelling, 309

Equilibrium Tide, 488

Errors in Chain Surveying, 140, 141

Errors in Compass Surveying, 263

Errors in Computation, 269

Errors in Levelling, 340–347

Errors in Subtense Method, 480

Errors in Tape Measurement, 167–171

Errors in Traversing, 233–237, 247

Euler Spiral Curve, 457

Extended Vernier, 39

Eyepiece of a Telescope, 23, 26

The numbers refer to Sections

Face Left: Face Right, 75, 94
False Origin of Coordinates, 284, 302
Field Book of Chain Survey, 181
Field Checking of Air Surveys, 562
Field Completion of Air Surveys, 561
Field of View of Telescope, 31
Fixed Needle Traversing, 258
Floats for Current Measurement, 518, 527–
Flying Traverse, 194, 243
Focal Length of a Lens, 11
Focus of a Lens, 11
Focusing a Telescope, 28
Foresight in Levelling, 312
Francis Formula for Discharge, 540
Free Needle Traversing, 257
French Curves, 187
Fungus on Lenses, 33

Galileo Telescope, 21
Geodimeter, 129, 195
Gradient, 384
Graduated Circle, 64, 65, 76
Graduation Errors of a Circle, 91
Graduation of Levelling Staff, 111, 113, 341
Graduation of Steel Tapes, 143, 158
Graphical Adjustment of Traverse, 293
Graphical Resection, 512
Grid Bearing, 201, 274
Grid Meridian, 201, 221
Grid North, 201
Ground Control for Air Survey, 546–
Ground Survey for Air Photography, 546–
Gunter's Chain, 43, 45
Gyro-theodolite, 464

Half-breadths of cross sections, 407
Hand Level, 119, 120, 335, 376
Harmonic Analysis of Tides, 488
Haulage of Excavation, 423, 424, 427
Height Control for Air Survey, 551
Height Correction of Linear Measurement, 165
Height of Instrument Method, 316, 317
High Water Mark, 496
Hook Gauge, 524
Horizontal Angle Measurement, 64, 85
Horizontal Circle of Theodolite, 64
Horizontal Collimation Adjustment, 94, 95
Horizontal Collimation of Theodolite Telescope, 101
Horizontal Line, 308
Horse Shoe Curve, 454
Hydraulic Radius, 535
Hydrographic Surveying, 486–
Hydrostatic Levelling, 122

Identification of Ground Control Points, 549
Illumination for theodolite reading, 78
Image formed by a Lens, 13
Impersonal Clamp of Theodolite, 67
Included angles in Traversing, 216–218
Indian Clinometer, 368
Inking of Plans, 191
Intermediate Sight in Levelling, 313
Internal Focusing Telescope, 23
Interpolation of Contours, 380, 383
Interpolation in Tables, 272
Intersection Angle of a Curve, 439
Intersection Computation, 296
Intersection with Plane Table, 363, 365
Interval between Contours, 372
Intervisibility of Points, 386
Invar Subtense Bar, 479
Inverted Staff in Levelling, 327
Isogonic Lines, 199

Jacob Staff, 103

Kepler Telescope, 21
Kutter & Ganguillet Formula, 535

Latitudes and Departures, 275, 276, 283
Leap Frog Method in Altimetry, 555
Least Count of a Vernier, 39
Lemniscate Curve, 459
Length Measurement in Traversing, 231
Lens Formula, 12
Lenses in Instruments, 10
Lettering on Plans, 192
Level Book, 316
Level Line, 308
Levelling, 307–
Levelling; field methods, 325–
Levelling Cams, 70
Levelling an Instrument, 36, 83
Levelling Screws, 66, 70, 83
Levelling Staff, 110–114
Levels, 105–109, 115, 120. 122
Line of Collimation of Telescope, 22
Line Ranger, 52
Line of Sight of Telescope, 22, 63, 93, 310
Linen Tapes, 50
Links (unit of Measurement), 45
Liquid Compass, 41, 104
Local Attraction in Compass Surveying, 262
Location Survey, 432
Longitudinal Sections, 324, 327
Loop Traverse, 194
Low Water Mark, 496
Lower Plate of Theodolite, 66, 72, 80

Lowest Astronomical Tide, 495
Lunitidal Interval, 488

Machine Computation, 273
Magnetic Declination, 41, 199, 222
Magnetic Dip, 41
Magnetic Meridian, 199, 222
Magnetic Needle, 41, 72
Magnetic Variation, 41
Magnification, optical, of a Lens, 13
Magnification, visual, of Telescope, 32
Map Projection, 298
Maskelyne's Rules for Small Angles, 272
Mass Diagram, 423–
Mathematical Tables, 266, 285
Mean Sea Level, 314, 489
Mekometer, 129
Meridians, 197–212
Metric Chains, 43
Microscope-reading Theodolite, 75
Mirrors in Instruments, 7
Mounting Paper on Plane Table, 359

National Physical Laboratory, 166
National Survey Systems of Co-ordinates, 298
Neap Tide, 488
Newlyn Datum for Heights, 314, 490
Non-repeating Theodolite, 80
Notched Weir, 537

Obstacle in Setting out a Curve, 449
Obstacles in Chain Surveying, 186
Obstacles in Theodolite Surveying, 252
Offset Scale, 188, 190
Offsets to a Curve, 447
Offsets, 175, 176, 180
Offsetting, 152, 183–
Oiling of Instruments, 127
Optical Centering, 74
Optical Distance Measurement, 470
Optical Magnification of a Lens, 13
Optical Square, 54–56
Optical-reading Theodolite, 75
Ordnance Survey Benchmarks, 315
Ordnance Survey Map Scales, 188
Ordnance Survey Methods, 193
Orthomorphic Projection, 298
Overhaul in Earthwork, 429

Pacing in Traverse Surveying, 259
Packing of Instruments, 128
Pappus's Theorem, 415
Parabolic Vertical Curve, 462
Parallax in Telescope, 4, 28
Parallel Plate, 9
Parallel Ruler, 300, 301
Pendulum Levels, 108

Permanent Adjustment of Instruments, 2
Pin-cushion Distortion of a Lens, 20
Plane Table, 351–353
Plane Table Surveying, 350–
Planimeter, 400–402
Plastic Tapes, 50
Plotting a Chain Survey, 187, 189, 190
Plotting Level Sections, 329
Plotting Machines, 190, 306
Plotting Soundings, 508
Plotting Traverses, 299–306
Plumb-bob, 74
Plumbing Fork, 357, 360
Plumbing rod of Tripod, 42
Plummet, 74
Plus Measurements in Chain Survey, 180
Polar Planimeter, 401
Poles; ranging, 51
Power of a Lens, 11
Precision of Computation, 267
Predicted Tide, 488, 495
Pre-marking of Control Points, 558, 559
Principal Axis of a Lens, 10
Principal Chord of a Curve, 443
Principal Focus of a Lens, 11
Prismatic Compass, 41, 104
Prismoid Volume, 411
Prismoidal Correction, 412
Prism Square, 54, 57
Prisms in Instruments, 8
Problems in Setting Out, 448
Profiles by Levelling, 324, 329
Prolonging a Line, 132
Propagation of Errors in Linear Measurement, 171
Propagation of Error in Traversing, 248–250
Protractors for Plotting, 300

Quadrantal Bearings, 203, 204

Radian Measure, 271
Radiation with Plane Table, 363, 364
Radius; Hydraulic, 535
Railway Curves, 187
Ramsden Eyepiece, 26
Random Errors, 170
Rangefinder, 482
Range of Tides, 488
Ranging a Line, 130–132
Ranging Poles, 51
Reciprocal Levelling, 348
Reconnaissance for Chain Survey, 177
Reconnaissance Survey, 431
Recording Chain Survey Measurements, 181
Recording a Compass Traverse, 261

Recording Length Measurements, 153, 157

Recording Levelling Observations, 316, 334

Recording Traverse Measurements, 217, 230

Rectangular Coordinates, 274, 284, 294, 295, 302

Reduced Bearings, 206, 275

Reduced Level, 308

Reduction of Scale, 192

Refraction, atmospheric, 91, 346

Refraction Effect in Levelling, 346

Refraction through Lenses, 11

Repetition Method for Horizontal Angles, 80, 86

Resection, 509–514

Resection Computation, 513

Resection on Plane Table, 362, 363, 367

Resolving Power of Telescope, 29

Reticules of Telescopes, 25

Reversal Principle in Adjustments, 3, 94

Reverse Bearing, 205

Right-angles: Setting out, 183, 184

Rise and Fall Method, 316, 318

Rod: Levelling, 110

Rolling up Steel Tape, 146

Rotation Axis of Theodolite, 63

Rounding Off Numbers, 268

Sag Correction, 162

Sag of a Chain, 138, 140

Scale-reading Theodolite, 75

Scales for Chain Surveys, 188

Scales of Ordnance Survey Maps, 188

Scotch Staff, 112

Sea-level, 488

Secondary Axis of a Lens, 10

Sections by Levelling, 323, 324

Sections drawn from Contours, 385

Self-reduction Tachymeter, 476, 477

Semi-graphical Computation, 296

Sensitivity of a Bubble, 37

Setting out a Horizontal Angle, 87

Setting out Cross Sections, 333

Setting out Curves, 440–

Setting out Gradients, 384

Setting out Levels, 338

Setting out Works, 430–

Sextant, 499–

Sextant; box, 58–62

Shift of a Transition Curve, 456

Shore Line Surveying, 496

Shrinkage of filled Material, 424

Signals for Traversing, 212

Simpson's Rule for Areas, 398

Sine Formulae, 273

Single Base Method for Altimetry, 554

Skew Bridge Setting Out, 433

Skew Surface of Prismoid, 413

Slope Correction, 164

Slope Measurement, 138, 149

Slope Method for Stream Discharge, 535

Slope Stakes, 463

Sloping Lines in Chain Surveying, 138, 185

Sloping Lines in Steel Tape Measuring, 149

Slow Motion movement of Theodolite, 67

Small Angles in Computations, 271, 272

Solution of Triangles, 273, 297

Sopwith Staff, 112

Soundings, 506–

Speed Limitations on Curvature, 436

Spherical Aberration of Lens, 16

Spirit Level: see 'Bubble Tube'.

Split Bubble, 35

Spot Heights, 370, 381, 416

Spring Balance, 47, 147, 166

Spring Tide, 488

Stadia Lines, 109, 318, 471

Staff: Levelling, 110–114

Standardisation of Tapes, 160, 166

Station Marks in Chain Survey, 178, 179

Station Marks in Traversing, 210, 211, 255

Station Pointer, 511

Steel Bands, 44, 48, 142, 143

Steel Tape; measuring methods, 144, 145

Stepping on Slope Lines, 138

Stops in Optical Systems, 20

Straight Edge, 187

Stream Gauging, 522–

Sub-chord of a Curve, 443

Subdivision of Area, 405, 406

Submerged Weir, 537

Subtense Method, 129, 479–

Sun-glass for theodolite, 79

Super-elevation of Curves, 436, 456–

Surveying Compass, 103

Surveying Telescope, 23, 77

Swinging the Staff in Levelling, 342

Symbols on Plans, 191

Systematic Errors, 168

Tables, Mathematical, 266, 285

Tacheometer, 468

Tachymeter, 468, 471, 477

Tachymetry, 129, 232, 468–

Tachymetric Contouring, 377, 381

Tachymetric Surveying, 468–

Tangent Length of a Curve, 439

Tangent Point of a Curve, 442, 445

Tangent Screw of Theodolite, 67

Tangential Method of Tachymetry, 478

Tapes; linen, 50
Tapes; plastic, 50
Tapes; steel, 44, 48, 142, 143
Tape Clips, 47, 151, 208
Target Staff, 114
Telemeter, 483
Telescopes, 21, 77
Telescopic Alidade, 355
Telescopic Tripod, 42
Tellurometer, 129, 195
Temperature of Steel Tape, 145, 150, 161
Temporary Adjustment of Instruments, 2
Temporary Adjustments of Level, 116
Temporary Adjustments of Theodolite, 82
Tension in Steel Tape, 145, 147, 148, 163
Tests of a Theodolite, 89, 91
Theodolite, 63, 66, 75
Theodolite; levelling up, 83
Thermometer, 47, 150, 166, 208
Three Point Problem, 510
Thrupp's Method for Stream Velocity, 534
Tidal Currents, 516–
Tidal Range, 488
Tide Gauge, 490, 491
Tides, 488–
Tide Tables, 488
Tie Lines in Chain Surveying, 174, 180
Tilting Level, 107
Tommy Bar, 4
Transformation of Coordinates, 294
Transit Axis of Theodolite, 63, 96, 100
Transit Theodolite, 75
Transition Curves, 437, 456–
Transport of Instruments, 124
Transverse Mercator Projection, 298
Trapezoidal Rule for Areas, 398
Trapezoidal Weir, 537
Traverse Base, 210, 254, 290
Traverse Computation, 280
Traverses: Plotting, 299–306
Traversing, 194–
Traversing with Plane Table, 363, 366
Traverse Tables, 285, 292
Triangulation with Plane Table, 365
Tribrach of Theodolite, 66
Tripods, for ranging poles, 51
Tripod Stand, 42, 69, 73, 82, 128
Tripod, telescopic, 42
Trough Compass, 72, 356
True Meridian, 198, 221
Tunnels; Setting out, 464
Turning Point in Levelling, 311, 328
Two-Peg Method for Adjusting Level, 118

Upper Plate of Theodolite, 66

Variation of the Compass, 199
Variation, magnetic, 41
Variations of the Magnetic Meridian, 200
Vernier, 39, 40
Vernier Theodolite, 75
Vertical Angle Measurement, 65, 68, 88
Vertical Angle Zero, 88, 90, 98
Vertical Curves, 438, 460–
Virtual Image by a Lens, 11
Visibility on Vertical Curves, 460, 461
Visual Magnification of Telescope, 32
Volume Measurement, 408–

Wasting Excavated Material, 427
Water Level, 122
Weight of a Tape, 166
Weir Method for Stream Discharge, 536–
Whole-Circle Bearings, 203, 204

Zero setting of Theodolite, 80